MODERN TRENDS IN CONTROLLED STOCHASTIC PROCESSES:

Theory and Applications

Volume II

Alexey B. Piunovskiy

Editor

MODERN TRENDS IN CONTROLLED STOCHASTIC PROCESSES:

Theory and Applications

Volume II

Luniver Press
2015

Published by Luniver Press
Frome BA11 3EY United Kingdom

British Library Cataloguing-in-Publication Data
A catalogue record for this book is available from the British Library

MODERN TRENDS IN CONTROLLED STOCHASTIC PROCESSES:
Theory and Applications, Volume II

ISBN-10: 1-905986-45-9
ISBN-13: 978-1-905986-45-3

Table of Contents

MODERN TRENDS IN CONTROLLED STOCHASTIC PROCESSES

Introduction

Alexey B. Piunovskiy

University of Liverpool
Dept. of Mathematical Sciences, Liverpool L69 7ZL, UK
piunov@liv.ac.uk

In July 2015, world class experts met in Liverpool to discuss the challenging problems of stochastic optimal control. This book contains extended reports from that workshop. Note that many presentations were about on-going research, so that the corresponding articles present preliminary results; the detailed proofs will be published later. The book will enable researchers, academics and research students to get a sense of novel results, concepts, models, methods, and applications of controlled stochastic processes. Below, we briefly describe the topics touched in the further chapters.

Models. Discrete-time processes still attract much attention [1, 4, 6, 9–13, 17–19]. Note that such models can serve as approximations to continuous-time models [11, §4.3], [19]; they also appear after preliminary investigation of continuous-time Markov decision processes [16]. The range of specific models is quite wide: optimal stopping [6, 12], long-run average [10, 17] and discounted [1, 4, 11, 13, 17, 19] costs, and so on. Some recent results on the total (undiscounted) cost can be found in [16]; article [4] is about stochastic games.

In [3, 7, 14, 15], different versions of controlled stochastic differential equations are studied, including networks of diffusions [7] and control of a group of (independent) agents [3]. The so called stochastic fluid models (close to piecewise deterministic) and continuous-time Markov decision processes are studied in [8] and [16] correspondingly.

It should be underlined that many authors investigate models with partial information [2, 6, 9, 11, 12] which are traditionally considered to be more challenging. Another important aspect is that [13, 15, 16, 19] consider constrained optimal control problems.

Methods. Dynamic programming remains powerful when investigating many different models [1, 4, 6, 7, 10, 11, 14, 15]; the linear programming approach, especially useful for the analysis of constrained problems, is developed in [10, 16, 19]. Additionally one can mention H_2-control [9], robust control [11] and the Whittle indexation [13]. As for the models with partial information, the Bayesian approach is essentially used in [6, 12]. Different approximate and numerical methods are presented in [1, 4, 17, 18].

No doubts, any active researcher in the field of stochastic optimal control will find useful material presented in this book.

Applications. Even the authors of purely theoretical articles find it necessary to illustrate their results by meaningful examples. Obviously, stochastic optimal control can be used for solving many important applied problems. One of traditional areas is Queueing Theory [1, 2, 13], another one is Financial Mathematics (including taxation, insurance etc) [4, 5, 14, 19]. Let us also mention control of water resources [11, 19], pollution control [11, 15], house selling [6], quality control [12], unmanned vehicle control [9] and power management control [8].

Structure of the book. The first eleven article [1, 4, 6, 9–13, 17–19] deal with discrete time. Roughly speaking, [6, 9, 10] are mainly theoretical, [1, 4, 11, 13, 17, 18] pay more attention to approximations and numerics, [12, 19] concentrate on special applied problems.

Articles [2, 3, 5, 7, 8, 14–16] deal with continuous time models. Again, the first five articles [3, 7, 14–16] provide theoretical/analytical investigation of the general enough models, and the last three [2, 5, 8] are devoted to specific applied problems.

Acknowledgements. All the authors are thankful to the Engineering and Physical Sciences Research Council (EPSRC, UK) and to the Research Centre in Mathematics and Modelling (RCMM, Uni. of Liverpool) for the financial support of the workshop "Modern Trends in Controlled Stochastic Processes: Theory and Applications" held at the Dept. of Mathematical Sciences of the University of Liverpool, 29 June - 04 July 2015.

References

1. Almudevar, A.: A functional analytic approach to approximate iterative algorithms. *This Book*, p.67-86.
2. Anselmi, J., Gaujal, B. and Nesti, T.: Control of parallel non-observable queues: asymptotic equivalence and optimality of periodic policies. *This Book*, p.286-306.
3. Anulova, S.: A multidimensional comparison theorem for SDE with monotone drift. Applications to control of a group of independent identical agents. *This Book*, p.256-262.
4. Avrachenkov, K. and Varava, A.: Completely mixed stochastic games with small unfixed discount factor. *This Book*, p.152-163.
5. Avram, F. and Minca, A.: Steps towards a management toolkit for central branch risk networks, using rational approximations and matrix scale functions. *This Book*, p.263-285.
6. Bäuerle, N. and Rieder, U.: Partially observable risk-sensitive stopping problems in discrete time. *This Book*, p.12-31.
7. Borkar, V.S. and Kumar, K.S.: Coordination in networked control processes through gossip-like local interactions. *This Book*, p.246-255.
8. Chernysh, K.: Average cost optimisation for a power supply management model. *This Book*, p.307-323.
9. Costa, O.L.V., Fragoso, M.D. and Todorov, M.G.: H_2 control and filtering of Markov jump linear systems with partial information. *This Book*, p.47-66.
10. Dufour, F. and Prieto-Rumeau, T.: Solving the average cost optimality equation for unichain Markov decision processes: a linear programming approach. *This Book*, p.32-46.
11. Gordienko, E. and Minjárez-Sosa, J.A.: Markov decision processes with state-dependent discount factors: stability with respect to the Prokhorov metric. *This Book*, p.112-129.
12. Horiguchi, M.: Bayesian inference in Markov decision processes. *This Book*, p.177-189.

13. Jacko, P. and Novák, V.: Whittle's indexation approach to and applications of bi-objective two-state binary-action Markov decision processes. *This Book*, p.140-151.
14. Lon, P.C. Rodosthenous, N. and Zervos, M.: On the optimal stopping of a skew geometric Brownian motion. *This Book*, p.231-245.
15. Mendoza-Pérez, A.F., Jasso-Fuentes, H. and Hernández-Lerma, O.: Ergodic control of pollution with cost constraints. *This Book*, p.213-230.
16. Piunovskiy, A.B.: Sufficient classes of strategies in continuous-time Markov decision processes with total expected cost. *This Book*, p.190-212.
17. Saldi, N., Yüksel,S. and Linder, T.: Finite approximations to Markov decision processes with Borel spaces. *This Book*, p.87-111.
18. Sonin, I.M. and Steinberg, C.: Elimination and insertion operations for finite Markov chains. *This Book*, p.130-139.
19. Zadorojniy, A.: Real life optimization problems: a CMDP-based approach. *This Book*, p.164-176.

Partially Observable Risk-Sensitive Stopping Problems in Discrete Time

Nicole Bäuerle*[1] and Ulrich Rieder[2]

[1] Karlsruhe Institute of Technology, 76128 Karlsruhe, Germany
nicole.baeuerle@kit.edu

[2] University of Ulm, 89069 Ulm, Germany
ulrich.rieder@uni-ulm.de

Abstract. In this paper we consider stopping problems with partial observation under a general risk-sensitive optimization criterion for problems with finite and infinite time horizon. Our aim is to maximize the certainty equivalent of the stopping reward. We develop a general theory and discuss the Bayesian risk-sensitive house selling problem as a special example. In particular we are able to study the influence of the attitude towards risk of the decision maker on the optimal stopping rule.
Keywords: optimal stopping, certainty equivalent, exponential utility, updating operator, value iteration.
AMS 2000 subject classification: Primary 62L15, Secondary 90C40

1 Introduction

In this paper we consider stopping problems with partial observation under a general risk-sensitive optimization criterion for problems with finite and infinite time horizon. More precisely our aim is to maximize the certainty equivalent of the stopping reward over the time horizon. In case of an infinite time horizon we assume that we have strictly negative cost when we do not stop. The certainty equivalent of a random variable is defined by $U^{-1}(\mathbb{E}U(X))$ where U is an increasing concave function. If $U(x) = x$ we obtain as a special case the classical risk-neutral decision maker. The case $U(x) = \frac{1}{\gamma}e^{\gamma x}$ is often referred to as 'risk-sensitive', however the risk-sensitivity is here only expressed in a special way through the risk-sensitivity parameter $\gamma \neq 0$. More general, the certainty equivalent may be written (assuming enough regularity of U) as

$$U^{-1}\Big(\mathbb{E}\big[U(X)\big]\Big) \approx \mathbb{E}X - \frac{1}{2}l_U(\mathbb{E}X)Var[X] \tag{1}$$

where

$$l_U(x) = -\frac{U''(x)}{U'(x)} \tag{2}$$

* Corresponding author

is the *Arrow-Pratt* function of absolute risk aversion. In case of an exponential utility, this absolute risk aversion is constant (for a discussion see [5]).

In addition we suppose that the process is partially observable. More precisely we suppose that we have a jointly Markovian process (X_n, Y_n) where only the first component is observable. Also the stopping reward depends only on the first component. This class of models includes in particular Bayesian models where the unobservable process reduces to an unknown parameter. Important applications are sequential probability ratio tests. A particular example can e.g. be found in [22]. The theory of risk-neutral stopping problems can e.g. be found in [8] or [18].

Stopping problems with general utility functions are rarely treated in the literature. [17] considers the classical house selling problem with general utility but with complete observation. In a separate section we show that most of the results in [17] also extend to the Bayesian case. [15] considers stopping problems with denumerable state space and arbitrary utility function. The authors there discuss the so-called *monotone case* and give conditions for the optimality of *one-step-look-ahead rules*. Of course the stopping problems we treat here can be seen as special partially observable risk-sensitive Markov Decision Processes. The general theory for these type of problems has been developed in [2]. The special case of an exponential utility combined with partial observation has been treated in [6, 11, 14, 19]. The first treatment of the exponential certainty equivalent as an optimization criterion can be traced back to [13]. See also [20] and [21] for other risk sensitive criteria like the variance and [3] for general certainty equivalents with complete observation. Applications to risk management can be found in [10]. In [7] the authors discuss the influence of the utility function on the optimal strategy.

Our paper is organized as follows. In the next section we present a short introduction to certainty equivalents and discuss their properties. In particular we show that they can serve as a reasonable criterion for risk-sensitive decision making. In Section 3 we introduce our risk-sensitive stopping problem with partial observation and with finite time horizon. We prove the existence of an optimal stopping time and give a recursive algorithm for the computation of the maximal expected utility. We are also able to show that the more risk averse a decision maker is, the later she will stop. In Section 4 we will consider the Bayesian, risk-sensitive house selling problem. Here we generalize the results in [17] to the Bayesian case. In particular we show the existence of so-called *reservation levels* which characterize the optimal stopping strategy. We also discuss the dependence of the reservation levels on the available information, the attitude towards risk of the decision maker and the time horizon. Finally in Section 5 we consider the general stopping problem with infinite time horizon.

2 Certainty Equivalents

In this section we briefly recall the properties of our objective function and demonstrate that it is very well-suited for risk-sensitive optimization. For a deeper investigation we refer the reader to [16]. In what follows we work with the space of all bounded real-valued random variables $\mathcal{L}^\infty$. We interpret the outcome of a random variable X as the realization of a risky position. The event $X < 0$ represents a loss whereas $X > 0$ represents a gain. For a strictly increasing function $U : \mathbb{R} \to \mathbb{R}$ we define the *certainty equivalent* $\rho_U : \mathcal{L}^\infty \to \mathbb{R}$ by

$$\rho_U(X) := U^{-1}\big(\mathbb{E}[U(X)]\big).$$

The quantity is called certainty equivalent since $U(\rho_U(X)) = \mathbb{E}[U(X)]$ which means that the utility of $\rho_U(X)$ is the same as the expected utility of X. Obviously ρ_U is *law-invariant*, i.e. $\rho_U(X) = \rho_U(Y)$ if X and Y have the same distribution. In what follows we summarize some properties of ρ_U. For this purpose we assume that U is *strictly increasing and concave.* Recall that for two real-valued random variable X and Y we say that

- $X \le_{st} Y$ if and only if $\mathbb{E}f(X) \le \mathbb{E}f(Y)$ for all increasing $f : \mathbb{R} \to \mathbb{R}$ for which the expectations exist,
- $X \le_{cv} Y$ if and only if $\mathbb{E}f(X) \le \mathbb{E}f(Y)$ for all concave $f : \mathbb{R} \to \mathbb{R}$ for which the expectations exist.

It is now easy to see that ρ_U is *monotone*, i.e.

- **(monotonicity)** $X \le_{st} Y$ implies that $\rho_U(X) \le \rho_U(Y)$.

Obviously we also have

- **(constancy)** $\rho_U(m) = m$ for any constant $m \in \mathbb{R}$.

Moreover, we have consistency w.r.t. the concave ordering. This property is also called *Schur-concavity* (see e.g. [9]), i.e.

- **(Schur-concavity)** $X \le_{cv} Y$ implies that $\rho_U(X) \le \rho_U(Y)$.

Note that Schur-concavity is not the same as *concavity of a risk measure* $\rho : \mathcal{L}^\infty \to \mathbb{R}$ which is defined by

$$\rho(\alpha X + (1-\alpha)Y) \ge \alpha\rho(X) + (1-\alpha)\rho(Y), \text{ for all } \alpha \in [0,1] \text{ and r.v. } X, Y \in \mathcal{L}^\infty.$$

It has been shown in [1] that any concave risk measure on a non-atomic probability space is Schur-concave. However, only when U is linear or exponential, the certainty equivalent ρ_U is also concave (see [16]).

Hence maximizing the certainty equivalent means to prefer risky positions with higher reward and lower variance. In the classical risk neutral setting, the variance of a position does not enter the decision criterion. This may sometimes lead to a very risky behavior (as e.g. demonstrated in [3] for a casino game). Finally let us consider the following axioms which are often imposed on a risk measure $\rho : \mathcal{L}^\infty \to \mathbb{R}$:

- **(translation-invariance)** $\rho(X + m) = \rho(X) + m$ for any constant $m \in \mathbb{R}$,
- **(positive homogeneity)** $\rho(\alpha X) = \alpha\rho(X)$ for any constant $\alpha > 0$.

It has been shown in [16] that ρ_U is translation-invariant essentially if and only if U is linear or exponential and ρ_U is positive homogeneous if and only if U is the logarithm or a power function (including the linear function).

More general there exists an axiomatic characterization of certainty equivalents which is know as *Nagumo-Kolmogorov-de Finetti theorem*. It states that a functional $\rho : \mathcal{L}^\infty \to \mathbb{R}$ is a certainty equivalent if and only if it is monotone, law invariant, quasi-linear and has the constancy property. For an early proof see [12]. For a more recent treatment see [16].

Another motivation for certainty equivalents is the fact that they may be written (assuming enough regularity of U) as in (1). Thus the certainty equivalent is approximately a weighted criterion of expectation and variance.

An important special case is obtained when

$$U(x) = \frac{1}{\gamma} e^{\gamma x}, \ x \in \mathbb{R}, \text{ with } \gamma < 0$$

is an exponential function. In this case

$$\rho_U(X) = \frac{1}{\gamma} \ln \mathbb{E} e^{\gamma X}$$

is the *entropic risk measure*. In this case if $X \sim \mathcal{N}(\mu, \sigma^2)$ then we obtain by direct calculation that $\rho_U(X) = \mu + \frac{1}{2}\gamma\sigma^2$. The entropic risk measure also has a dual representation given by

$$\rho_U(X) = \inf_{\mathbb{Q} \ll \mathbb{P}} \Big\{ \mathbb{E}_{\mathbb{Q}}[X] - \frac{1}{\gamma} \mathbb{E}_{\mathbb{Q}}\Big[\ln \frac{d\mathbb{Q}}{d\mathbb{P}}\Big] \Big\},$$

where the infimum is attained at

$$\mathbb{Q}^*(dz) = \frac{e^{\gamma z} \mathbb{P}(dz)}{\int e^{\gamma y} \mathbb{P}(dy)}.$$

Thus we can interpret the optimization problem in this case as a game against nature where the second player (nature) chooses the probability measure.

3 Risk-Sensitive Stopping Problems

We suppose that a *partially observable Markov Process* is given which we introduce as follows: We denote this process by $(X_n, Y_n)_{n \in \mathbb{N}_0}$ and assume a Borel state space $E_X \times E_Y$. The x-component will be the *observable* part, the y-component *cannot be observed* by the controller. There is a stochastic transition kernel Q from $E_X \times E_Y$ to $E_X \times E_Y$ which determines the distribution of the new state pair given the current state. So $Q(B|x, y)$ is the probability that the next state pair is in $B \in \mathcal{B}(E_X \times E_Y)$, given the current state is (x, y). In what follows we assume that the transition kernel Q has a density q with respect to some σ-finite measures λ and ν, i.e.

$$Q(B|x, y) = \int_B q(x', y'|x, y) \lambda(dx') \nu(dy'), \quad B \in \mathcal{B}(E_X \times E_Y).$$

For convenience we introduce the marginal transition kernel density by

$$q^X(x'|x,y) := \int_{E_Y} q(x',y'|x,y)\nu(dy').$$

We assume that the initial distribution Q_0 of Y_0 is known. For a fixed (observable) initial state $x \in E_X$, the initial distribution Q_0 together with the transition probability Q define by the Theorem of Ionescu Tulcea a probability measure $\mathbb{P}_x$ on $(E_X \times E_Y)^{N+1}$ endowed with the product σ-algebra. More precisely $\mathbb{P}_{xy}$ is the probability measure given $X_0 = x$ and $Y_0 = y$ and we define $\mathbb{P}_x(\cdot) := \int P_{xy}(\cdot)Q_0(dy)$.

Next we have a measurable one-stage reward function $c : E_X \to \mathbb{R}$ which depends only on the observable part of the process. Since $c < 0$ is possible this could also be a cost. This reward (cost) is obtained as long as the process is not stopped. A reward $g : E_X \to \mathbb{R}$ is obtained when the process is stopped. Stopping-times are taken w.r.t. the observable filtration $(\mathcal{F}_n)$ where $\mathcal{F}_n = \sigma(X_0, \ldots, X_n)$. We first consider problems with a finite time horizon N and maximize the certainty equivalent of the stopping reward. Since U^{-1} is increasing we can skip it from the optimization problem. Thus, we define

$$J_N(x) := \sup_{0\leq\tau\leq N} \mathbb{E}_x\Big[U\Big(\sum_{k=0}^{\tau-1} c(X_k) + g(X_\tau)\Big)\Big]. \tag{3}$$

In order to have a well-defined problem we assume that for all $x \in E_X$

$$\sup_{0\leq\tau\leq N} \sup_{y\in E_Y} \mathbb{E}_{xy}\Big[\sum_{k=0}^{\tau-1} c^+(X_k) + g^+(X_\tau)\Big] < \infty. \tag{4}$$

Note that since U is concave it can be bounded by a linear function and thus (4) implies that $J_N(x) < \infty$ for all $x \in E_X$. Unfortunately (3) cannot be solved with classical MDP techniques directly. Thus, we introduce the following auxiliary problems. For a probability measure $\mu \in \mathbb{P}(E_Y)$ and a constant $s \in \mathbb{R}$ define for $n = 1, \ldots, N$:

$$V_n(x,\mu,s) := \sup_{0\leq\tau\leq n} \int_{E_Y} \mathbb{E}_{xy}\left[U\Big(\sum_{k=0}^{\tau-1} c(X_k) + s + g(X_\tau)\Big)\right]\mu(dy) \tag{5}$$

In view of (4) $V_n(x,\mu,s) < \infty$ for all (x,μ,s). Obviously we have by this embedding technique that $J_N(x) = V_N(x, Q_0, 0)$. We claim now that (5) can be solved with MDP techniques. In order to do so, we first have to cope with the fact that (Y_n) is not observable. Hence we define the operator $\Phi : E_X \times E_X \times \mathbb{P}(E_Y) \to \mathbb{P}(E_Y)$ by

$$\Phi(x,x',\mu)(B) := \frac{\int_B q^X(x'|x,y)\mu(dy)}{\int_{E_Y} q^X(x'|x,y)\mu(dy)}, \quad B \in \mathcal{B}(E_Y).$$

Note that Φ is exactly the usual updating (Bayesian) operator which appears in classical POMDP (see e.g. [4], Section 5.2). It updates the conditional probability of the unobservable state. In what follows denote by (μ_n) the sequence of probability measures on E_Y generated by Φ with $\mu_0 := Q_0$. I.e. for $n \in \mathbb{N}$, a history $h_n := (x_0, \dots, x_n) \in E_X^{n+1}$ and $B \in \mathcal{B}(E_Y)$ we define

$$\begin{aligned} \mu_0(B|h_0) &:= Q_0(B), \\ \mu_{n+1}(B|h_n, x') &:= \Phi\big(x_n, x', \mu_n(\cdot|h_n)\big)(B) \end{aligned} \tag{6}$$

Then it is well-known (see e.g. [4], Theorem 5.2.1) that

$$\mu_n(B|X_0, \dots, X_n) = \mathbb{P}_x\big(Y_n \in B|X_0, \dots, X_n\big), \quad B \in \mathcal{B}(E_Y).$$

We consider now a stopping problem with a process on the state space $E := E_X \times \mathbb{P}(E_Y) \times \mathbb{R}$. The running reward is zero and the stopping reward is $g(x, \mu, s) := U\big(g(x)+s\big)$. The transition law of the process is given by $\tilde{Q}(\cdot|x, \mu, s)$ which is for $(x, \mu, s) \in E$ and a measurable function $v : E \to \mathbb{R}$ defined by

$$\begin{aligned} &\int_{E_X} \int_{\mathbb{P}(E_Y)} \int_{\mathbb{R}} v(x', \mu', s') \tilde{Q}(d(x', \mu', s')|x, \mu, s) \\ &\quad = \int_{E_X} v\Big(x', \Phi(x, x', \mu), s + c(x)\Big) Q^X(dx'|x, \mu) \end{aligned}$$

where

$$Q^X(B|x, \mu) := \int_B \int_{E_Y} q^X(x'|x, y) \mu(dy) \lambda(dx'). \tag{7}$$

Stopping times are now considered w.r.t. the filtration $(\mathcal{G}_n)$ which is defined by $\mathcal{G}_n = \sigma(X_0, \mu_0, S_0, \dots, X_n, \mu_n, S_n)$ with $S_n := \sum_{k=0}^{n-1} c(X_k)$. However note that obviously by construction of the sequences (μ_n) and (S_n) we obtain $\mathcal{F}_n = \mathcal{G}_n$.

Theorem 1. *It holds:*

a) For $n = 1, \dots, N$ and $(x, \mu, s) \in E_X \times \mathbb{P}(E_Y) \times \mathbb{R}$, the functions V_n are given by

$$\begin{aligned} &V_0(x, \mu, s) = U\big(g(x) + s\big) \\ &V_n(x, \mu, s) = \\ &\max\Big\{U\big(g(x) + s\big), \int_{E_X} V_{n-1}\Big(x', \Phi(x, x', \mu), s + c(x)\Big) Q^X(dx'|x, \mu)\Big\} \end{aligned} \tag{8}$$

The value function of (3) is then given by $J_N(x) = V_N(x, Q_0, 0)$.

b) For every $n = 1, \dots, N$ and $(x, \mu, s) \in E_X \times \mathbb{P}(E_Y) \times \mathbb{R}$, let $f_n^*(x, \mu, s) = 1$ if the maximum in the recursion (8) is taken at $U\big(g(x) + s\big)$ and define $(g_0^*, \dots, g_{N-1}^*)$ by

$$g_n^*(h_n) := f_{N-n}^*\Big(x_n, \mu_n(\cdot|h_n), \sum_{k=0}^{n-1} c(x_k)\Big), \quad n = 0, \dots, N-1.$$

Then the optimal stopping time is given by

$$\tau^* := \inf\{n \in \mathbb{N}_0 : g_n^*(h_n) = 1\} \wedge N.$$

Note that τ^* *depends on the history* h_n *of the process.*

Proof. The proof follows from Theorem 3.3 in [2]. ■

Remark 1. By definition we have $U\big(g(x)+s\big) \leq V_n(x,\mu,s) \leq V_{n+1}(x,\mu,s)$ for all $n \in \mathbb{N}$ and $(x,\mu,s) \in E_X \times \mathbb{P}(E_Y) \times \mathbb{R}$ i.e. the value of the stopping problem increases with the time horizon. This observation is of course obvious, since all stopping times which are feasible for a smaller time horizon are also feasible for a larger time horizon. Note that we use increasing and decreasing in a non-strict sense throughout.

Example 1. Let $U(x) = \frac{1}{\gamma}e^{\gamma x}$ with $\gamma < 0$. In this case Theorem 1 simplifies due to the positive homogeneity of ρ_U and it is easy to see that

$$V_n(x,\mu,s) = e^{\gamma s} h_n(x,\mu)$$

and the h_n satisfy the recursion

$$h_0(x,\mu) = \frac{1}{\gamma}e^{\gamma g(x)}$$
$$h_n(x,\mu) = \max\Big\{\frac{1}{\gamma}e^{\gamma g(x)}, e^{\gamma c(x)} \int_{E_X} h_{n-1}\Big(x', \Phi(x,x',\mu)\Big) Q^X(dx'|x,\mu)\Big\}.$$

As a result, in the exponential case the state space is reduced by one variable and the state of the auxiliary problem consists only of x and the conditional distribution of y given the history h_n. In particular the optimal stopping rule depends on the history only through the conditional probability of the unobservable random variable. The same situation arises in the classical risk-neutral stopping problem.

Finally we discuss the influence of the risk attitude of the decision maker on the optimal stopping time. We use the Arrow-Pratt function of absolute risk aversion (2) to measure the risk sensitivity. A utility function U is said to be *more risk averse* than a utility function W if $l_U(x) \geq l_W(x)$ for all $x \in \mathbb{R}$. For our purpose it is crucial to note that a utility function U is more risk averse than a utility function W if and only if, there exits an increasing concave function $r : \mathbb{R} \to \mathbb{R}$ such that $U = r \circ W$. In what follows we denote all quantities which refer to utility function U by $g_n^*(h_n, U), V_n(x,\mu,s,U)$ etc. and similar for W.

Theorem 2. *Suppose that the utility function* U *is more risk averse than the utility function* W*. For all* $n = 0, 1, \ldots, N-1$ *and histories* h_n *we obtain that* $g_n^*(h_n, W) = 1$ *implies* $g_n^*(h_n, U) = 1$*, i.e. a more risk-averse decision maker will not stop later.*

Proof. Let r be such that $U = r \circ W$. We first prove by induction on n that $V_n(x,\mu,s,U) \le r \circ V_n(x,\mu,s,W)$ for all (x,μ,s) and n. First for $n = 0$ we have

$$V_0(x,\mu,s,U) = U\big(g(x)+s\big) = r \circ W\big(g(x)+s\big) = r \circ V_0(x,\mu,s,W).$$

Using the Jensen inequality, the induction hypothesis and the fact that r is increasing and concave we obtain

$$\begin{aligned}
&V_n(x,\mu,s,U) = \\
&= \max\Big\{U\big(g(x)+s\big), \int_{E_X} V_{n-1}\Big(x', \Phi(x,x',\mu), s+c(x), U\Big) Q^X(dx'|x,\mu)\Big\} \\
&\le \max\Big\{r \circ W\big(g(x)+s\big), \int_{E_X} r \circ V_{n-1}\Big(x', \Phi(x,x',\mu), s+c(x), W\Big) Q^X(dx'|x,\mu)\Big\} \\
&\le \max\Big\{r \circ W\big(g(x)+s\big), r \circ \int_{E_X} V_{n-1}\Big(x', \Phi(x,x',\mu), s+c(x), W\Big) Q^X(dx'|x,\mu)\Big\} \\
&= r \circ \max\Big\{W\big(g(x)+s\big), \int_{E_X} V_{n-1}\Big(x', \Phi(x,x',\mu), s+c(x), W\Big) Q^X(dx'|x,\mu)\Big\} \\
&= r \circ V_n(x,\mu,s,W).
\end{aligned}$$

This implies in particular that

$$W\big(g(x)+s\big) \ge \int_{E_X} V_{n-1}\Big(x', \Phi(x,x',\mu), s+c(x), W\Big) Q^X(dx'|x,\mu)$$

leads to

$$\begin{aligned}
&U\big(g(x)+s\big) = \\
&= r \circ W\big(g(x)+s\big) \ge r \circ \int_{E_X} V_{n-1}\Big(x', \Phi(x,x',\mu), s+c(x), W\Big) Q^X(dx'|x,\mu) \\
&\ge \int_{E_X} r \circ V_{n-1}\Big(x', \Phi(x,x',\mu), s+c(x), W\Big) Q^X(dx'|x,\mu) \\
&\ge \int_{E_X} V_{n-1}\Big(x', \Phi(x,x',\mu), s+c(x), U\Big) Q^X(dx'|x,\mu).
\end{aligned}$$

By definition this means that $f_n^*(x,\mu,s,W) = 1$ implies $f_n^*(x,\mu,s,U) = 1$. Transferring this observation to g_n^* implies the statement. ■

4 Risk-Sensitive Bayesian House Selling Problem

Suppose someone wants to sell her house. We assume that offers $X_0, \dots, X_N$ arrive independently and are identically distributed with distribution Q_θ. Here $\theta \in \Theta$ is an unknown parameter. We assume that Q_θ has a density $q(x|\theta)$ and some prior distribution Q_0 for θ is given. As long as offers are rejected a maintenance cost of $c > 0$ has to be paid. When an offer is accepted, the price is obtained and the game ends. The aim is to find the maximal risk-sensitive stopping reward

$$J_N(x) := \sup_{0 \le \tau \le N} \int_\Theta \mathbb{E}_{x\theta}\Big[U\Big(X_\tau - c\tau\Big)\Big] Q_0(d\theta). \tag{9}$$

This is a special case of our general model with $Y_n \equiv \theta$, $c(x) = -c$ and $g(x) = x$. The integrability condition (4) reduces to

$$\sup_\theta \mathbb{E}_\theta[X_1^+] < \infty.$$

Note that this problem without partial observation has been investigated in [17]. According to Theorem 1 we obtain J_N by computing the functions V_n. These are given by

$$\begin{aligned} V_0(x, \mu, s) &= U\big(x + s\big) \\ V_n(x, \mu, s) &= \max\Big\{U\big(x + s\big), c_n(\mu, s)\Big\} \end{aligned}$$

with $c_n(\mu, s) := \int_{\mathbb{R}} V_{n-1}\Big(x', \Phi(x', \mu), s - c\Big)Q(dx'|\mu)$ and we have that $J_N(x) = V_N(x, Q_0, 0)$. Note that $Q(dx'|\mu)$ corresponds to $Q^X(dx'|x, \mu)$ in (7) and is thus given by

$$Q(\cdot|\mu) = \int_\Theta Q_\theta(\cdot)\mu(d\theta)$$

and the updating operator simplifies to

$$\Phi(x, \mu)(B) := \frac{\int_B q(x|\theta)\mu(d\theta)}{\int_\Theta q(x|\theta)\mu(d\theta)}, \quad B \in \mathcal{B}(\Theta).$$

Moreover, when we define $f_n^*(x, \mu, s) = 1$ if $U\big(x+s\big) \ge c_n(\mu, s)$ and $(g_0^*, \ldots, g_{N-1}^*)$ by

$$g_n^*(h_n) := f_{N-n}^*\big(x_n, \mu_n(\cdot|h_n), -nc\big), \quad n = 0, \ldots, N-1,$$

then the optimal stopping time for problem (9) is given by

$$\tau^* := \inf\{n \in \mathbb{N}_0 : g_n^*(h_n) = 1\} \wedge N.$$

Let us now further investigate the optimal stopping time. We have

$$g_n^*(h_n) = 1 \quad \Leftrightarrow \quad U(x_n - nc) \ge c_{N-n}(\mu_n(\cdot|h_n), -nc).$$

When we define

$$\begin{aligned} U_n(x) &:= U(x - nc) \\ d_k(\mu, U_n) &:= c_k(\mu, -nc) = \int_{\mathbb{R}} V_{k-1}\Big(x', \Phi(x', \mu), -(n+1)c\Big)Q(dx'|\mu), \end{aligned}$$

then we obtain

$$g_n^*(h_n) = 1 \quad \Leftrightarrow \quad x_n \geq U_n^{-1}\big(d_{N-n}(\mu_n(\cdot|h_n), U_n)\big) =: x_{n,N}^*(\mu_n(\cdot|h_n)).$$

Note that $U_n^{-1}(x) = nc + U^{-1}(x)$. We call $x_{n,N}^*(\cdot)$ *reservation level.* The reservation levels depend on μ_n and U. The optimal stopping time is hence the first time, the offer exceeds the corresponding, history dependend reservation level.

Theorem 3. *a) The optimal stopping time for the Bayesian house selling problem is given by*

$$\tau^* = \inf\big\{n \in \mathbb{N}_0 : X_n \geq x_{n,N}^*(\mu_n(\cdot|h_n))\big\} \wedge N.$$

b) The reservation levels can recursively be computed by

$$x_{N-1,N}^*(\mu) = U_{N-1}^{-1} \circ \int_{\mathbb{R}} U_N(x) Q(dx|\mu)$$

$$x_{n,N}^*(\mu) = U_n^{-1} \circ \int_{\mathbb{R}} U_{n+1}\Big(\max\big\{x, x_{n+1,N}^*(\Phi(x,\mu))\big\}\Big) Q(dx|\mu).\square$$

Proof. Part a) is clear from the definition and the previous results. Part b) can be shown by inserting the correct definitions. For $n = N-1$ we obtain from the definition of $x_{N-1,N}^*(\mu)$ that

$$x_{N-1,N}^*(\mu) = U_{N-1}^{-1}\big(d_1(\mu, U_{N-1})\big)$$

with

$$d_1(\mu, U_{N-1}) = \int V_0\big(x, \Phi(x,\mu), -Nc\big) Q(dx|\mu) = \int U_N(x) Q(dx|\mu).$$

For $x_{n,N}^*$ we obtain by definition:

$$x_{n,N}^*(\mu) = U_n^{-1}\big(d_{N-n}(\mu, U_n)\big).$$

Further $d_{N-n}(\mu, U_n)$ can be written as

$$\begin{aligned}
d_{N-n}(\mu, U_n) &= \int_{\mathbb{R}} V_{N-n-1}\Big(x, \Phi(x,\mu), -(n+1)c\Big) Q(dx|\mu) \\
&= \int_{\mathbb{R}} \max\Big\{U(x-(n+1)c), \\
&\qquad \int V_{N-n-2}\Big(x', \Phi(x', \Phi(x,\mu)), -(n+2)c\Big) Q(dx'|\Phi(x,\mu))\Big\} Q(dx|\mu) \\
&= \int_{\mathbb{R}} \max\Big\{U_{n+1}(x), d_{N-n-1}(\Phi(x,\mu), U_{n+1})\Big\} Q(dx|\mu) \\
&= \int_{\mathbb{R}} U_{n+1}\Big(\max\big\{x, U_{n+1}^{-1} \circ d_{N-n-1}(\Phi(x,\mu), U_{n+1})\big\}\Big) Q(dx|\mu)
\end{aligned}$$

and the statement follows from the definition of $x_{n+1,N}^*$. ■

Remark 2. It obviously holds that

$$d_N(\mu, U) = \int_{\mathbb{R}} V_{N-1}\Big(x, \Phi(x,\mu), -c\Big) Q(dx|\mu)$$
$$= \sup_{1\le\tau\le N} \int_{\Theta} \int_{\mathbb{R}} \mathbb{E}_{x\theta}\big[U(X_\tau - c\tau)\big] Q_\theta(dx)\mu(d\theta),$$

i.e. $d_N(Q_0, U)$ is the value of the stopping problem when we start without known initial offer.

Example 2. In case $U(x) = \frac{1}{\gamma}e^{\gamma x}$ with $\gamma < 0$ the recursion for the reservation levels simplifies. In order to see this, note that $U_n(x) = U(x - nc) = \frac{1}{\gamma}e^{\gamma x} \cdot e^{-\gamma nc}$ and $U_n^{-1}(x) = U^{-1}(x) + nc = \frac{1}{\gamma}\ln(\gamma x) + nc$. With these observation we obtain

$$x^*_{N-1,N}(\mu) = -c + \frac{1}{\gamma}\ln\Big(\int_{\mathbb{R}} e^{\gamma x} Q(dx|\mu)\Big)$$
$$x^*_{n,N}(\mu) = -c + \frac{1}{\gamma}\ln\Big(\int_{\mathbb{R}} e^{\gamma \max\left\{x, x^*_{n+1,N}(\Phi(\mu,x))\right\}} Q(dx|\mu)\Big).$$

Note that in contrast to the $x^*_{n,N}$ with general utility function, the reservation levels in the exponential utility case depend only on the time difference to the planning horizon. More precisely we could also define $x^*_n(\mu) := x^*_{N-n,N}(\mu)$ and obtain

$$x^*_1(\mu) = -c + \frac{1}{\gamma}\ln\Big(\int_{\mathbb{R}} e^{\gamma x} Q(dx|\mu)\Big)$$
$$x^*_n(\mu) = -c + \frac{1}{\gamma}\ln\Big(\int_{\mathbb{R}} e^{\gamma \max\left\{x, x^*_{n-1}(\Phi(\mu,x))\right\}} Q(dx|\mu)\Big).$$

In particular we can start computing the reservation levels without fixing a planning horizon in advance.

We did this in the following *numerical example*, where $Q_\theta = B(1, \theta)$ is a Bernoulli distribution with unknown 'success' probability θ, the prior distribution of θ is uniform on $[0, 1]$ and $c = 0.1$. For a planning horizon of $N = 10$ we computed the optimal stopping rule. Of course an offer of 1 will always be accepted. An offer of 0 may be accepted when we fear the accumulation of cost. It turns out that this decision heavily depends on the risk aversion parameter $\gamma < 0$. The smaller γ, the more risk averse the decision maker is. **Figure 1** has to be read as follows: When $\gamma < -2.2$, the decision maker will stop immediately. For $-2.2 < \gamma < -1.51$ she will reject at least the first offer when it is zero. For $-1.51 < \gamma < -1.1$ she will reject the first two offers when they are zero. And so on. The switch from rejecting 8 to 9 zeros is for γ smaller than -10^{-8}.

4.1 Influence of the Filter on the Reservation Levels

In order to discuss the influence of the filter on the reservation levels we make some further simplifying assumptions. Indeed it is often the case that the filter

$-\infty < \gamma < -2.2$	$-2.2 < \gamma < -1.51$	$-1.51 < \gamma < -1.1$	$-1.1 < \gamma < -0.8$
0	1	2	3
$-0.8 < \gamma < -0.56$	$-0.56 < \gamma < -0.34$	$-0.34 < \gamma < -0.18$	$-0.18 < \gamma < -0.03$
4	5	6	7

Fig. 1. Optimal number of zeros which are rejected.

$\mu_n(\cdot|h_n)$ does only depend on a part of the history h_n or on a certain function of it. In what follows we assume that there is a (Borel) information set I endowed with a σ-algebra such that there exist a measurable function $t_n : H_n \to I$ and a transition kernel $\hat{\mu}$ from I to Θ such that

$$\mu_n(B|h_n) = \hat{\mu}(B|t_n(h_n)), \quad \text{for } B \in \mathcal{B}(E_Y).$$

The function t_n is sometimes called *sufficient statistics.* Further we assume that there is a measurable mapping $\hat{\Phi} : I \times E_X$ such that

$$t_{n+1}(h_{n+1}) = \hat{\Phi}(x_{n+1}, t_n(h_n))$$

and that $\hat{\mu}$ has a density $\hat{p}(\cdot|i)$. Thus, we have to replace μ_n by the current information state and obtain in particular for the reservation levels:

$$x^*_{N-1,N}(i) := U^{-1}_{N-1} \circ \int_{\mathbb{R}} U_N(x) Q(dx|i)$$
$$x^*_{n,N}(i) := U^{-1}_n \circ \int_{\mathbb{R}} U_{n+1}\Big(\max\left\{x, x^*_{n+1,N}(\hat{\Phi}(x,i))\right\}\Big) Q(dx|i)$$

where

$$Q(B|i) := \int_B \int_\Theta q(x|\theta)\hat{\mu}(d\theta|i)\lambda(dx).$$

The next step is to introduce an order relation on the set I where we assume now that $\Theta \subset \mathbb{R}$. We define here for $i, i' \in I$

$$i \le i' \quad :\Leftrightarrow \quad \hat{\mu}(\cdot|i) \le_{lr} \hat{\mu}(\cdot|i')$$

where $\le_{lr}$ is the likelihood ratio ordering which is defined by

$$\hat{\mu}(\cdot|i) \le_{lr} \hat{\mu}(\cdot|i') \quad \Leftrightarrow \quad \frac{\hat{p}(\theta|i')}{\hat{p}(\theta|i)} \quad \text{is increasing in } \theta.$$

Note that $\hat{p}$ is the density of $\hat{\mu}$. The likelihood ratio ordering implies the stochastic ordering. Now we are able to formulate the main result of this subsection

Theorem 4. *Suppose that $q(x|\theta)$ is MTP_2, i.e. $q(\cdot|\theta) \le_{lr} q(\cdot|\theta')$ for all $\theta \le \theta'$, then the reservation levels $x^*_{n,N}(i)$ are increasing in i.*

Proof. We prove the statement by induction on n. First consider

$$x^*_{N-1,N}(i) = U_{N-1}^{-1} \circ \int_{\mathbb{R}} U_N(x) \int_{\Theta} q(x|\theta)\hat{\mu}(d\theta|i)\lambda(dx).$$

Since $x \mapsto U_N(x)$ is increasing and the $\leq_{lr}$ implies the $\leq_{st}$ order we obtain that

$$\theta \mapsto \int_{\mathbb{R}} U_N(x)q(x|\theta)\lambda(dx) =: f(\theta)$$

is increasing in θ. Thus by the definition of $i \leq i'$ we obtain that $\int_{\Theta} f(\theta)\hat{\mu}(d\theta|i)$ is increasing in i. Now suppose the statement is true for $x^*_{n+1,N}(i)$. Next note that by our assumption on $q(x|\theta)$ we have that

$$(i, x) \mapsto \hat{\Phi}(x, i)$$

is increasing. This follows from Lemma 5.4.9 in [4]. Hence

$$(i, x) \mapsto U_{n+1}\Big(\max \big\{x, x^*_{n+1,N}(\hat{\Phi}(x, i))\big\}\Big) =: f(i, x)$$

is increasing. Next by our assumption on $q(x|\theta)$ we have that

$$(i, \theta) \mapsto \int_{\mathbb{R}} f(i, x)q(x|\theta)\lambda(dx) =: \hat{f}(i, \theta)$$

is increasing. And finally we obtain that

$$x^*_{n,N}(i) = U_n^{-1} \circ \int_{\Theta} \hat{f}(i, \theta)\hat{\mu}(d\theta|i)$$

is increasing which completes the induction. ■

For further details and examples we refer the reader to Section 5.4 in [4].

Example 3. In this example we consider the special case of exponentially distributed random variables (offers)

$$q(x|\theta) = \frac{1}{\theta}e^{-\frac{1}{\theta}x}, \quad x \geq 0,\ \theta \in \Theta := (0, \infty).$$

According to our definition of μ_n we get by recursion for $h_n = (x_1, \dots, x_n)$ and $B \in \mathcal{B}(\Theta)$

$$\mu_n(B|h_n) = \frac{\int_B \frac{1}{\theta^n} \exp(-\frac{1}{\theta}\sum_{k=1}^n x_k)Q_0(d\theta)}{\int_{\Theta} \frac{1}{\theta^n} \exp(-\frac{1}{\theta}\sum_{k=1}^n x_k)Q_0(d\theta)}.$$

Thus $t_n(x_1, \dots, x_n) = \left(\sum_{k=1}^n x_k, n\right)$ is a sufficient statistic. Thus, we have $I := \mathbb{R}_+ \times \mathbb{N}_0$ and denote $i = (s, n) \in I$. Moreover, $\hat{\Phi}(x, (s, n)) = (s + x, n + 1)$ and the conditional distribution of the unknown parameter has the form

$$\hat{\mu}(d\theta|s, n) \propto \left(\frac{1}{\theta}\right)^n e^{-\frac{s}{\theta}}Q_0(d\theta)$$

if the information (s, n) is given. With this representation it is not difficult to verify that

$$i = (s, n) \leq i' = (s', n') \quad \Leftrightarrow \quad s \leq s' \text{ and } n \geq n'.$$

Further, the family of densities $q(x|\theta)$ is MTP_2 in $x \geq 0$ and $\theta \geq 0$, thus Theorem 4 applies.

If we assume now a special prior distribution of the unknown parameter then we obtain an explicit distribution for $Q(\cdot|i)$. We assume that the prior distribution Q_0 is a so-called *Inverse Gamma distribution*, i.e. the density is given by

$$Q_0(d\theta) = \frac{b^a}{\Gamma(a)}\left(\frac{1}{\theta}\right)^{a+1} e^{-\frac{b}{\theta}} d\theta, \quad \theta > 0$$

where $a > 1$ and $b > 0$ are fixed. Then the distribution Q is given by

$$Q\big(dx|s, n\big) = \int q(dx|\theta)\hat{\mu}\big(d\theta|s, n\big) = (n + a)\frac{(s + b)^{n+a}}{(x + s + b)^{n+a+1}}dx.$$

Hence Q is a special *Second Order Beta distribution.* In a risk-neutral situation a similar setting has been considered in [22].

4.2 Influence of the Utility Function on the Reservation Levels

Here we proceed as in [17] (Theorem 3.3) in order to study the impact of the risk aversion on the reservation levels. As in section 3 we use the Arrow-Pratt function of absolute risk aversion (2) to measure the risk attitude in the sense that a utility function U is more risk averse than a utility function W if $l_U(x) \geq l_W(x)$ for all $x \in \mathbb{R}$. In what follows we denote by $x^*_{n,N}(\mu, U)$ the reservation levels which belong to the utility function U. Then we obtain

Theorem 5. *If U is a more risk averse utility function than W, then the reservation levels satisfy $x^*_{n,N}(\mu, U) \leq x^*_{n,N}(\mu, W)$ for all $n = 0, 1 \ldots, N-1$ and all $\mu \in \mathbb{P}(E_Y)$.*

Proof. The proof follows from Theorem 2. ■

Remark 3. Theorem 5 includes as a special case the comparison to the risk neutral stopping problem: Suppose U is an arbitrary increasing concave utility function, then obviously $U = U \circ \mathrm{id}$. Thus, we choose $r = U$ in this context and see that the reservation levels of a risk neutral decision maker will always be above the reservation levels of a risk averse decision maker.

Example 4. In case $U(x) = \frac{1}{\gamma}e^{\gamma x}$ with $\gamma < 0$ we obtain from Theorem 5 that the reservation levels $x^*_n(\mu)$ are increasing in γ.

4.3 Influence of the Time Horizon on the Reservation Levels

In a risk neutral setting it is often the case that the reservation levels are decreasing as time goes by, i.e. the decision maker becomes less selective when she approaches the time horizon. However in [17] it has been shown that this is no longer true in a risk averse setting. Indeed he gave some examples where the reservation levels are increasing.

Without additional assumptions it is difficult to determine how the reservation levels behave in time. However in general the reservation levels satisfy the following relation.

Theorem 6. *For all $n = 0, 1, \dots, N-2$ and $\mu \in \mathbb{P}(E_Y)$ it holds that*

$$x^*_{n,N}(\mu) \geq nc + \rho_U\Big(x^*_{n+1,N}(\Phi(X_1,\mu)) - (n+1)c\Big).$$

In case $U(x) = \frac{1}{\gamma}e^{\gamma x}$ with $\gamma < 0$ we obtain

$$x^*_{n,N}(\mu) + c \geq \rho_U\Big(x^*_{n+1,N}(\Phi(X_1,\mu))\Big).$$

Proof. From Theorem 3 b) we obtain

$$\begin{aligned} x^*_{n,N}(\mu) &= U_n^{-1} \circ \int_{\mathbb{R}} U_{n+1}\Big(\max\big\{x, x^*_{n+1,N}(\Phi(x,\mu))\big\}\Big)Q(dx|\mu) \\ &\geq U_n^{-1} \circ \int_{\mathbb{R}} U_{n+1}\Big(x^*_{n+1,N}(\Phi(x,\mu))\Big)Q(dx|\mu) \\ &= nc + U^{-1} \circ \int_{\mathbb{R}} U\Big(x^*_{n+1,N}(\Phi(x,\mu)) - (n+1)c\Big)Q(dx|\mu)\big). \end{aligned}$$

Using the definition of the certainty equivalent yields the first statement. For the second statement note that ρ is translation invariant in the case of exponential utility. ■

In order to obtain decreasing reservation levels further assumptions are necessary. For example the property that the utility function has a *decreasing absolute risk aversion* (DARA). This means that $x \mapsto l_U(x)$ is decreasing, i.e. the decision maker becomes more risk averse with decreasing wealth.

Theorem 7. *a) The reservation levels $x^*_{n,N}(\mu)$ are increasing in N for all $\mu \in \mathbb{P}(E_Y)$.*

b) If the utility function U is DARA, then $x^*_{n,N}(\mu)$ are decreasing in n for all $\mu \in \mathbb{P}(E_Y)$.

Proof.

a) It is obvious that $x^*_{N-1,N}$ is increasing in N. Now suppose that $x^*_{n+1,N}$ is increasing in N. Then due to the recursion of the reservation levels in Theorem 3 and the fact that U is increasing we obtain that $x^*_{n,N}$ is increasing in N.

b) Now suppose the utility function U is DARA. Thus, in particular $U_1(x) = U(x-c)$ is more risk averse than U. Then we obtain with Theorem 5 that

$$x^*_{n,N}(\mu, U) \geq x^*_{n,N}(\mu, U_1) = x^*_{n+1,N}(\mu)$$

which implies the result. ■

Example 5. In case $U(x) = \frac{1}{\gamma}e^{\gamma x}$ with $\gamma < 0$ we obtain from Theorem 7 that the reservation levels $x^*_n(\mu)$ are increasing in n. This effect can also be seen in the numerical example 2.

5 Risk-Sensitive Stopping Problems with Infinite Time Horizon

Let us now consider the risk-sensitive stopping problem from Section 3 with infinite time horizon. Here we assume that the stopping reward g is bounded, i.e. $\underline{g} \leq g \leq \bar{g}$ and cost are strictly negative, i.e. $\sup_{x \in E_X} c(x) =: \bar{c} < 0$. Thus we consider

$$J_\infty(x) := \sup_{\tau < \infty} \mathbb{E}_x \Big[U\Big(\sum_{k=0}^{\tau-1} c(X_k) + g(X_\tau) \Big) \Big] \tag{10}$$

where the supremum is taken over all $(\mathcal{F}_n)$-stopping times τ with $\mathbb{P}_x(\tau < \infty) = 1$ for all $x \in E_X$. This problem can be seen as the limiting problem of stopping problems with bounded horizon.

Theorem 8. *The sequence (V_n) of value functions defined in* (5) *has a limit V and this limit satisfies*

$$V(x, \mu, s) = \max\Big\{ U\big(g(x) + s\big), \int_{E_X} V\big(x', \Phi(x, x', \mu), s + c(x)\big) Q(dx'|\mu) \Big\}. \tag{11}$$

Proof. We have $V_n \leq V_{n+1}$ and the sequence V_n is bounded from above by our assumptions, thus the limit $V = \lim_{n\to\infty} V_n$ exists. Moreover, we can take the limit on both sides in the recursion (8) for V_n to obtain the fixed point property of V. ■

Next we show the relation between J_∞ and V.

Theorem 9. *a) It holds that $V(x, Q_0, 0) = J_\infty(x)$.*

b) Let $f^*(x, \mu, s) = 1$ if the maximum in (11) *is attained at $U\big(g(x) + s\big)$ and define $(g^*_0, g^*_1 \ldots)$ by*

$$g^*_n(h_n) := f^*\Big(x_n, \mu_n(\cdot|h_n), \sum_{k=0}^{n-1} c(x_k) \Big), \quad n \in \mathbb{N}_0.$$

Then the optimal stopping time τ^* is given by

$$\tau^* := \inf\{ n \in \mathbb{N}_0 : g^*_n(h_n) = 1 \}.$$

Proof.

a) First note that we have $V_n(x,Q_0,0) \le J_\infty(x)$ for all $n \in \mathbb{N}$ which implies that $V(x,Q_0,0) \le J_\infty(x)$. Next for every admissible stopping time τ with $\mathbb{P}_x(\tau < \infty) = 1$ and $\mathbb{E}_x\Big[\big|U\Big(\sum_{k=0}^{\tau-1} c(X_k) + g(X_\tau)\Big)\big|\Big] < \infty$ it holds for any $n \in \mathbb{N}$ that

$$\begin{aligned} V(x,Q_0,0) &\ge V_n(x,Q_0,0) \ge \\ &\ge \mathbb{E}_x\Big[U\Big(\sum_{k=0}^{(\tau\wedge n)-1} c(X_k) + g(X_{\tau\wedge n})\Big)\Big] \\ &\ge \mathbb{E}_x\Big[U\Big(\sum_{k=0}^{\tau-1} c(X_k) + g(X_{\tau\wedge n})\Big)\Big] \\ &= \mathbb{E}_x\Big[U\Big(\sum_{k=0}^{\tau-1} c(X_k) + g(X_\tau)\Big)1_{[\tau\le n]}\Big] + \mathbb{E}_x\Big[U\Big(\sum_{k=0}^{\tau-1} c(X_k) + g(X_n)\Big)1_{[\tau> n]}\Big] \end{aligned}$$

Then letting $n \to \infty$ yields with dominated convergence and the fact that $\mathbb{P}_x(\tau < \infty) = 1$

$$V(x,Q_0,0) \ge \mathbb{E}_x\Big[U\Big(\sum_{k=0}^{\tau-1} c(X_k) + g(X_\tau)\Big)\Big].$$

Taking the supremum over all admissible stopping times and combining the result with the first inequality implies the result.

b) Iterating the fixed point equation n-times and using the definition of τ^* we obtain

$$\begin{aligned} V(x,Q_0,0) = \mathbb{E}_x\Big[U\Big(\sum_{k=0}^{\tau^*-1} c(X_k) + g(X_{\tau^*})\Big)1_{[\tau^*\le n]}\Big] + \\ + \mathbb{E}_x\Big[V\Big(X_n, \mu_n(\cdot|H_n), \sum_{k=0}^{n-1} c(X_k)\Big)1_{[\tau^*> n]}\Big]. \end{aligned} \tag{12}$$

First we show that $\mathbb{P}_x(\tau^* < \infty) = 1$ for all $x \in E_X$. We can extend part a) easily to arbitrary states (x,μ,s), i.e.

$$V(x,\mu,s) = \sup_{\tau<\infty} \int_{E_Y} \mathbb{E}_{xy}\Big[U\Big(\sum_{k=0}^{\tau-1} c(X_k) + g(X_\tau) + s\Big)\Big]\mu(dy).$$

Then we obtain

$$V\Big(x_n, \mu_n(\cdot|h_n), \sum_{k=0}^{n-1} c(x_k)\Big) \le U\big(n\bar{c} + \bar{g}\big)$$

and from (12)

$$\begin{aligned} U\big(g(x)\big) &\le V(x, Q_0, 0) \le \\ &\le U(\bar g)\mathbb{P}_x(\tau^* \le n) + U(n\bar c + \bar g)\mathbb{P}_x(\tau^* > n), \end{aligned}$$

i.e.

$$\begin{aligned} U\big(g(x)\big) &\le U(\bar g) + \Big(U(n\bar c + \bar g) - U(\bar g)\Big)\mathbb{P}_x(\tau^* > n) \\ &= U(\bar g) + a_n \mathbb{P}_x(\tau^* > n) \end{aligned}$$

where $a_n := U(n\bar c + \bar g) - U(\bar g)$. It holds that $a_n < 0$ for all $n \in \mathbb{N}$ large enough and $\lim_{n\to\infty} a_n = -\infty$. Hence in total

$$\mathbb{P}_x(\tau^* > n) \le \frac{U\big(g(x)\big) - U(\bar g)}{a_n}$$

for all $n \in \mathbb{N}$ large enough which implies $\mathbb{P}_x(\tau^* < \infty) = 1$. Since V is bounded from above, letting $n \to \infty$ in (12) implies that

$$J_\infty(x) = V(x, Q_0, 0) \le \mathbb{E}_x\Big[U\Big(\sum_{k=0}^{\tau^*-1} c(X_k) + g(X_{\tau^*})\Big)\Big]$$

and τ^* is optimal.

■

5.1 Risk-Sensitive Bayesian House Selling Problem

Let us consider the risk-sensitive Bayesian house selling problem with infinite time horizon.When we assume that the offers are bounded, i.e. $X_i \in [m, M]$ our general assumptions of this section are satisfied. As in the finite horizon case we can see from the fixed point equation that the optimal stopping time is characterized by *reservation levels.* Indeed Theorem 8 and Theorem 9 apply directly and the fixed point equation reads

$$V(x, \mu, s) = \max\Big\{U(x+s), \int_{E_X} V\big(x', \Phi(x', \mu), s - c\big)Q(dx'|\mu)\Big\}.$$

Then we obtain

$$\begin{aligned} g_n^*(h_n) = 1 \quad \Leftrightarrow \quad x_n &\ge U_n^{-1}\Big(\int V\big(x', \Phi(x', \mu), -c(n+1)\big)Q(dx'|\mu)\Big) \\ &=: x_{n,\infty}^*(\mu). \end{aligned}$$

Note that the reservation levels in general still depend on the time stage, in contrast to the problem in the risk neutral setting because we have to memorize the cost which has accumulated so far. As for the case of finite time horizon we obtain a similar recursion for the reservation levels.

Theorem 10. *a) The optimal stopping time for the Bayesian house selling problem with infinite time horizon is given by*

$$\tau^* = \inf\big\{n \in \mathbb{N}_0 : X_n \geq x^*_{n,\infty}(\mu_n(\cdot|h_n))\big\}.$$

b) It holds that $x^*_{n,\infty}(\mu) = \lim_{N\to\infty} x^*_{n,N}(\mu)$ *and the reservation levels satisfy the following recursion*

$$x^*_{n,\infty}(\mu) = U_n^{-1} \circ \int_{\mathbb{R}} U_{n+1}\Big(\max\big\{x, x^*_{n+1,\infty}(\Phi(x,\mu))\big\}\Big) Q(dx|\mu).$$

Example 6. In case $U(x) = \frac{1}{\gamma} e^{\gamma x}$ with $\gamma < 0$ the recursion for the reservation levels simplifies to a fixed point equation

$$x^*_\infty(\mu) = -c + \frac{1}{\gamma} \ln\Big(\int_{\mathbb{R}} e^{\gamma \max\{x, x^*_\infty(\Phi(x,\mu))\}} Q(dx|\mu)\Big).$$

Moreover, it holds $x^*_\infty(\mu) = \lim_{n\to\infty} x^*_n(\mu)$. From Example 4 it follows that the reservation levels $x^*_\infty(\mu)$ are increasing in γ.

Also it is possible to discuss the influence of the attitude towards risk as before. Here we use the notation $x^*_{n,\infty}(\mu, U)$ when the reservation level belongs to utility function U.

Theorem 11. *a) Suppose that utility function U is more risk averse than utility function W. Then for all $n \in \mathbb{N}$ we obtain that* $x^*_{n,\infty}(\mu, U) \leq x^*_{n,\infty}(\mu, W)$.
b) If the utility function U is DARA, then $x^*_{n,\infty}(\mu)$ is decreasing in n.

Proof. For part a) we proceed as in the proof of Theorem 5 and show first that $V_n(x,\mu,s,z,U) \leq r \circ V_n(x,\mu,s,z,W)$ for $n \in \mathbb{N}$ where r is such that $U = r \circ W$. Taking the limit yields $V(x,\mu,s,z,U) \leq r \circ V(x,\mu,s,z,W)$ and the statement follows as in the proof of Theorem 5. Part b) follows from part a). ■

6 Conclusion

We have seen that the theory for partially observable risk-sensitive stopping problems is only slightly more complicated than the theory for the risk neutral case. We were also able to show in general that the more risk averse a decision maker is, the later she will stop. Though the numerical algorithms are more demanding than in the risk neutral case, the setting with exponential utility is still feasible.

References

1. Bäuerle, N. and Müller, A.: Stochastic orders and risk measures: consistency and bounds. *Insurance: Mathematics and Economics*, **38** (2006) 132-148.

2. Bäuerle, N. and Rieder, U. *Partially Observable Risk-Sensitive Markov Decision Processes.* Preprint (2015).
3. Bäuerle, N. and Rieder, U.: More risk-sensitive Markov decision processes. *Mathematics of Operations Research*, **39** (1) (2014) 105-120.
4. Bäuerle, N. and Rieder, U. *Markov Decision Processes with Applications to Finance.* Springer-Verlag, Berlin Heidelberg, 2011.
5. Bielecki, T. and Pliska, S.: Economic properties of the risk sensitive criterion for portfolio management. *Review of Accounting and Finance*, **2** (2003) 3-17.
6. Cavazos-Cadena, R. and Hernández-Hernández, D.: Successive approximations in partially observable controlled Markov chains with risk-sensitive average criterion. *Stochastics*, **77** (2005) 537-568.
7. Cavazos-Cadena, R. and Hernández-Hernández, D.: A Characterization of the Optimal Certainty Equivalent of the Average Cost via the Arrow-Pratt Sensitivity Function. *Mathematics of Operations Research* (to appear), 2015.
8. Chow, Y.S., Robbins, H. and Siegmund, D. *Great Expectations: The Theory of optimal Stopping.* houghton Mifflin, Boston, 1971.
9. Dana, R.A..: A representation result for concave Schur concave functions. *Mathematical Finance*, **15** (2005) 613-634.
10. Davis, M.A.H. and Lleo, S. *Risk-Sensitive Investment Management.* World Scientific, 2014.
11. Di Masi, G., and Stettner, L.: Risk sensitive control of discrete time partially observed Markov processes with infinite horizon. *Stochastics* **67**(3-4) (1999) 309-322.
12. Hardy, G.H., Littlewood, J.E. and Pólya, G. *Inequalities.* Cambridge University Press, Cambridge, 1934.
13. Howard R.A. and Matheson J.E.: Risk-sensitive Markov Decision Processes. *Management Science*, **18** (1972) 356–369.
14. James M.R., Baras, J.S. and Elliott, R.J.: Risk-sensitive control and dynamic games for partially observed discrete-time nonlinear systems. *IEEE Transactions on Automatic Control*, **39** (1994) 780-792.
15. Kadota, Y., Kurano M. and Yasuda M.: Utility-optimal stopping in a denumerable Markov chain. *Bulletin of Informatics and Cybernetics*, **28** (1996) 15-21.
16. Müller, A.: Certainty equivalents as risk measures. *Brazilian Journal of Probability and Statistics*, **21** (2007) 1-12.
17. Müller, A.: Expected utility maximization of optimal stopping problems. *European Journal of Operational Research*, **122** (2000) 101-114.
18. Shiryaev, A.N. *Optimal Stopping Rules.* Springer-Verlag, Berlin Heidelberg, 2008.
19. Stettner, L.: Risk sensitive portfolio optmization with completely and partially observed factors. *IEEE Transactions on Automatic Control*, **49** (2004) 457-464.
20. White, D.J.: Utility, probabilistic constraints, mean and varaince of discounted rewards in Markov processes. *OR Spektrum*, **9** (1987) 13-22.
21. Whittle, P. *Risk Sensitive Optimal Control.* Wiley, Chichester, 1990.
22. Tamaki, M.: Optimal selection from a gamma distribution with unknown parameter. *Z. Oper. Res. Ser. A-B*, **28** (1984) 47-57.

Solving the Average Cost Optimality Equation for Unichain Markov Decision Processes: a Linear Programming Approach

François Dufour[1] and Tomás Prieto-Rumeau[2]

[1] INRIA Bordeaux Sud-Ouest, France, dufour@math.u-bordeaux1.fr

[2] UNED, Madrid, Spain, tprieto@ccia.uned.es

Abstract. We study a unichain Markov decision process with finite state and action space, under the average cost optimality criterion. The usual linear programming formulation of the control problem does not necessarily yield a solution to the average cost dynamic programming optimality equation. In this paper we show that the solution to this optimality equation can be obtained by solving two linear programming problems.
Keywords: Markov decision processes, Unichain models, Linear programming.
AMS 2010 subject classification: Primary 90C40, Secondary 90C05.

1 Introduction

We deal with a finite state and action Markov decision process under the long-run expected average cost optimality criterion. This problem has been investigated since the early developments of Markov decision processes. In particular, the dynamic programming optimality equation for such models has been established under various hypotheses, in the multichain, unichain, and irreducible settings. The linear programming formulation of the control problem is also well known for the multichain, unichain, and irreducible cases. In general, the linear programming approach allows to obtain the optimal average cost and to determine, as well, an average cost optimal policy. Here, we are interested in studying the relation between the average cost optimality equation and the linear programming formulation of a unichain Markov decision process.

For such a basic problem as the one studied here, we believe that it is useless to give an extensive list of references. The interested reader can consult, for instance, Chapters 5 and 6 in Kallenberg's lecture notes [1] or Chapter 9 in Puterman's book [3], in which he will find the most important results on average cost unichain Markov decision processes.

The link between the dynamic programming optimality equation and the linear programming formulation of a unichain Markov decision process is not straightforward. To fix ideas, we introduce the average cost optimality equation (although we will formally introduce our model in Section 2, we believe that the notation below is sufficiently standard to be understood), which takes the form

$$g = \min_{a \in A(x)} \left\{ c(x,a) + \sum_{y \in X} Q(x,a,y)h(y) \right\} \quad \text{for all } x \in X. \tag{1}$$

If the pair (g, h) is a solution to this optimality equation, then g equals the constant optimal average cost $g = g^*$ of the control problem. The primal linear program **P** for the control problem is to minimize $\sum_{x,a} c(x,a)z(x,a)$ subject to

$$\sum_{a} z(x,a) = \sum_{y,b} z(y,b)Q(y,b,x) \quad \text{for all } x \in X,$$

with $\sum_{x,a} z(x,a) = 1$ and all $z(x,a) \geq 0$. Its optimal value is g^* and an average optimal policy can be easily determined from an optimal solution of **P**. The dual linear program **D** is to maximize g subject to

$$g + h(x) \leq \min_{a \in A(x)} \left\{ c(x,a) + \sum_{y \in X} Q(x,a,y)h(y) \right\} \quad \text{for each } x \in X,$$

where $g \in \mathbb{R}$ and the $h(x) \in \mathbb{R}$ are unconstrained. Obviously, its optimal value is again g^* but, at optimality, *we do not necessarily obtain a solution of the average optimality equation* (1). This issue is closely related to the existence of transient states for deterministic stationary policies and the existence of average optimal policies that are not canonical. Indeed, if each deterministic stationary policy is recurrent (with no transient states) then the sets of average optimal and canonical policies coincide, and **D** indeed yields a solution of the optimality equation (1). Therefore, our interest mainly focus on unichain non recurrent Markov decision processes.

Notwithstanding that there exist algorithms that yield solutions to the optimality equation (1), such as the policy iteration algorithm and variants of the value iteration algorithm, the above remark is the starting point of our research: *Can the average optimality equation of a unichain Markov decision process be solved using linear programming?* Apparently, the linear problem **D** does not suffice for that purpose. In this paper we show that the answer to that question is yes, but that in order to reach a solution of (1) we must solve *two* linear programs. Namely, our main result shows that we must solve **P** and then a variant of the dual problem **D**, which is somehow parametrized by the optimal solution of the problem **P**.

The rest of the paper is organized as follows. Section 2 settles the notation and gives the definition of the average control problem. Unichain models are studied in Section 3, while our main results on linear programming are given in Section 4. We illustrate our results with an example in Section 5.

2 Preliminaries

2.1 Definition of the Control Model

Consider a Markov decision process model $\mathcal{M}$ with finite state and action spaces given by the following elements:

- The state space X is a finite set.
- The finite action space is A. For each $x \in X$, let $A(x) \subseteq A$ be the nonempty set of actions available when the system is in state x. Define

$$\mathbb{K} = \{(x,a) \in X \times A : a \in A(x)\}.$$

- For each $(x,a) \in \mathbb{K}$, let $Q(x,a,\cdot)$ be the transition probability measure, i.e., $Q(x,a,y) \geq 0$ for all $y \in X$ and $\sum_{y \in X} Q(x,a,y) = 1$.
- The cost function is $c : \mathbb{K} \to \mathbb{R}$.

Consider the canonical sample space $\Omega = \mathbb{K}^\infty$ endowed with the discrete sigma-algebra. A generic element of Ω will be written $\omega = (x_0, a_0, x_1, a_1, \ldots)$. The projection functions are $X_n : \Omega \to X$ and $A_n : \Omega \to A$ given by $X_n(\omega) = x_n$ and $A_n(\omega) = a_n$, respectively, for $n \geq 0$.

2.2 Control Policies

A control policy $\pi \in \Pi$ is a sequence $\{\pi_n\}_{n \geq 0}$ of probability distributions on A given $\mathbb{K}^n \times X$ such that $\pi_n\{A(x_n)|x_0, a_0, \ldots, x_n\} = 1$ for all $n \geq 0$ and $\omega \in \Omega$. For each initial state $x \in X$ and every policy $\pi \in \Pi$ there exists a unique probability measure P_x^π on Ω such that $P_x^\pi\{X_0 = x\} = 1$, and

$$P_x^\pi\{A_n = a | X_0 = x_0, A_0 = a_0, \ldots, X_n = x_n) = \pi_n(a|x_0, a_0, \ldots, x_n)$$
$$P_x^\pi\{X_{n+1} = x | X_0 = x_0, A_0 = a_0, \ldots, X_n = x_n, A_n = a_n\} = Q(x_n, a_n, x)$$

for all $n \geq 0$ and $\omega \in \Omega$. Its expectation operator is denoted by E_x^π.

We say that the policy $\pi \in \Pi$ is (randomized) stationary if there exist a family $\varphi = \{\varphi(\cdot|x)\}_{x \in X}$ of probability distributions on $A(x)$, for $x \in X$, such that

$$\pi_n(\cdot|x_0, a_0, \ldots, x_n) = \varphi(\cdot|x_n) \quad \text{for all } n \geq 0 \text{ and } \omega \in \Omega.$$

In this case, we will identify π with φ, and the set of stationary policies will be denoted by $\Phi \subseteq \Pi$. Under the probability measure P_x^φ, for the stationary policy $\varphi \in \Phi$ and for any initial state $x \in X$, the process $\{X_n\}_{n \geq 0}$ is a homogeneous Markov chain on X with transition matrix Q_φ given by

$$Q_\varphi(x,y) = \sum_{a \in A(x)} Q(x,a,y)\varphi(a|x) \quad \text{for } x, y \in X.$$

The corresponding expected cost function will be written

$$c_\varphi(x) = \sum_{a \in A(x)} c(x,a)\varphi(a|x) \quad \text{for } x \in X.$$

If the stationary policy $\varphi \in \Phi$ is such that $\varphi(\cdot|x)$ is concentrated on a single action in $A(x)$, then we will say that φ is a deterministic stationary policy. Such policies will be identified with the class $\mathbb{F}$ of functions $f : X \to A$ such that $f(x) \in A(x)$ for all $x \in X$, by means of $\varphi(\cdot|x) = \delta_{f(x)}(\cdot)$ for $x \in X$, with δ the Dirac distribution. Thus, we have $\mathbb{F} \subseteq \Phi \subseteq \Pi$. For a deterministic stationary policy, the corresponding transition matrix and cost function are

$$Q_f(x,y) = Q(x,f(x),y) \quad \text{and} \quad c_f(x) = c(x,f(x))$$

for $x, y \in X$, respectively.

2.3 The Long-Run Average Cost Criterion and the ACOE

The long-run expected average cost of the policy $\pi \in \Pi$ for the initial state $x \in X$ is defined as

$$J(\pi,x) = \overline{\lim_{N\to\infty}} \frac{1}{N} \sum_{n=0}^{N-1} E_x^{\pi}\big[c(X_n,A_n)\big].$$

The optimal value function is $J^*(x) = \sup_{\pi\in\Pi} J(\pi,x)$ for $x \in X$, and the policy $\pi^* \in \Pi$ is said to be average optimal when $J(\pi,x) = J^*(x)$ for each $x \in X$.

The set of functions from X to $\mathbb{R}$ is denoted by $\mathbb{R}^X$. We say that the pair $(g,h) \in X \times \mathbb{R}^X$ is a solution to the average cost optimality equation (ACOE) for $\mathcal{M}$ when

$$g = \min_{a\in A(x)} \Big\{c(x,a) + \sum_{y\in X} Q(x,a,y)h(y)\Big\} \quad \text{for all } x \in X.$$

Given a solution (g,h) of the ACOE, we say that the deterministic stationary policy $f \in \mathbb{F}$ is canonical when it "attains the minimum" in the ACOE, that is,

$$g = c(x,f(x)) + \sum_{y\in X} Q(x,f(x),y)h(y) \quad \text{for all } x \in X.$$

Notation. In the sequel, we will interpret Q_φ and Q_f as square matrices of order $|X|$. The restriction of these matrices to transitions from the states in $S \subseteq X$ to $S' \subseteq X$ will be written $Q_\varphi(S,S')$ and $Q_f(S,S')$. The identity matrix is $\mathbf{I}$. The zero matrix, of whatever dimension, is $\mathbf{0}$. Let $\mathbf{1}$ be the constant function equal to one.

Any function on X will be interpreted, unless otherwise mentioned, as a column vector. The restriction of a (row or column) vector $u \in \mathbb{R}^X$ to the states of some subset $S \subseteq X$ will be denoted by $u(S)$.

3 Unichain Control Models

3.1 Unichain Stationary Policies

We say that a stationary policy $\varphi \in \Phi$ is unichain if, for the homogeneous Markov chain with transition matrix Q_φ, the state space X can be partitioned into a unique closed irreducible class R_φ plus a (possibly empty) class of transient states T_φ.

This means that if the initial state of the Markov chain with transition matrix Q_φ is in R_φ, then the Markov chain does not leave R_φ and, moreover, for every pair of states x, y in R_φ, the state x communicates with y, which will be written $x \overset{\varphi}{\rightsquigarrow} y$. In particular, all the states in R_φ are positive recurrent. If the initial state is in T_φ, then the Markov chain will be eventually absorbed by the set R_φ.

Write the transition matrix Q_φ of a unichain policy in block form:

$$Q_\varphi = \begin{pmatrix} Q_\varphi(R_\varphi, R_\varphi) & \mathbf{0} \\ Q_\varphi(T_\varphi, R_\varphi) & Q_\varphi(T_\varphi, T_\varphi) \end{pmatrix}, \tag{2}$$

where we highlight the transitions between the states in R_φ and T_φ. It is well known (provided that T_φ is nonempty) that

$$\big(\mathbf{I} - Q_\varphi(T_\varphi, T_\varphi)\big)^{-1} = \mathbf{I} + \sum_{k=1}^{\infty} \big(Q_\varphi(T_\varphi, T_\varphi)\big)^k, \tag{3}$$

where the term (x, y) of $\big(\mathbf{I} - Q_\varphi(T_\varphi, T_\varphi)\big)^{-1}$ is the expected number of visits to the state $y \in T_\varphi$ when the initial state is $x \in T_\varphi$. Also, the term (x, y) of

$$\big(\mathbf{I} - Q_\varphi(T_\varphi, T_\varphi)\big)^{-1} Q_\varphi(T_\varphi, R_\varphi)$$

is the probability that, starting from $x \in T_\varphi$, absorption by R_φ occurs through $y \in R_\varphi$. Consequently,

$$\big(\mathbf{I} - Q_\varphi(T_\varphi, T_\varphi)\big)^{-1} Q_\varphi(T_\varphi, R_\varphi)\mathbf{1} = \mathbf{1} \tag{4}$$

because we reach R_φ with probability one for any initial state in T_φ.

3.2 The Poisson Equation

We say that a probability distribution μ on X, interpreted as a row vector, is invariant for the stationary policy $\varphi \in \Phi$ when $\mu = \mu Q_\varphi$.

The system of linear equations in Lemma 1(iii) below is called the Poisson equation for φ.

Lemma 1. *Let $\varphi \in \Phi$ be a unichain stationary policy.*

(i) *The policy φ has a unique invariant distribution μ_φ. This invariant distribution satisfies $\mu_\varphi(x) > 0$ when $x \in R_\varphi$ and $\mu_\varphi(x) = 0$ when $x \in T_\varphi$.*
(ii) *For every $x \in X$, the average cost of φ is $J(\varphi, x) = \mu_\varphi c_\varphi$.*
(iii) *There exist $g \in \mathbb{R}$ and $h \in \mathbb{R}^X$ with*

$$g + h(x) = c_\varphi(x) + \sum_{y \in X} Q_\varphi(x, y)h(y) \quad \textit{for all } x \in X$$

or, in matrix notation, $g\mathbf{1} + h = c_\varphi + Q_\varphi h$. Moreover, $g = \mu_\varphi c_\varphi$ and h is unique up to additive constants.

Proof. (i). Recalling the expression of Q_φ in block form in (2), it follows that an invariant probability measure μ satisfies the equations

$$\mu(R_\varphi) = \mu(R_\varphi)Q_\varphi(R_\varphi, R_\varphi) + \mu(T_\varphi)Q_\varphi(T_\varphi, R_\varphi)$$
$$\mu(T_\varphi) = \mu(T_\varphi)Q_\varphi(T_\varphi, T_\varphi).$$

Iterating the last equation, we have $\mu(T_\varphi) = \mu(T_\varphi)\big(Q_\varphi(T_\varphi, T_\varphi)\big)^n$ for all $n \geq 1$. The series $\sum_n \big(Q_\varphi(T_\varphi, T_\varphi)\big)^n$ converges (recall (3)), and so $\big(Q_\varphi(T_\varphi, T_\varphi)\big)^n \to \mathbf{0}$. Hence, we necessarily have $\mu(T_\varphi) = \mathbf{0}$. Therefore, $\mu(R_\varphi) = \mu(R_\varphi)Q_\varphi(R_\varphi, R_\varphi)$ has a unique solution (as a probability vector), with strictly positive terms, because the states in R_φ are positive recurrent.

(ii). Given an initial state $x \in X$, noting that $E_x^\varphi\big[c(X_n, A_n)\big]$ is the x-term of the column vector $Q_\varphi^n c_\varphi$ for all $n \geq 1$, and since the matrix $\mathbf{1}\mu_\varphi$ is the Cesaro limit of the sequence $\{Q_\varphi^n\}_{n\geq 0}$, the results follows.

(iii). Let H_φ be the deviation matrix of Q_φ, defined as the Drazin inverse of $\mathbf{I} - Q_\varphi$, given by

$$H_\varphi = \big(\mathbf{I} - Q_\varphi + \mathbf{1}\mu_\varphi\big)^{-1}(\mathbf{I} - Q_\varphi).$$

The deviation matrix satisfies

$$\mathbf{1}\mu_\varphi + H_\varphi = \mathbf{I} + Q_\varphi H_\varphi.$$

By right-multiplying this equation by c_φ, we get that $g = \mu_\varphi c_\varphi$ and $h = H_\varphi c_\varphi$ is a solution to the equation in (iii).

Uniqueness of g in the equation $g\mathbf{1} + h = c_\varphi + Q_\varphi h$ follows easily by left-multiplying this equation by μ_φ. If $h, h' \in \mathbb{R}^X$ are solutions to the equation in (iii) then we have $h - h' = Q_\varphi(h - h')$. The restriction of these equations to R_φ yields $(h-h')(R_\varphi) = Q_\varphi(R_\varphi, R_\varphi)(h-h')(R_\varphi)$. Since $Q_\varphi(R_\varphi, R_\varphi)$ is irreducible, the eigenvalue 1 is simple and $(h - h')(R_\varphi)$, which is a right-eigenvector, is constant, say, $(h-h')(R_\varphi) = v\mathbf{1}$ for some $v \in \mathbb{R}$. On the transient states we have

$$(h - h')(T_\varphi) = v \cdot Q_\varphi(T_\varphi, R_\varphi)\mathbf{1} + Q_\varphi(T_\varphi, T_\varphi)(h - h')(T_\varphi),$$

and so

$$(h - h')(T_\varphi) = v \cdot \big(\mathbf{I} - Q_\varphi(T_\varphi, T_\varphi)\big)^{-1} Q_\varphi(T_\varphi, R_\varphi)\mathbf{1}.$$

As a consequence of (4), we have $(h - h')(T_\varphi) = v\mathbf{1}$. We have thus proved that $h - h'$ is indeed constant. ■

The above result shows that the Poisson equation for a unichain stationary policy admits a solution, with g the expected average cost of the policy φ and h unique up to additive constants.

3.3 Unichain Control Models and the ACOE

Now we give the definition of a unichain control model.

Definition 1. *The control model $\mathcal{M}$ is unichain when every deterministic stationary policy in $\mathbb{F}$ is unichain.*

The next result result gives some insight on the unichain property for (randomized) stationary policies.

Lemma 2. *If the control model $\mathcal{M}$ is unichain then every stationary policy in Φ is unichain.*

Proof. It should be clear that, given $\varphi \in \Phi$, there exist distinct $f_1, \ldots, f_k \in \mathbb{F}$ and strictly positive $\beta_1, \ldots, \beta_k$ with $\sum \beta_i = 1$ such that

$$\varphi = \beta_1 f_1 + \ldots + \beta_k f_k.$$

As a consequence, $Q_\varphi(x,y) = \sum_{i=1}^k \beta_i Q_{f_i}(x,y)$ for all states $x, y \in X$ and, therefore, if $x \overset{f_i}{\rightsquigarrow} y$ for some $1 \leq i \leq k$ then $x \overset{\varphi}{\rightsquigarrow} y$.

In the sequel, to avoid trivial situations, we will assume that $k \geq 2$. Suppose that the transition matrix Q_φ possesses at least two disjoint closed irreducible classes C_1 and C_2. Given a state $x \in C_1$ and the policy, say f_1, then $x \overset{f_1}{\rightsquigarrow} z$ for all $z \in R_{f_1}$ (regardless x is recurrent or transient for f_1), and so $x \overset{\varphi}{\rightsquigarrow} z$ for all $z \in R_{f_1}$. Consequently, $R_{f_1} \subseteq C_1$. Repeat the same argument for any $y \in C_2$, and then $R_{f_1} \subseteq C_2$, which is a contradiction.

Therefore, the transition matrix Q_φ has a unique closed irreducible class, hence, it is unichain. ■

Remark 1. The above proof shows that $\bigcup_{i=1}^k R_{f_i} \subseteq R_\varphi$, that is, the recurrent states under any f_i are recurrent under φ. Inclusion may be strict, however, as shown by the following example. Let $X = \{1,2,3,4\}$ and $A = \{1,2\}$. The transition probabilities when taking action 1 are $Q(x,1,y)$, given by the matrix

$$\begin{pmatrix} \star & \star & 0 & 0 \\ \star & \star & 0 & 0 \\ 0 & \star & 0 & 0 \\ 0 & 0 & \star & 0 \end{pmatrix},$$

where the $\star$ denote the strictly positive terms. If $f_1 \in \mathbb{F}$ is the policy that takes action 1 at every state, then $R_{f_1} = \{1,2\}$. The transition probabilities when taking action 2 are $Q(x,2,y)$, given by the matrix

$$\begin{pmatrix} 0 & 0 & 0 & \star \\ 0 & \star & \star & 0 \\ 0 & \star & \star & 0 \\ 0 & 0 & \star & 0 \end{pmatrix}.$$

If $f_2 \in \mathbb{F}$ is the policy that takes action 2 at every state, then $R_{f_2} = \{2,3\}$. The transition matrix of the randomized strategy $\varphi = \frac{1}{2} f_1 + \frac{1}{2} f_2$ is

$$\begin{pmatrix} \star & \star & 0 & \star \\ \star & \star & \star & 0 \\ 0 & \star & \star & 0 \\ 0 & 0 & \star & 0 \end{pmatrix},$$

and so $R_\varphi = X$, which is strictly larger than $R_{f_1} \cup R_{f_2}$.

The next result gives the existence of solutions to the ACOE.

Theorem 1. *Suppose that the control model $\mathcal{M}$ is unichain.*

(i) *There exist solutions $(g, h) \in \mathbb{R} \times \mathbb{R}^X$ to the ACOE for $\mathcal{M}$. Moreover, g is unique and $g = J^*(x)$ for all $x \in X$, while h is unique up to additive constants.*

(ii) *Any canonical policy is average optimal.*

Proof. The existence of solutions to the ACOE is a well known fact. For completeness, we give a sketch of the proof. There exists a policy $f \in \mathbb{F}$ that is α-discount optimal for some sequence of discount factors $\alpha \uparrow 1$. Let $V_\alpha^* \in \mathbb{R}^X$ be the optimal discounted value function. For this policy, the series expansion of the resolvent yields (with o the Landau notation)

$$(\mathbf{I} - \alpha Q_f)^{-1} = \frac{1}{1-\alpha}\mathbf{1}\mu_f + H_f + \mathrm{o}(1-\alpha)$$

as $\alpha \uparrow 1$. Therefore, the total expected discounted cost of the policy f is

$$\begin{aligned} V_\alpha^*(x) &= \sum_{k=0}^{\infty} \alpha^k E_x^f[c(X_k, A_k)] = (\mathbf{I} - \alpha Q_f)^{-1} c_f \\ &= \frac{1}{1-\alpha}\mathbf{1}\mu_f c_f + H_f c_f + \mathrm{o}(1-\alpha). \end{aligned}$$

Choose a state $x_0 \in X$ and note that for the above mentioned sequence of $\alpha \uparrow 1$ we have $(1-\alpha)V_\alpha^*(x_0) \to g_0$ for some $g_0 \in \mathbb{R}$ and $V_\alpha^* - V_\alpha^*(x_0)\mathbf{1} \to h_0$ for some $h_0 \in \mathbb{R}^X$. Taking the limit through this subsequence in the discounted cost optimality equation

$$V_\alpha^*(x) = \min_{a \in A(x)} \{c(x,a) + \alpha \sum_{y \in X} Q(x,a,y) V_\alpha^*(y)\} \quad \text{for } x \in X$$

shows that (g_0, h_0) is a solution to the ACOE. The fact that $g_0 = \sup_{\pi \in \Pi} J(\pi, x)$ for all $x \in X$, and that f is average cost optimal, follows from the standard "verification" theorem in Markov decision processes.

It remains to show that, if h and h' are taken from two solutions of the ACOE, then they differ by a constant. Let the policies f and f' in $\mathbb{F}$ attain the minimum in the ACOEs for (g, h) and (g, h'), respectively. Therefore, for all $x \in X$ we have

$$g + h(x) = c_f(x) + \sum_{y \in X} Q_f(x,y)h(y) \tag{5}$$

$$\leq c_{f'}(x) + \sum_{y \in X} Q_{f'}(x,y)h(y) \tag{6}$$

$$g + h'(x) = c_{f'}(x) + \sum_{y \in X} Q_{f'}(x,y)h'(y) \tag{7}$$

$$\leq c_f(x) + \sum_{y \in X} Q_f(x,y)h'(y). \tag{8}$$

With R_f and T_f be the sets of recurrent and transient states for Q_f, we write

$$Q_f = \begin{pmatrix} Q_f(R_f, R_f) & \mathbf{0} \\ Q_f(T_f, R_f) & Q_f(T_f, T_f) \end{pmatrix};$$

recall (2). By (5) and (8) we have that the function $h - h'$ on X is superharmonic for Q_f, meaning that $h - h' \geq Q_f(h - h')$. As a consequence, $h - h' \geq \mu_f \cdot (h - h')$, and so $h - h'$ is constant on R_f because μ_f is strictly positive on R_f. Similarly, using (6) and (7) we obtain that $h - h'$ is subharmonic for $Q_{f'}$, and it follows that $h - h'$ is constant on $R_{f'}$. Summarizing, there exist γ and $\gamma' \in \mathbb{R}$ such that

$$h(x) - h'(x) = \gamma \quad \text{for all } x \in R_f \quad \text{and} \quad h(x) - h'(x) = \gamma' \quad \text{for all } x \in R_{f'}. \tag{9}$$

Now, by (8), for every $x \in T_f$,

$$h'(x) - \sum_{y \in T_f} Q_f(x, y) h'(y) \leq -g + c_f(x) + \sum_{y \in R_f} Q_f(x, y) h'(y)$$

or, in matrix notation,

$$(\mathbf{I} - Q_f(T_f, T_f)) h'(T_f) \leq -g\mathbf{1} + c_f(T_f) + Q_f(T_f, R_f) h'(R_f). \tag{10}$$

Also, from (5),

$$(\mathbf{I} - Q_f(T_f, T_f)) h(T_f) = -g\mathbf{1} + c_f(T_f) + Q_f(T_f, R_f) h(R_f). \tag{11}$$

Since the matrix $\big(\mathbf{I} - Q_f(T_f, T_f)\big)^{-1}$ is nonnegative (recall (3)), it follows from (10) that

$$h'(T_f) \leq \big(\mathbf{I} - Q_f(T_f, T_f)\big)^{-1} \cdot \big(-g\mathbf{1} + c_f(T_f) + Q_f(T_f, R_f) h'(R_f)\big).$$

However, we have $h(R_f) = h'(R_f) + \gamma\mathbf{1}$, and so

$$\begin{aligned} h'(T_f) &\leq \big(\mathbf{I} - Q_f(T_f, T_f)\big)^{-1} \cdot \big(-g\mathbf{1} + c_f(T_f) + Q_f(T_f, R_f)(h(R_f) - \gamma\mathbf{1})\big) \\ &= h(T_f) - \gamma\mathbf{1}, \end{aligned}$$

where we have applied (4) and (11). By following a similar argument, we obtain that $h(T_{f'}) - h'(T_{f'}) \leq \gamma'\mathbf{1}$. So far, we have established that

$$\begin{cases} h(x) - h'(x) = \gamma & \text{on } R_f \\ h(x) - h'(x) \geq \gamma & \text{on } T_f \end{cases} \quad \text{and} \quad \begin{cases} h(x) - h'(x) = \gamma' & \text{on } R_{f'} \\ h(x) - h'(x) \leq \gamma' & \text{on } T_{f'}. \end{cases}$$

In particular, $\gamma \leq h(x) - h'(x) \leq \gamma'$ for all $x \in X$.

Suppose for a moment that $\gamma < \gamma'$. Then we necessarily have $R_f \cap R_{f'} = \emptyset$. Define the policy $\tilde{f} \in \mathbb{F}$ as follows: $\tilde{f}(x) = f(x)$ if $x \in R_f$, and $\tilde{f}(x) = f'(x)$ if $x \in R_{f'}$ (the definition of $\tilde{f}$ outside $R_f \cup R_{f'}$ is not relevant). We have that R_f and $R_{f'}$ are two disjoint recurrent classes for $\tilde{f}$, which leads to a contradiction, and so $\gamma = \gamma'$.

(ii). The proof that any canonical policy is average cost optimal is standard and we omit it. ■

In the sequel, we will write $g^* = J^*(x)$, for $x \in X$. We will refer to g^* as to the optimal gain. Note also that the set of canonical policies does not depend on the particular solution h in the ACOE for $\mathcal{M}$.

4 Solving the ACOE with Linear Programming

Next we address the linear programming formulation of the control model $\mathcal{M}$ which, in what follows, is assumed to be unichain.

Primal problem. Consider the following (primal) linear programming problem **P**:

$$\text{minimize} \quad \sum_{(x,a)\in\mathbb{K}} c(x,a)z(x,a)$$

subject to

$$\sum_{a\in A(x)} z(x,a) = \sum_{(x',a')\in\mathbb{K}} z(x',a')Q(x',a',x) \quad \text{for all } x \in X,$$

$$\sum_{(x,a)\in\mathbb{K}} z(x,a) = 1, \qquad \text{and} \qquad z(x,a) \geq 0 \quad \text{for all } (x,a) \in \mathbb{K}.$$

It is well known that the optimal value of **P** equals g^*, the optimal gain of $\mathcal{M}$. Indeed, given $\varphi \in \Phi$ and letting $z(x,a) = \mu_\varphi(x)\varphi(a|x)$ for all $(x,a) \in \mathbb{K}$ we obtain a feasible solution of **P**. The objective function at such z equals $\mu_\varphi c_\varphi$, which is the (constant) average cost of φ. Conversely, given a feasible solution $\{z(x,a)\}$ of **P**, consider the (nonempty) set

$$X_z = \{x \in X : \sum_{a\in A(x)} z(x,a) > 0\}. \tag{12}$$

Define $\varphi \in \Phi$ arbitrarily on $X - X_z$ and by

$$\varphi(a|x) = \frac{z(x,a)}{\sum_{a'\in A(x)} z(x,a')} \quad \text{for } x \in X_z \text{ and } a \in A(x). \tag{13}$$

Then the relation $z(x,a) = \mu_\varphi(x)\varphi(a|x)$ holds for all $(x,a) \in \mathbb{K}$ and, moreover, $\sum c(x,a)z(x,a) = \mu_\varphi c_\varphi$. Therefore, starting from an optimal solution $\{z(x,a)\}$ of **P** the so constructed policy $\varphi \in \Phi$ is average cost optimal. Besides, $X_z = R_\varphi$ is the set of recurrent states of φ.

Dual problem. The dual problem of **P** is the linear programming problem **D** given by:

$$\text{maximize} \quad g \quad \text{subject to}$$

$$g + h(x) \leq c(x,a) + \sum_{y \in X} Q(x,a,y)h(y) \quad \text{for all } (x,a) \in \mathbb{K},$$

$$g \in \mathbb{R} \qquad \text{and} \qquad h(x) \in \mathbb{R} \quad \text{for } x \in X.$$

Its optimal value is g^*, the optimal gain of $\mathcal{M}$. If $(g^*, h) \in \mathbb{R} \times \mathbb{R}^X$ is an optimal solution of **D**, the following inequalities hold

$$g^* + h(x) \leq \min_{a \in A(x)} \Big\{ c(x,a) + \sum_{y \in X} Q(x,a,y)h(y) \Big\} \quad \text{for all } x \in X, \tag{14}$$

and it should be clear that at least one of these inequalities holds with equality (otherwise, g^* could be increased). However, by solving **D** we might not obtain a solution to the ACOE for the control model $\mathcal{M}$, that is, at optimality, some of the inequalities in (14) may be strict, although we know from Theorem 1 that solutions to the ACOE indeed exist.

Next we show how we can find a solution to the ACOE for the control model $\mathcal{M}$ by solving two linear programs. The next lemma uses the notation X_z, defined in (12) for a feasible solution z of **P**.

Lemma 3. *Let* (g^*, h) *be a solution of the ACOE for* $\mathcal{M}$ *and let* $(g^*, \overline{h})$ *be an optimal solution of the dual linear program* **D**. *Then, for any state* $x^* \in X_{z^*}$ *with* z^* *an optimal solution of the primal linear program* **P**, *we have*

$$\min_{x \in X} \{ h(x) - \overline{h}(x) \} = h(x^*) - \overline{h}(x^*).$$

Proof. Let z^* be an optimal solution of **P** and fix any state $x^* \in X_{z^*}$. Observe that

$$h_0 = h - h(x^*)\mathbf{1} \quad \text{and} \quad \overline{h}_0 = \overline{h} - \overline{h}(x^*)\mathbf{1}$$

are also such that (g^*, h_0) and $(g^*, \overline{h}_0)$ are solutions to the ACOE and an optimal solution of **D**, respectively. Therefore, to establish the lemma we must prove that $h_0 \geq \overline{h}_0$.

Let $\varphi^* \in \Phi$ be the average cost optimal stationary policy constructed, starting from z^*, as in (13). We deduce from (14) that

$$g^* + \overline{h}_0(x) \leq c_{\varphi^*}(x) + \sum_{y \in X} Q_{\varphi^*}(x,y)\overline{h}_0(y) \quad \text{for all } x \in X.$$

Since φ^* is average cost optimal, i.e., $\mu_{\varphi^*} c_{\varphi^*} = g^*$, the above inequality necessarily holds with equality for the recurrent states $R_{\varphi^*} = X_{z^*}$ of φ^*:

$$g^* + \overline{h}_0(x) = c_{\varphi^*}(x) + \sum_{y \in X} Q_{\varphi^*}(x,y)\overline{h}_0(y) \quad \text{for all } x \in R_{\varphi^*}.$$

On the other hand, we deduce from the ACOE that

$$g^* + h_0(x) \leq c_{\varphi^*}(x) + \sum_{y \in X} Q_{\varphi^*}(x,y) h_0(y) \quad \text{for all } x \in R_{\varphi^*}.$$

Notice that the above summations range, in fact, over $y \in R_{\varphi^*}$. Therefore, the function $h_0 - \overline{h}_0$, when restricted to the set of recurrent states R_{φ^*}, is subharmonic for Q_{φ^*} and, hence, constant on R_{φ^*}. Since $\overline{h}_0(x^*) = h_0(x^*) = 0$, we conclude that $\overline{h}_0(x) = h_0(x)$ for all $x \in R_{\varphi^*}$.

Now, let $f^* \in \mathbb{F}$ be a canonical policy for $\mathcal{M}$, that is, it attains the minimum in the ACOE:

$$g^* + h_0(x) = c_{f^*}(x) + \sum_{y \in X} Q_{f^*}(x,y) h_0(y) \quad \text{for all } x \in X. \tag{15}$$

Since we also have, by (14),

$$g^* + \overline{h}_0(x) \leq c_{f^*}(x) + \sum_{y \in X} Q_{f^*}(x,y) \overline{h}_0(y) \quad \text{for all } x \in X, \tag{16}$$

we obtain that $\overline{h}_0 - h_0$ is subharmonic for the kernel Q_{f^*} and hence constant on the set R_{f^*} of recurrent states for f^*. Arguing as in the proof of Theorem 1(i), we have that $R_{f^*} \cap R_{\varphi^*}$ is not empty, and so $\overline{h}_0(x) = h_0(x)$ for all $x \in R_{f^*}$.

Let us now write (16) in matrix form for the transient states T_{f^*} of f^*:

$$\overline{h}_0(T_{f^*}) \leq c_{f^*}(T_{f^*}) - g^* \mathbf{1} + Q_{f^*}(T_{f^*}, T_{f^*}) \overline{h}_0(T_{f^*}) + Q_{f^*}(T_{f^*}, R_{f^*}) \overline{h}_0(R_{f^*}).$$

This implies that

$$\begin{aligned}
\overline{h}_0(T_{f^*}) &\leq \big(\mathbf{I} - Q_{f^*}(T_{f^*}, T_{f^*})\big)^{-1} \big(c_{f^*}(T_{f^*}) - g^* \mathbf{1} + Q_{f^*}(T_{f^*}, R_{f^*}) \overline{h}_0(R_{f^*})\big) \\
&= \big(\mathbf{I} - Q_{f^*}(T_{f^*}, T_{f^*})\big)^{-1} \big(c_{f^*}(T_{f^*}) - g^* \mathbf{1} + Q_{f^*}(T_{f^*}, R_{f^*}) h_0(R_{f^*})\big) \\
&= h_0(T_{f^*}),
\end{aligned}$$

where equalities follow from the fact that $\overline{h}_0 = h_0$ on R_{f^*} and from (15). This completes the proof that $\overline{h}_0(x) \leq h_0(x)$ for all $x \in X$. ∎

Note that the proof that $\overline{h}_0 \leq h_0$ mainly relies on the properties of the canonical policy f^*. Since our goal is in fact to solve the ACOE, such a canonical policy is not therefore "available". That is why this proof uses the policy φ^*, that can be explicitly determined by solving $\mathbf{P}$, and then uses the link between them: $R_{f^*} \cap R_{\varphi^*} \neq \emptyset$, deduced from the unichain nature of the control model $\mathcal{M}$.

The modified dual problem. We define now a linear programming problem which is parameterized by a constant $g_0 \in \mathbb{R}$ and a state $x_0 \in X$. The linear programming problem $\mathbf{D}'(g_0, x_0)$ is given by

$$\text{maximize} \quad \sum_{x \in X} h(x) \quad \text{subject to}$$

$$g_0 + h(x) \leq c(x,a) + \sum_{y \in X} Q(x,a,y)h(y) \quad \text{for all } (x,a) \in \mathbb{K},$$

$$h(x_0) = 0 \qquad \text{and} \qquad h(x) \in \mathbb{R} \quad \text{for } x \in X.$$

We call this the modified dual linear programming problem because it is the same problem as **D** except for the objective function, the additional constraint that $h(x_0) = 0$, and the fact that g_0 is a data of the problem, and not a variable as in the dual problem **D**.

By adequately choosing $g_0 \in \mathbb{R}$ and $x_0 \in X$, the modified linear programming problem finds a solution to the ACOE for $\mathcal{M}$. This is our main result next in the paper.

Theorem 2. *The following procedure allows to solve the ACOE for a unichain control model* $\mathcal{M}$.

(a) Solve **P**, *let* $g^* \in \mathbb{R}$ *be its optimal value, and let* $\{z^*(x,a)\}$ *be an optimal solution. Determine a state* $x^* \in X$ *with* $\sum_{a \in A(x)} z^*(x^*,a) > 0$.
(b) For $g^* \in \mathbb{R}$ *and* $x^* \in X$ *as in (a), solve* $\mathbf{D}'(g^*,x^*)$ *and let* $\overline{h} \in \mathbb{R}^X$ *be an optimal solution.*

Then $(g^*,\overline{h}) \in \mathbb{R} \times \mathbb{R}^X$ *is a solution to the ACOE for* $\mathcal{M}$.

Proof. Let $\overline{h} \in \mathbb{R}^X$ be an optimal solution of $\mathbf{D}'(g^*,x^*)$ and let $h \in \mathbb{R}^X$ be such that (g^*,h) is the unique solution of the ACOE for $\mathcal{M}$ that vanishes at x^*. Observe that h is in fact a feasible solution of $\mathbf{D}'(g^*,x^*)$ and so

$$\sum_{x \in X} \overline{h}(x) \geq \sum_{x \in X} h(x). \tag{17}$$

Note also that $(g^*,\overline{h})$ is an optimal solution of the dual linear program **D**. As a consequence of Lemma 3 we have that $h(x) \geq \overline{h}(x)$ for all $x \in X$. Combined with (17), we deduce that all these inequalities hold with equality, that is, $h = \overline{h}$, which shows that $(g^*,\overline{h})$ solves the ACOE for $\mathcal{M}$. ■

Therefore, we can find the solutions to the ACOE for $\mathcal{M}$ by solving two "connected" linear programming problems **P** and **D**′. These linear programming problems are connected in the sense that, first of all, we must solve **P** and then, with some data obtained from this solution, we solve **D**′, which yields the solution to the ACOE.

5 An Example

Consider the following example taken from [2, Section 4.2.2]. The state and action spaces are $X = \{x_1, x_2\}$ and $A = \{a_1, a_2\}$, and the action sets are

$$A(x_1) = \{a_1, a_2\} \quad \text{and} \quad A(x_2) = \{a_1\}.$$

The transition probabilities are

$$Q(x_1, a_1, x_1) = Q(x_1, a_1, x_2) = 1/2, \; Q(x_1, a_2, x_2) = 1, \text{ and } Q(x_2, a_1, x_2) = 1.$$

The cost function is

$$c(x_1, a_1) = -5, \quad c(x_1, a_2) = -10, \quad \text{and} \quad c(x_2, a_1) = 1.$$

There are two deterministic stationary policies: $\mathbb{F} = \{f_1, f_2\}$, with $f_i(x_1) = a_i$ for $i = 1, 2$. This control model is unichain, with $R_{f_1} = R_{f_2} = \{2\}$. The ACOE is

$$g + h(x_1) = \min\{-5 + \frac{1}{2}h(x_1) + \frac{1}{2}h(x_2), -10 + h(x_2)\} \tag{18}$$

$$g + h(x_2) = 1 + h(x_2). \tag{19}$$

The equation (19) yields that the optimal gain is $g^* = 1$, while the solutions of (18) are of the form $h(x_2) - h(x_1) = 12$, the minimum being attained in the first term, corresponding to action a_1. The solutions of the ACOE are therefore

$$\big(1, (\lambda, 12 + \lambda)\big) \quad \text{for } \lambda \in \mathbb{R}. \tag{20}$$

We also have that the unique canonical policy is f_1, although both f_1 and f_2 are average optimal.

We solve the dual linear programming problem **D** with the `linprog` function from Matlab and it returns the optimal solution

$$g^* = 1, \quad h(x_1) = -82.5037, \quad h(x_2) = 82.5037,$$

which does not solve the ACOE because

$$g^* + h(x_1) < \min\{-5 + \frac{1}{2}h(x_1) + \frac{1}{2}h(x_2), -10 + h(x_2)\}.$$

This simple example show that an optimal solution of **D** might not be a solution for the ACOE.

Now we use the procedure described in Theorem 2. We solve the primal problem **P** and we obtain the optimal solution

$$z^*(x_1, a_1) = z^*(x_1, a_2) = 0 \quad \text{and} \quad z^*(x_2, a_1) = 1$$

with optimal objective function $g^* = 1$. Therefore, $X_{z^*} = \{x_2\}$, and we choose $x^* = x_2$. We solve the modified linear program $\mathbf{D}'(1, x_2)$ with `linprog`, which returns the optimal solution $h(x_1) = -12$ and $h(x_2) = 0$. In this way, we have indeed obtained a solution for the ACOE, namely, the solution

$$\big(1, (-12, 0)\big)$$

(cf. (20)), which is the unique solution h of the ACOE that vanishes at x_2.

6 Acknowledgement

This research was supported by grant MTM2012-31393 from the Ministerio de Economía y Competitividad, Spain.

References

1. Kallenberg, L.C.M. *Markov Decision Processes.* Lecture notes (2010 version) available at **http://www.math.leidenuniv.nl/~kallenberg/**
2. Piunovskiy, A.B. *Examples in Markov Decision Processes.* Imperial College Press, London, 2013.
3. Puterman, M.L. *Markov Decision Processes: Discrete Stochastic Dynamic Programming.* John Wiley & Sons, New York, 1994.

H_2 Control and Filtering of Markov Jump Linear Systems with Partial Information

Oswaldo L.V. Costa[1], Marcelo D. Fragoso[2] and Marcos G. Todorov[2]

[1] Universidade de São Paulo
Departamento de Engenharia de Telecomunicações e Controle, 05508 900 São Paulo SP, Brazil
oswaldo@lac.usp.br

[2] National Laboratory for Scientific Computing - LNCC/CNPq
Systems and Control Department, 25651-070 Petrópolis - RJ, Brazil
frag@lncc.br; todorov@lncc.br

Abstract. The goal of this paper is to consider the H_2-control and filtering for discrete-time Markov Jump Linear Systems (MJLS) assuming partial information on some variables. Three partial information scenarios are considered, named as Problems 1, 2 and 3. For Problem 1 the Markov jump parameter is assumed to be known, but the state variable not, and the goal is to design a dynamic Markov jump controller in order to stabilize the closed loop system in the mean square sense, and minimize the H_2 norm. In Problem 2 it is assumed that both the Markov jump parameter and the state variable are not available, and it is desired to design the optimal linear minimum mean square filter (LMMSE) for MJLS as well as to analyze the stationary solution of the associated Riccati like equation. For Problem 3 we assume that there is a detector that emits signals which provides information on the Markov parameter, and that the state variable is available to the controller. The idea is to use the information provided by this detector in order to design a feedback linear control that stochastically stabilizes the closed loop system with guaranteed H_2-cost. A Linear Matrix Inequalities (LMI) formulation is provided in order to achieve this goal.
Keywords: H_2-control, filtering problem, Markov Jump Linear Systems, partial information.
AMS 2000 subject classification: Primary 93E20, Secondary 93E11

1 Introduction

This paper deals with the H_2 control and filtering problems for Markov jump linear systems (MJLS) under partial information. Regarding the MJLS, the partial information problem may be associated either with the state variable $x(k)$, the Markov chain $\theta(k)$, or yet with both variables, which is of course the hardest problem. For the control problem with partial observations of the Markov chain the readers are referred, for instance, to [2, 10, 11]. The case with partial information of the state and perfect measurement of the Markov chain (including the

H_2 control problem) is treated, for instance, in [3, 6, 7, 9, 13]. The case in which both the state variable and the Markov chain are only partially observable was also studied in [9]. The H_2-norm control problem of discrete-time Markov jump linear systems when part of, or the total of the Markov states is not accessible to the controller was addressed in [8]. In this case the non-observed part of the Markov states is grouped in a number of clusters of observations, with the case of a single cluster retrieving the situation when no Markov state is observed. In this paper we consider the following problems.

1) Problem 1: Perfect information of the mode of operation $\theta(k)$ and output observation $y(k)$ of the state variable $x(k)$.
2) Problem 2: No knowledge of the mode of operation $\theta(k)$ and output observation $y(k)$ of the state variable $x(k)$.
3) Problem 3: There is a detector that emits signals $\widehat{\theta}(k)$ providing information on the Markov parameter $\theta(k)$, and perfect observation of the state variable $x(k)$.

For Problems 1 and 3 the idea is to design a feedback linear controller, using the information provided by the detector $\widehat{\theta}(k)$ $(= \theta(k)$ for Problem 1) in order to stochastically stabilize the closed loop system and minimize the H_2 norm (infinite horizon quadratic cost) among the controllers of this linear feedback class. For Problem 1 a separation principle, based on two coupled algebraic Riccati equations, is derived. For Problem 3 we first analyze the stochastic stabilizability problem through a feedback control for the MJLS, using the signal from the detector instead of the unknown Markov parameter. We can show that the existence of a solution to a set of linear matrix inequalities (LMIs) provides a stochastically stabilizing feedback gain for the MJLS. In the sequel it is provided an LMI optimization formulation in order to design a stochastically stabilizing feedback control with guaranteed H_2-cost. Notice that following this approach it is possible to get explicit numerical tools for the H_2 control problem of MJLS with partial information, unlike other approaches as, for instance, in [2, 10–12]. We also present conditions (one of them always satisfied for the limit case in which the detector provides perfect information on the Markov parameter), under which our results recast the usual results for the H_2 control of MJLS as presented in [4]. For Problem 2 the idea is to design a time-invariant filter (independent of the mode of operation) in order to minimize the stationary expected least square error. A Riccati like equation and its associated stationary solution are considered. A key feature of the approach adopted in this paper is that the linear controller depends only on the information from the detector, which is $\theta(k)$ for Problem 1, $\widehat{\theta}(k)$ for Problem 3. In Problem 2, which is the worst scenario, information comes only from the output of the system.

The paper is organized as follows. In section 2 we present the notation adopted in this paper. In section 3 we introduce the problem formulation, assumptions, the definition of stochastic stability, and some auxiliary results. In section 4 we present the main results related to Problem 1, in section 5 the results related to Problem 2, and in section 6 those related to Problem 3. In subsection

6.2 we present in Theorem 8 the first main result of Problem 3, dealing with the stochastic stabilizability problem. In subsection 6.4 we present in Theorem 10 the second main result of Problem 3, dealing with the guaranteed cost H_2-control problem. In subsection 6.5 we present a numerical example related to a fault-tolerant problem.

2 Notation

For $\mathbb{X}$ and $\mathbb{Y}$ complex Banach spaces we set $\mathbb{B}(\mathbb{X}, \mathbb{Y})$ the Banach space of all bounded linear operators of $\mathbb{X}$ into $\mathbb{Y}$, with the uniform induced norm represented by $\|.\|$. For simplicity we set $\mathbb{B}(\mathbb{X}) = \mathbb{B}(\mathbb{X}, \mathbb{X})$. The spectral radius of an operator $\mathcal{T} \in \mathbb{B}(\mathbb{X})$ is denoted by $r_\sigma(\mathcal{T})$. If $\mathbb{X}$ is a Hilbert space then the inner product is denoted by $\langle .;. \rangle$, and for $\mathcal{T} \in \mathbb{B}(\mathbb{X})$, $\mathcal{T}^*$ denotes the adjoint operator of $\mathcal{T}$. As usual, $\mathcal{T} \geq 0$ ($\mathcal{T} > 0$ respectively) will denote that the operator $\mathcal{T} \in \mathbb{B}(\mathbb{X})$ is positive-semi-definite (positive-definite). In particular, we denote respectively by $\mathbb{R}^n$ and $\mathbb{C}^n$ the n dimensional real and complex Euclidean spaces and $\mathbb{B}(\mathbb{C}^n, \mathbb{C}^m)$ ($\mathbb{B}(\mathbb{R}^n, \mathbb{R}^m)$ respectively) the normed bounded linear space of all $m \times n$ complex (real) matrices, with $\mathbb{B}(\mathbb{C}^n) = \mathbb{B}(\mathbb{C}^n, \mathbb{C}^n)$ ($\mathbb{B}(\mathbb{R}^n) = \mathbb{B}(\mathbb{R}^n, \mathbb{R}^n)$). Unless otherwise stated, $\|.\|$ will denote the standard norm in $\mathbb{C}^n$, and for $M \in \mathbb{B}(\mathbb{C}^n, \mathbb{C}^m)$, $\|M\|$ denotes the induced uniform norm in $\mathbb{B}(\mathbb{C}^n, \mathbb{C}^m)$. The superscript * indicates the conjugate transpose of a matrix, while $'$ indicates the transpose. Clearly for real matrices * and $'$ will have the same meaning. The identity matrix will be denoted by I and the trace operator by tr(.). For N integer set $\mathbb{N} = \{1, \ldots, N\}$. Define $\mathbb{H}^{n,m}$ as the linear space made up of all N-sequences of complex matrices $V = (V_1, \ldots, V_N)$ with $V_i \in \mathbb{B}(\mathbb{C}^n, \mathbb{C}^m)$, $i \in \mathbb{N}$. For simplicity, we set $\mathbb{H}^n = \mathbb{H}^{n,n}$ and $\mathbb{H}^{n+}$ such that $V = (V_1, \ldots, V_N) \in \mathbb{H}^{n+}$ if $V \in \mathbb{H}^n$ and $V_i \geq 0$ for each $i \in \mathbb{N}$. It is easy to verify that $\mathbb{H}^{n,m}$ is a Hilbert space when equipped with the inner product $\langle .;. \rangle$ given, for $V = (V_1, \ldots, V_N)$ and $S = (S_1, \ldots, S_N)$ in $\mathbb{H}^{n,m}$, by $\langle V; S \rangle = \sum_{i=1}^N \operatorname{tr}(V_i^* S_i)$. For $M_i \in \mathbb{B}(\mathbb{C}^n, \mathbb{C}^m)$, $i \in \mathbb{N}$, we set diag$[M_i]$ the $Nm \times Nn$ block diagonal matrix formed with $M_1, \ldots, M_N$ in the diagonal and zero elsewhere. We define the operators φ and $\hat{\varphi}$ in the following way: for $V = (V_1, \ldots, V_N) \in \mathbb{H}^{n,m}$, considering $V_i = \begin{bmatrix} v_{i1} & \cdots & v_{in} \end{bmatrix} \in \mathbb{B}(\mathbb{C}^n, \mathbb{C}^m), v_{ij} \in \mathbb{C}^m$

$$\varphi(V_i) = \begin{bmatrix} v_{i1} \\ \vdots \\ v_{in} \end{bmatrix} \in \mathbb{C}^{mn} \quad \text{and} \quad \hat{\varphi}(V) = \begin{bmatrix} \varphi(V_1) \\ \vdots \\ \varphi(V_N) \end{bmatrix} \in \mathbb{C}^{Nmn}.$$

For two matrices A and B in $\mathbb{B}(\mathbb{C}^n)$ we consider $A \otimes B \in \mathbb{B}(\mathbb{C}^{n^2})$ as the Kronecker product between these matrices (see [1]). Consider the stochastic basis $(\Omega, \mathcal{P}, \mathcal{F}, \{\mathcal{F}_k\})$ and denote by $E(.)$ the expected value operator, and by $E(.|.)$ the conditional expected value. We denote by L_2^n the Hilbert space of sequences of random vectors $z(k)$, $z(k) : \Omega \to \mathbb{R}^n$, with $z(k)$ $\mathcal{F}_k$-measurable, such that $\|z\|_2^2 = \sum_{k=0}^{\infty} E(\|z(k)\|^2) < \infty$. For $A \in \mathcal{F}$ we set $\mathbf{1}_A$ as the Dirac measure or equivalently, the indicator function of the event A (thus $\mathbf{1}_A(\omega) = 1$ if $\omega \in A$, and 0 otherwise).

3 Problem Formulation, Assumptions and Definitions

3.1 Preliminaries

We will consider throughout the paper the following controlled discrete-time linear system with Markov jumps on a probability space $(\Omega, P, \mathcal{F})$:

$$\mathcal{G}_S = \begin{cases} x(k+1) = A_{\theta(k)}x(k) + B_{\theta(k)}u(k) + E_{\theta(k)}w(k), \\ \quad x(0) = x_0, \;\; \theta(0) = \theta_0, \\ \quad y(k) = H_{\theta(k)}x(k) + G_{\theta(k)}w(k), \\ \quad z(k) = C_{\theta(k)}x(k) + D_{\theta(k)}u(k). \end{cases} \tag{1}$$

Here $\theta(k)$ is a homogeneous Markov chain with state space $\mathbb{N} = \{1, \ldots, N\}$ and transition probability $\mathbf{P} = [p_{ij}]$, and $\pi_i(k) = P(\theta(k) = i)$. It is assumed that x_0 and θ_0 are independent random variables with $E(x_0) = \mu_0$ and $E(x_0x_0') = \mathbb{Q}_0$. We have that the state variable is given by $x(k) \in \mathbb{R}^n$ and the control variable by $u(k) \in \mathbb{R}^m$; $y(k) \in \mathbb{R}^p$ is the output observable variable, $w(k) \in \mathbb{R}^r$ is the external disturbance, and $z(k) \in \mathbb{R}^q$ is the controlled output. We assume that $C_i'D_i = 0$, $E_iG_i' = 0$, $D_i'D_i > 0$ and $G_iG_i' > 0$. We need the following definition.

Definition 1. *Consider $w(k) = 0$ and $u(k) = 0$ in (1). We say that the system $\mathcal{G}_S$ is mean square stable (MSS) if $\lim_{k\to\infty} E(\|x(k)\|^2) = 0$ and stochastic stable (SS) if $\sum_{k=0}^{\infty} E(\|x(k)\|^2) < \infty$ for any initial condition x_0, θ_0.*

It was shown in [4] that for the finite dimensional Markov chain case, MSS and SS are equivalent concepts, and moreover they are equivalent to the spectral radius of an augmented matrix $\mathcal{A}$ being less than one or to the existence of a unique solution to a set of coupled Lyapunov equations, which can be written in four equivalent forms. This augmented matrix $\mathcal{A}$ is defined as $\mathcal{A} = \mathcal{C}\mathcal{N}$ where $\mathcal{C} = \mathbf{P}' \otimes I$ and $\mathcal{N} = \text{diag}[A_i \otimes A_i]$ (recall that $\otimes$ represents the Kronecker operator). We present next 3 examples to illustrate the application of the equivalence between MSS and the spectral radius of $\mathcal{A}$, unveiling some subtleties of MJLS. In all examples consider the homogeneous case ($w(k) = 0$ and $u(k) = 0$ in (1)).

Example 1: Consider the following system with two operation modes $A_1 = \frac{4}{3}$, $A_2 = \frac{1}{3}$ (note that mode 1 is unstable and mode 2 is stable). The transition probability matrix is $\mathbf{P} = \begin{bmatrix} \frac{1}{2} & \frac{1}{2} \\ \frac{1}{2} & \frac{1}{2} \end{bmatrix}$. It is easy to verify that $\mathcal{A} = \frac{1}{2}\begin{bmatrix} \frac{16}{9} & \frac{1}{9} \\ \frac{16}{9} & \frac{1}{9} \end{bmatrix}$ and $r_\sigma(\mathcal{A}) = \frac{17}{18} < 1$, and so the system is MSS. Suppose now that we have a different transition probability matrix, say $\bar{\mathbf{P}} = \begin{bmatrix} 0.9 & 0.1 \\ 0.9 & 0.1 \end{bmatrix}$, so that the system will most likely stay longer in mode 1, which is unstable. Then $\mathcal{A} = \begin{bmatrix} \frac{144}{90} & \frac{1}{10} \\ \frac{16}{90} & \frac{1}{90} \end{bmatrix}$, $r_\sigma(\mathcal{A}) = 1.61 > 1$ and the system is no longer MSS. This evinces a connection between MSS and the probability of visits to the unstable modes, which is translated in the expression for $\mathcal{A}$.

The next 2 examples deal with the fact that an MJLS composed only of unstable modes can be MSS and, alternatively, an MJLS composed only of stable modes can be unstable in the mean square sense.

Example 2: [A Non MSS System with Stable Modes] Consider a system with two operation modes, defined by matrices $A_1 = \begin{bmatrix} 0 & 2 \\ 0 & 0.5 \end{bmatrix}$ and $A_2 = \begin{bmatrix} 0.5 & 0 \\ 2 & 0 \end{bmatrix}$ and the transition probability matrix $\mathbf{P} = \begin{bmatrix} 0.5 & 0.5 \\ 0.5 & 0.5 \end{bmatrix}$. Note that both modes are stable. Curiously, we have that $r_\sigma(\mathcal{A}) = 2.125 > 1$, which means that the system is not MSS.

Example 3: [A MSS System with Unstable Modes] Consider the following system: $A_1 = \begin{bmatrix} 2 & -1 \\ 0 & 0 \end{bmatrix}$ and $A_2 = \begin{bmatrix} 0 & 1 \\ 0 & 2 \end{bmatrix}$ and the transition probability matrix $\mathbf{P} = \begin{bmatrix} 0.1 & 0.9 \\ 0.9 & 0.1 \end{bmatrix}$. Note that both modes are unstable, but we have that $r_\sigma(\mathcal{A}) = 0.4 < 1$.

The general conclusion one extracts from these examples is that stability of each operation mode is neither a necessary nor sufficient condition for mean square stability of the system. MSS depends upon a balance between the transition probability of the Markov chain and the operation modes.

3.2 Problem 1 - a Separation Principle

In this case $\theta(k)$ is known in system (1) and the controller is of the form:

$$\mathcal{G}_K = \begin{cases} \widehat{x}(k+1) = \widehat{A}_{\theta(k)}\widehat{x}(k) + \widehat{B}_{\theta(k)}y(k) \\ u(k) = \widehat{C}_{\theta(k)}\widehat{x}(k). \end{cases} \tag{2}$$

The goal is to obtain the mean square stability of the closed loop system and minimize the closed-loop H_2 norm of channel $w \to z$ (to be defined in Definition 3, and which can be seen as "energy of impulse response").

3.3 Problem 2 - an Optimal Filter

In this case $\theta(k)$ is unknown, and the system of interest is given by

$$\begin{cases} x(k+1) = A_{\theta(k)}x(k) + E_{\theta(k)}w(k) \\ y(k) = H_{\theta(k)}x(k) + G_{\theta(k)}w(k). \end{cases}$$

It is well known that the optimal nonlinear filter for this problem is obtained from a bank of filters, which requires exponentially increasing memory and computation with time. To limit the computational requirements, we consider the optimal linear minimum mean square filter (LMMSE) for MJLS. It will be shown that it has dimension Nn (n dimension of the state, N number of states of the Markov chain), which leads to a time-varying linear filter which is easy to implement, with all calculations performed off-line. Conditions to guarantee the convergence of the error covariance matrix to the stationary solution of an Nn dimensional algebraic Riccati equation will be presented, as well as the stability of the stationary filter.

3.4 Problem 3 - a Detector-Based Approach

In this case $\theta(k)$ is unknown and $x(k)$ known. We assume that only a partial observation $\widehat{\theta}(k)$ is available (like a hidden Markov model). It is assumed that a detector $\theta(k) \mapsto \widehat{\theta}(k)$ emits symbols from $\mathbb{N}$ to other discrete set $\mathbb{M}$. Formally:

$$P\big(\widehat{\theta}(k) = \ell \,\big|\, \mathcal{F}_k\big) = P\big(\widehat{\theta}(k) = \ell \,\big|\, \theta(k)\big) = \alpha_{\theta(k)\ell} \tag{3}$$

where $\mathcal{F}_k = \sigma\big\{x(0), u(0), \theta(0), \widehat{\theta}(0), \ldots, x(k), u(k), \theta(k)\big\}$ is the σ-algebra of the information up to time k, excluding $\widehat{\theta}(k)$, and $[\alpha_{i\ell}] \in \mathbb{R}^{N\times M}$ is dubbed the *emission matrix* in the HMM literature. We *do not* employ nonlinear filtering to estimate $\theta(k)$ but, instead, controls are derived directly from detected symbols, $\widehat{\theta}(k)$. We consider controls of the form $u(k) = K_{\widehat{\theta}(k)}x(k)$, so that the closed-loop system is given by

$$\begin{cases} x(k+1) = \big(A_{\theta(k)} + B_{\theta(k)}K_{\widehat{\theta}(k)}\big)x(k) + E_{\theta(k)}w(k) \\ \quad\;\; z(k) = \big(C_{\theta(k)} + D_{\theta(k)}K_{\widehat{\theta}(k)}\big)x(k) \end{cases} \tag{4}$$

The main problems dealt with are to obtain $K_{\widehat{\theta}(k)}$ such that the closed loop system is stochastic stable and it minimizes the H_2 cost using tools via LMIs.

4 Main Results for Problem 1

4.1 Preliminaries

In this section we consider Problem 1 (see subsection 3.2), in which $\theta(k)$ is assumed to be known for the controller but not $x(k)$. Instead the controller has access to the output $y(k)$ as in (1). We also assume in this section that $\{\theta(k), k = 0, 1, \ldots\}$ is an ergodic Markov chain and thus we have that $P(\theta(k) = i) \overset{k\uparrow\infty}{\to} \pi_i > 0$. We consider dynamic Markov controllers given by (2). The closed loop system is

$$\begin{bmatrix} x(k+1) \\ \widehat{x}(k+1) \end{bmatrix} = \begin{bmatrix} A_{\theta(k)} & B_{\theta(k)}\widehat{C}_{\theta(k)} \\ \widehat{B}_{\theta(k)}H_{\theta(k)} & \widehat{A}_{\theta(k)} \end{bmatrix} \begin{bmatrix} x(k) \\ \widehat{x}(k) \end{bmatrix} + \begin{bmatrix} E_{\theta(k)} \\ \widehat{B}_{\theta(k)}G_{\theta(k)} \end{bmatrix} w(k)$$

$$z(k) = \begin{bmatrix} C_{\theta(k)} & D_{\theta(k)}\widehat{C}_{\theta(k)} \end{bmatrix} \begin{bmatrix} x(k) \\ \widehat{x}(k) \end{bmatrix}.$$

Writing

$$\Gamma_i = \begin{bmatrix} A_i & B_i\widehat{C}_i \\ \widehat{B}_iH_i & \widehat{A}_i \end{bmatrix};\ \Psi_i = \begin{bmatrix} E_i \\ \widehat{B}_iG_i \end{bmatrix};\ \Phi_i = [C_i \;\; D_i\widehat{C}_i];\ \mathbf{v}(k) = \begin{bmatrix} x(k) \\ \widehat{x}(k) \end{bmatrix}$$

we have that the Markov jump closed loop system $\mathcal{G}_{cl}$ is given by

$$\mathcal{G}_{cl} = \begin{cases} \mathbf{v}(k+1) = \Gamma_{\theta(k)}\mathbf{v}(k) + \Psi_{\theta(k)}w(k) \\ \quad z(k) = \Phi_{\theta(k)}\mathbf{v}(k) \end{cases}$$

with $\mathbf{v}(k)$ of dimension n_{cl}.

Definition 2. *We say that the controller $\mathcal{G}_K$ is admissible if the closed loop MJLS $\mathcal{G}_{cl}$ is MSS.*

Definition 3. *The H_2- norm of the closed loop system $\mathcal{G}_{cl}$ is defined as:*

$$\|\mathcal{G}_{cl}\|_2^2 = \sum_{s=1}^{n_{cl}} \sum_{i=1}^{N} \|z_s\|_2^2 \pi_i \tag{5}$$

where $x(0) = 0$, $\widehat{x}(0) = 0$, $\|z_s\|_2^2 = \sum_{k=1}^{\infty} E(\|z_s(k)\|^2)$ and $z_s = (z_s(0), z_s(1), \ldots)$ is the output under the following conditions:

a) The input is δ_s, defined as: $w_s(0) = e_s$, $w_s(k) = 0$ for all $k > 0$, with $e_s \in \mathbb{R}^p$ the unitary vector formed by 1 at the sth position, 0 elsewhere; and
b) $\theta(0) = i$ with probability π_i.

Let $P = (P_1, \ldots, P_N)$ and $S = (S_1, \ldots, S_N)$ be the only solution (see [4]) of the observability and controllability gramians

$$S_i = \Gamma_i'\mathcal{E}_i(S)\Gamma_i + \Phi_i'\Phi_i \qquad \text{Observability Gramian}$$

$$P_j = \sum_{i=1}^{N} p_{ij}[\Gamma_i P_i \Gamma_i' + \pi_i \Psi_i \Psi_i'] \quad \text{Controllability Gramian}$$

where $\mathcal{E}(X) = (\mathcal{E}_1(X), \ldots, \mathcal{E}_N(X))$ is defined as $\mathcal{E}_i(X) = \sum_{j=1}^{N} p_{ij}X_j$ for $X = (X_1, \ldots, X_N)$. We have the following theorem, whose proof can be found in [4]:

Theorem 1. *We have that*

$$\|\mathcal{G}_{cl}\|_2^2 = \sum_{i=1}^{N}\sum_{j=1}^{N} \pi_i p_{ij} tr(\Psi_i' S_j \Psi_i) = \sum_{i=1}^{N} tr(\Phi_i P_i \Phi_i').$$

Therefore, the optimal H_2 control (OC for short) problem we want to study is: find $(\widehat{A}, \widehat{B}, \widehat{C})$, where $\widehat{A} = (\widehat{A}_1, \ldots, \widehat{A}_N)$, $\widehat{B} = (\widehat{B}_1, \ldots, \widehat{B}_N)$, $\widehat{C} = (\widehat{C}_1, \ldots, \widehat{C}_N)$, such that the closed loop MJLS $\mathcal{G}_{cl}$ is MSS and minimize $\|\mathcal{G}_{cl}\|_2^2$.

4.2 Filtering Problem

For the optimal filtering (OF for short) problem consider that $\{w(0), w(1), \ldots\}$ is a wide sense white noise sequence (that is, $E(w(k)) = 0$, $E(w(k)w(k)') = I$, $E(w(k)w(l)') = 0$ for $k \neq l$), and $x(0) = 0$. In the OF problem we want to find

$\widehat{A}, \widehat{B}, \widehat{C}$ such that the closed loop system $\mathcal{G}_{cl}$ with $\widehat{x}(0) = 0$ is MSS and minimize $\lim_{k\to\infty} E(\|z(k)\|^2)$. Let $P_i(k) = E(\mathbf{v}(k)\mathbf{v}(k)'\mathbf{1}_{\{\theta(k)=i\}})$, $i = 1, \ldots, N$. Recalling that $\pi_i(k) = P(\theta(k) = i)$, we have that

$$P_j(k+1) = \sum_{i=1}^{N} p_{ij}[\Gamma_i P_i(k)\Gamma_i' + \pi_i(k)\Psi_i\Psi_i'].$$

Moreover, since the closed loop system is assumed to be MSS, we have that $P(k) \overset{k\uparrow\infty}{\longrightarrow} P$ where $P = (P_1, \ldots, P_N)$ is the unique solution of the controlability gramian. Notice that

$$\begin{aligned} E(\|z(k)\|^2) &= E(tr(z(k)z(k)')) = tr(E(\Phi_{\theta(k)}\mathbf{v}(k)\mathbf{v}(k)'\Phi_{\theta(k)}')) \\ &= \sum_{i=1}^{N} tr(E(\Phi_i[\mathbf{v}(k)\mathbf{v}(k)'\mathbf{1}_{\{\theta(k)=i\}}]\Phi_i')) \\ &= \sum_{i=1}^{N} tr(\Phi_i P_i(k)\Phi_i') \ \overset{k\uparrow\infty}{\longrightarrow} \sum_{i=1}^{N} tr(\Phi_i P_i \Phi_i') = \|\mathcal{G}_{cl}\|_2^2 \end{aligned}$$

and thus problems OC and OF are equivalent. It will be convenient to consider the OF problem for the following system:

$$\mathcal{G}_v = \begin{cases} x(k+1) = A_{\theta(k)}x(k) + B_{\theta(k)}u(k) + E_{\theta(k)}w(k), & x(0) = 0 \\ y(k) = H_{\theta(k)}x(k) + G_{\theta(k)}w(k) \\ v(k) = R_{\theta(k)}^{1/2}[F_{\theta(k)}x(k) + u(k)] \end{cases}$$

with $u(k)$ given by (2), and where we assume that $F = (F_1, \ldots, F_N)$ stabilizes (A, B), in the mean square sense, and that $R_i \geq 0$, $i = 1, \ldots, N$. Suppose that there exists $Y = (Y_1, \ldots, Y_N), Y_i \geq 0$ the stabilizing solution of the following filtering coupled Algebraic Riccati equations (CARE)

$$Y_j = \sum_{i=1}^{N} p_{ij}[A_i Y_i A_i' + \pi_i E_i E_i' - A_i Y_i H_i'(G_i G_i' \pi_i + H_i Y_i H_i')^{-1} H_i Y_i A_i']. \tag{6}$$

Define also

$$M_i = A_i Y_i H_i'(G_i G_i' \pi_i + H_i Y_i H_i')^{-1} \tag{7}$$

and

$$\widehat{x}_e(k+1) = A_{\theta(k)}\widehat{x}_e(k) + B_{\theta(k)}u(k) + M_{\theta(k)}(k)(Y(k) - H_{\theta(k)}\widehat{x}_e(k)), \ \widehat{x}_e(0) = 0$$

where $u(k)$ is an admissible controller, and

$$\begin{aligned} Y_j(k+1) &= \sum_{i=1}^{N} p_{ij}[A_i Y_i(k) A_i' + \pi_i(k) E_i E_i' \\ &\quad - A_i Y_i(k) H_i'(G_i G_i' \pi_i(k) + H_i Y_i(k) H_i')^{-1} H_i Y_i(k) A_i'] \\ M_i(k) &= A_i Y_i(k) H_i'(G_i G_i' \pi_i(k) + H_i Y_i(k) H_i')^{-1}. \end{aligned}$$

Define $\widetilde{x}_e(k) = x(k) - \widehat{x}_e(k)$, so that

$$\widetilde{x}_e(k+1) = [A_{\theta(k)} - M_{\theta(k)}(k)H_{\theta(k)}]\widetilde{x}_e(k) + [E_{\theta(k)} - M_{\theta(k)}(k)G_{\theta(k)}]w(k).$$

It is easy to check that $Y_j(k) = E(\widetilde{x}_e(k)\widetilde{x}_e(k)'\mathbf{1}_{\{\theta(k)=j\}})$. We have the following theorems, whose proofs can be found in [4].

Theorem 2. *$E(\widetilde{x}_e(k)\widehat{x}_e(k)'\mathbf{1}_{\{\theta(k)=i\}}) = 0$ and $E(\widetilde{x}_e(k)\widehat{x}(k)'\mathbf{1}_{\{\theta(k)=i\}}) = 0$ for $i = 1, \ldots, N$ and $k = 0, 1, \ldots$.*

Theorem 3. *Setting $\Phi_i = R_i^{1/2}[F_i \quad \widehat{C}_i]$ we have for every $k = 0, 1, \ldots$ that*

$$E(\|v(k)\|^2) = \sum_{i=1}^{N} tr(\Phi_i P_i(k)\Phi_i') \geq \sum_{i=1}^{N} tr(R_i^{1/2} F_i Y_i(k) F_i' R_i^{1/2}).$$

Theorem 4. *An optimal solution for the OF problem posed above is:*

$$\widehat{A}_i^{op} = A_i - M_i H_i - B_i F_i; \quad \widehat{B}_i^{op} = M_i; \quad \text{and} \quad \widehat{C}_i^{op} = -F_i$$

and the optimal cost is $\min \|\mathcal{G}_{cl}\|_2^2 = \|\mathcal{G}_{cl}^{op}\|_2^2 = \sum_{i=1}^N tr(\overline{F}_i Y_i \overline{F}_i')$, where $\overline{F}_i = R_i^{1/2} F_i$.

Remark 1. Since $Y(k) \overset{k\uparrow\infty}{\longrightarrow} Y$ we have from Theorem 3 that $\lim_{k\to\infty} E(\|v(k)\|^2) = \sum_{i=1}^N tr(\Phi_i P \Phi_i') \geq \lim_{k\to\infty} \sum_{i=1}^N tr(\overline{F}_i Y_i(k) \overline{F}_i') = \sum_{i=1}^N tr(\overline{F}_i Y_i \overline{F}_i')$.

4.3 The Separation Principle for H_2-Control of MJLS

Suppose that there exists $X = (X_1, \ldots, X_N), X_i \geq 0$ the stabilizing solution of the following optimal control CARE

$$X_i = A_i'\mathcal{E}_i(X)A_i - A_i'\mathcal{E}_i(X)B_i(D_i'D_i + B_i'\mathcal{E}_i(X)B_i)^{-1}B_i'\mathcal{E}_i(X)A_i + C_i'C_i \quad (8)$$

and let

$$F_i = (D_i'D_i + B_i'\mathcal{E}_i(X)B_i)^{-1}B_i'\mathcal{E}_i(X)A_i, \quad \text{and} \quad R_i = (D_i'D_i + B_i'\mathcal{E}_i(X)B_i). \quad (9)$$

Define $\widetilde{A}_i = A_i - B_iF_i$, $\widetilde{C}_i = C_i - D_iF_i$ and consider the following change of variables: $u(k) = v(k) - F_{\theta(k)}x(k)$, so that,

$$\begin{cases} x(k+1) = \widetilde{A}_{\theta(k)}x(k) + B_{\theta(k)}v(k) + E_{\theta(k)}w(k) \\ z(k) = \widetilde{C}_{\theta(k)}x(k) + D_{\theta(k)}v(k) \\ y(k) = H_{\theta(k)}x(k) + G_{\theta(k)}w(k) \end{cases}$$

We can decompose the above system such that $x(k) = x_1(k) + x_2(k)$, $z(k) = z_1(k) + z_2(k)$ and

$$\mathcal{G}_c = \begin{cases} x_1(k+1) = \widetilde{A}_{\theta(k)}x_1(k) + E_{\theta(k)}w(k) \\ z_1(k) = \widetilde{C}_{\theta(k)}x_1(k) \end{cases}$$

$$\mathcal{G}_U = \begin{cases} x_2(k+1) = \widetilde{A}_{\theta(k)}x_2(k) + B_{\theta(k)}R_{\theta(k)}^{-1/2}R_{\theta(k)}^{1/2}v(k) \\ z_2(k) = \widetilde{C}_{\theta(k)}x_2(k) + D_{\theta(k)}R_{\theta(k)}^{-1/2}R_{\theta(k)}^{1/2}v(k) \end{cases}$$

so that $z(k) = \mathcal{G}_c(w)(k) + \mathcal{G}_U(R^{1/2}v)(k)$. We have the following result (see [4] for the proof).

Proposition 1. $\mathcal{G}_U^* \mathcal{G}_U = I$.

Setting

$$\mathcal{G}_v = \begin{cases} \mathbf{v}(k+1) = \Gamma_{\theta(k)} \mathbf{v}(k) + \Psi_{\theta(k)} w(k) \\ \quad v(k) = R_{\theta(k)}^{1/2} [F_{\theta(k)} \;\; \widehat{C}_{\theta(k)}] \begin{bmatrix} x(k) \\ \widehat{x}(k) \end{bmatrix} = \Phi_{\theta(k)} \mathbf{v}(k) \end{cases}$$

we have that $z(k) = \mathcal{G}_S(w)(k) = \mathcal{G}_c(w)(k) + \mathcal{G}_U(\mathcal{G}_v(w))(k)$. The norm of system $\mathcal{G}_S$ can be written as

$$\begin{aligned} \|\mathcal{G}_S(w)\|_2^2 &= \langle \mathcal{G}_c(w) + \mathcal{G}_U(\mathcal{G}_v(w)); \mathcal{G}_c(w) + \mathcal{G}_U(\mathcal{G}_v(w)) \rangle \\ &= \|\mathcal{G}_c(w)\|_2^2 + \langle \mathcal{G}_U^* \mathcal{G}_c(w); \mathcal{G}_v(w) \rangle + \langle \mathcal{G}_v(w); \mathcal{G}_U^* \mathcal{G}_c(w) \rangle \\ &\quad + \langle \mathcal{G}_U^* \mathcal{G}_U \mathcal{G}_v(w); \mathcal{G}_v(w) \rangle. \end{aligned}$$

We can show that $\langle \mathcal{G}_U^* \mathcal{G}_c(\delta_r); \mathcal{G}_v(\delta_r) \rangle = 0$. Furthermore from Proposition 1 we have that $\mathcal{G}_U^* \mathcal{G}_U = I$, and thus, $\langle \mathcal{G}_U^* \mathcal{G}_U \mathcal{G}_v(w); \mathcal{G}_v(w) \rangle = \|\mathcal{G}_v(w)\|_2^2$. This leads to

$$\|\mathcal{G}_S\|_2^2 = \sum_r \|\mathcal{G}_S(\delta_r)\|_2^2 = \|\mathcal{G}_c\|_2^2 + \|\mathcal{G}_v\|_2^2$$

and since $\mathcal{G}_c$ does not depends on u,

$$\min_u \|\mathcal{G}_S\|_2^2 = \|\mathcal{G}_c\|_2^2 + \min_u \|\mathcal{G}_v\|_2^2.$$

Moreover, from Theorems 1 and 4

$$\|\mathcal{G}_c\|_2^2 = \sum_{i=1}^N \pi_i tr(E_i^* \mathcal{E}_i(X) E_i), \quad \min_u \|\mathcal{G}_v\|_2^2 = \sum_{i=1}^N tr(R_i^{1/2} F_i Y_i F_i' R_i^{1/2}).$$

The main result can now be presented, for the proof see [4].

Theorem 5. *Consider system $\mathcal{G}_S$ and Markov jump mean square stabilizing controllers $\mathcal{G}_K$. Suppose that there exist the mean square stabilizing solutions $Y = (Y_1, \ldots, Y_N)$ and $X = (X_1, \ldots, X_N)$ for the filtering and control CARE and let $M = (M_1, \ldots, M_N)$ and $F = (F_1, \ldots, F_N)$ be the control and filtering gains obtained from (6),(7),(8),(9). Then an optimal solution for the H_2-control problem is given by $\widehat{A}^{op}$, $\widehat{B}^{op}$, $\widehat{C}^{op}$ as in $\widehat{A}_i^{op} = A_i - M_i H_i - B_i F_i$, $\widehat{B}_i^{op} = M_i$ and $\widehat{C}_i^{op} = -F_i$. Moreover the value of the H_2-norm for this control is*

$$\min_{\mathcal{G}_K} \|\mathcal{G}_S\|_2^2 = \sum_{i=1}^N \pi_i \operatorname{tr}(E_i^* \mathcal{E}_i(X) E_i) + \sum_{i=1}^N \operatorname{tr}(R_i^{1/2} F_i Y_i F_i^* R_i^{1/2}).$$

5 Main Results for Problem 2

Bearing in mind subsection 3.3, consider the following MJLS:

$$\mathcal{G}_S = \begin{cases} x(k+1) = A_{\theta(k)}x(k) + E_{\theta(k)}w(k) \\ y(k) = H_{\theta(k)}x(k) + G_{\theta(k)}w(k) \\ x(0) = x_0,\ \theta(0) = \theta_0 \end{cases}$$

assuming that $\theta(k)$ **is not known** at each time k, and that $\{w(0), w(1), \ldots\}$ is a wide sense white noise sequence. We adopt the following notation: for $r(k)$ a random vector, $\widehat{r}(k|t)$ is the best affine estimator of $r(k)$ given $\{y(0), \ldots, y(t)\}$, and $\widetilde{r}(k|t) = r(k) - \widehat{r}(k|t)$. The first goal is to obtain $\widehat{x}(k|k)$. Set

$$z_j(k) = x(k)\mathbf{1}_{\{\theta(k)=j\}} \in \mathbb{R}^n \quad z(k) = \begin{pmatrix} z_1(k) \\ \vdots \\ z_N(k) \end{pmatrix} \in \mathbb{R}^{Nn}$$

and notice that $\widehat{x}(k|k) = \sum_{i=1}^N \widehat{z}_i(k|k)$. The second-moment matrices associated to the above variables are

$$\begin{aligned} Q_i(k) &= E(z_i(k)z_i(k)'),\ i \in \mathbb{N},\ \ Z(k) = E(z(k)z(k)') = \operatorname{diag}[Q_i(k)], \\ \widehat{Z}(k|l) &= E(\widehat{z}(k|l)\widehat{z}(k|l)'),\ 0 \le l \le k, \\ \widetilde{Z}(k|l) &= E(\widetilde{z}(k|l)\widetilde{z}(k|l)'),\ 0 \le l \le k. \end{aligned}$$

We consider the following augmented matrices

$$\begin{aligned} \mathsf{A} &= \begin{bmatrix} p_{11}A_1 & \cdots & p_{N1}A_N \\ \vdots & \ddots & \vdots \\ p_{1N}A_1 & \cdots & p_{NN}A_N \end{bmatrix} \\ \mathsf{G}(k) &= [G_1\pi_1(k)^{1/2} \cdots G_N\pi_N(k)^{1/2}], \\ \mathsf{H} &= [H_1 \cdots H_N], \\ \mathsf{E}(k) &= \operatorname{diag}[[(p_{1j}\pi_1(k))^{1/2}E_1 \cdots (p_{Nj}\pi_N(k))^{1/2}E_N]]. \end{aligned}$$

We have the following theorem, for the proof see [4].

Theorem 6. *The LMMSE* $\widehat{x}(k|k)$ *is given by* $\widehat{x}(k|k) = \sum_{i=1}^N \widehat{z}_i(k|k)$ *where* $\widehat{z}(k|k)$ *satisfies the recursive equation*

$$\begin{aligned} \widehat{z}(k|k) =& \widehat{z}(k|k-1) + \widetilde{Z}(k|k-1)\mathsf{H}'\big(\mathsf{H}\widetilde{Z}(k|k-1)\mathsf{H}' \\ &+ \mathsf{G}(k)\mathsf{G}(k)'\big)^{-1}(y(k) - \mathsf{H}\widehat{z}(k|k-1)) \\ \widehat{z}(k|k-1) =& \mathsf{A}\widehat{z}(k-1|k-1),\ k \ge 1,\ \widehat{z}(0|-1) = q(0) = \begin{pmatrix} \mu_0\pi_1(0) \\ \vdots \\ \mu_0\pi_N(0) \end{pmatrix}. \end{aligned}$$

The positive semi-definite matrices $\widetilde{Z}(k|k-1)$ *are obtained from* $\widetilde{Z}(k|k-1) = Z(k) - \widehat{Z}(k|k-1)$ *where* $Z(k) = \text{diag}[Q_j(k)]$ *are given by the recursive equation*

$$Q_j(k+1) = \sum_{i=1}^{N} p_{ij} A_i Q_i(k) A_i' + \sum_{i=1}^{N} p_{ij} \pi_i(k) E_i E_i', \; Q_j(0) = \mathbb{Q}_0 \pi_j(0), \quad j \in \mathbb{N}$$

and $\widehat{Z}(k|k-1)$ *are given by the recursive equation*

$$\begin{aligned}\widehat{Z}(k|k) =& \widehat{Z}(k|k-1) + \widehat{Z}(k|k-1)\mathsf{H}'\big(\mathsf{H}\widetilde{Z}(k|k-1)\mathsf{H}' \\ &+ \mathsf{G}(k)\mathsf{G}(k)'\big)^{-1}\mathsf{H}\widehat{Z}(k|k-1) \\ \widehat{Z}(k|k-1) =& \mathsf{A}\widehat{Z}(k-1|k-1)\mathsf{A}', \; \widehat{Z}(0|-1) = q(0)q(0)'.\end{aligned}$$

Notice that in Theorem 6 the inverse of $\mathsf{H}\widetilde{Z}(k|k-1)\mathsf{H}' + \mathsf{G}(k)\mathsf{G}(k)'$ is well defined since for each $k = 0, 1, \ldots$ there exists $\iota(k) \in \mathbb{N}$ such that $\pi_{\iota(k)}(k) > 0$ and thus $\mathsf{H}\widetilde{Z}(k|k-1)\mathsf{H}' + \mathsf{G}(k)\mathsf{G}(k)' \geq \mathsf{G}(k)\mathsf{G}(k)' = \sum_{i=1}^{N} \pi_i(k) G_i G_i' > 0$. In Theorem 6 the term $\widetilde{Z}(k|k-1)$ is expressed as the difference between $Z(k)$ and $\widehat{Z}(k|k-1)$, which are obtained from recursive equations. In the next lemma we shall write $\widetilde{Z}(k|k-1)$ directly as a recursive Riccati equation, with an additional term that depends on the second moment matrices $Q_i(k)$. Notice that this extra term would be zero for the case in which there are no jumps ($N = 1$).

Define the linear operator $\mathfrak{V}(\cdot)$ as follows: for $\Upsilon = (\Upsilon_1, \ldots, \Upsilon_N) \in \mathbb{H}^n$,

$$\mathfrak{V}(\Upsilon) = \text{diag}\left[\sum_{i=1}^{N} p_{ij} A_i \Upsilon_i A_i'\right] - \mathsf{A}(\text{diag}[\Upsilon_i])\mathsf{A}'.$$

Notice that if $\Upsilon_i \geq 0$ for all $i = 1, \ldots, N$, then $\mathfrak{V}(\Upsilon) \geq 0$. We have the following lemma (see [4] for the proof):

Lemma 1. $\widetilde{Z}(k|k-1)$ *satisfies the following recursive Riccati equation*

$$\begin{aligned}&\widetilde{Z}(k+1|k) = \mathsf{A}\widetilde{Z}(k|k-1)\mathsf{A}' + \mathfrak{V}(Q(k)) + \mathsf{E}(k)\mathsf{E}(k)' \\ &\quad - \mathsf{A}\widetilde{Z}(k|k-1)\mathsf{H}'(\mathsf{H}\widetilde{Z}(k|k-1)\mathsf{H}' + \mathsf{G}(k)\mathsf{G}(k)')^{-1}\mathsf{H}\widetilde{Z}(k|k-1)\mathsf{A}'.\end{aligned}$$

where $Q(k) = (Q_1(k), \ldots, Q_N(k))$ *are given by the recursive equations in Theorem 6.*

As mentioned before, for the case with no jumps ($N = 1$) we would have $\mathfrak{V}(Q(k)) = 0$ and therefore the equation above would reduce to the standard recursive Ricatti equation for the Kalman filter.

Stationary Linear Filter

In Lemma 1 we have a recursive Riccati equation for $\widetilde{Z}(k|k-1)$. We shall establish now its convergence when $k \to \infty$. We assume that

1. System $\mathcal{G}_S$ is mean square stable (MSS) and that

2. the Markov chain $\{\theta(k)\}$ is ergodic so that

$$0 < \pi_i = \lim_{k \to \infty} \pi_i(k).$$

Re-define the matrices $\mathsf{G}(k)$, $\mathsf{E}(k)$ as

$$\mathsf{G} = [G_1 \pi_1^{1/2} \cdots G_N \pi_N^{1/2}],\ \mathsf{E} = \mathrm{diag}[[(p_{1j}\pi_1)^{1/2} E_1 \cdots (p_{Nj}\pi_N)^{1/2} E_N]].$$

From MSS and ergodicity of the Markov chain, we have that $Q(k) \to Q$ as $k \to \infty$, where $Q = (Q_1, \ldots, Q_N)$ is the unique solution that satisfies:

$$Z_j = \sum_{i=1}^{N} p_{ij} (A_i Z_i A_i' + \pi_i E_i E_i'),\quad j \in \mathbb{N}. \tag{10}$$

In what follows we define for any matrix $Z \geq 0$, $\mathsf{T}(Z)$ as: $\mathsf{T}(Z) = -\mathsf{A}Z\mathsf{H}'(\mathsf{H}Z\mathsf{H}' + \mathsf{G}\mathsf{G}')^{-1}$. As before, we have that $\mathsf{H}Z\mathsf{H}' + \mathsf{G}\mathsf{G}' > 0$ and thus the above inverse is well defined. The following theorem establishes the asymptotic convergence of $\widetilde{Z}(k|k-1)$. The idea of the proof is first to show that there exists a unique positive semi-definite solution P for the algebraic Riccati equation, and then prove that for some positive integer $\kappa > 0$, there exists lower and upper bounds, $R(k)$ and $P(k+\kappa)$ respectively, for $\widetilde{Z}(k+\kappa|k+\kappa-1)$, such that it squeezes asymptotically $\widetilde{Z}(k+\kappa|k+\kappa-1)$ to P (see [4] for the proof).

Theorem 7. *Consider the algebraic Riccati equation given by:*

$$Z = \mathsf{A}Z\mathsf{A}' + \mathsf{E}\mathsf{E}' - \mathsf{A}Z\mathsf{H}'(\mathsf{H}Z\mathsf{H}' + \mathsf{G}\mathsf{G}')^{-1}\mathsf{H}Z\mathsf{A}' + \mathfrak{V}(Q) \tag{11}$$

where $Q = (Q_1, \ldots, Q_N)$ *satisfies (10). Then there exists a unique positive semi-definite solution* $Nn \times Nn$ P *to (11). Moreover,* $r_\sigma(\mathsf{A} + \mathsf{T}(P)\mathsf{H}) < 1$ *and for any* $Q(0) = (Q_1(0), \ldots, Q_N(0))$ *with* $Q_i(0) \geq 0$, $i \in \mathbb{N}$, *and* $\widetilde{Z}(0|-1) = \mathrm{diag}[Q_i(0)] - q(0)q(0)' \geq 0$ *we have that* $\widetilde{Z}(k+1|k)$ *given as in Theorem 6 satisfies* $\widetilde{Z}(k+1|k) \stackrel{k\to\infty}{\longrightarrow} P$.

6 Main Results for Problem 3

6.1 Preliminaries

We consider in this section that the transition probability matrix $\mathbf{P} = [p_{ij}]$ satisfies:

Assumption 1. *For each* $j \in \mathbb{N}$, $\sum_{i=1}^{N} p_{ij} > 0$.

We assume that $\theta(k)$ *is not directly observed* but, instead, there is a finite set $\mathbb{M} = \{1, \ldots M\}$ such that a signal $\widehat{\theta}(k) \in \mathbb{M}$ is emitted associated to the Markov chain $\theta(k)$, independently of all previous and present values of the other processes, satisfying (3). Notice that $\sum_{\ell=1}^{M} \alpha_{i\ell} = 1$ for each $i \in \mathbb{N}$. We define for each $i \in \mathbb{N}$, $\mathcal{I}_i = \{\ell \in \mathbb{M}; \alpha_{i\ell} > 0\} = \{k_1^i, \ldots, k_{\tau^i}^i\}$, $\mathcal{I} = \bigcup_{i=1}^{N} \mathcal{I}_i \subset \mathbb{M}$. We have 2 extreme situations:

a) $M = N$ and $\alpha_{ii} = 1$, for $i \in \mathbb{N}$, which would correspond to the situation in which $\widehat{\theta}(k) = \theta(k)$, that is, $\theta(k)$ is known. In this case $\mathcal{I}_i = \{i\}$ and $\mathcal{I} = \mathbb{N}$.
b) $M = 1$ and $\alpha_{i1} = 1$ for all $i \in \mathbb{N}$, which corresponds to the situation in which $\widehat{\theta}(k)$ doesn't provide any information about $\theta(k)$, that is, $\theta(k)$ is unknown.

We will consider state-feedback controls using the observed emitted signal $\widehat{\theta}(k)$ instead of the unknown variable $\theta(k)$, that is, $u(k)$ will be of the following form:

$$u(k) = K_{\widehat{\theta}(k)}x(k), \tag{12}$$

for $K_\ell \in \mathbb{B}(\mathbb{R}^n, \mathbb{R}^m)$, $\ell \in \mathcal{I}$. Associated to a control as in (12) set for $i \in \mathbb{N}$, $\ell \in \mathcal{I}_i$,

$$A_{i\ell} = A_i + B_i K_\ell. \tag{13}$$

We define for each $i \in \mathbb{N}$ the following operators $\mathcal{T}$, $\mathcal{L}$ in $\mathbb{B}(\mathbb{H}^n)$. For $V = (V_1, \ldots, V_N) \in \mathbb{H}^n$, and $i, j \in \mathbb{N}$,

$$\mathcal{T}_j(V) = \sum_{i=1}^{N} \sum_{\ell \in \mathcal{I}_i} p_{ij}\alpha_{i\ell} A_{i\ell} V_i A'_{i\ell}, \tag{14}$$

$$\mathcal{L}_i(V) = \sum_{\ell \in \mathcal{I}_i} \alpha_{i\ell} A'_{i\ell} \mathcal{E}_i(V) A_{i\ell}. \tag{15}$$

It is easy to see that $\mathcal{T}^* = \mathcal{L}$ (see [5]). We recall the following definition of stochastic stabilizability.

Definition 4. *We say that system (1) is stochastically stabilizable if there exists* $K_\ell \in \mathbb{B}(\mathbb{R}^n, \mathbb{R}^m)$, $\ell \in \mathcal{I}$, *such that for* $u(k)$ *as in (12) we have, for every initial condition* x_0 *with finite second moment and every initial Markov state* θ_0, *that* $\|x\|_2^2 = \sum_{k=0}^{\infty} E(\|x(k)\|^2) < \infty$. *We denote by* $\mathcal{K}$ *the set of feedback gains* $K = \{K_\ell; \ell \in \mathcal{I}\}$, *such that stochastically stabilizes system (1).*

Define the matrices

$$\Phi_{ij} = \begin{bmatrix} p_{ij}\alpha_{ik_1^i} I \cdots p_{ij}\alpha_{ik_{\tau^i}^i} I \end{bmatrix}, \quad \Phi = \begin{bmatrix} \Phi_{11} & \cdots & \Phi_{N1} \\ \vdots & \ddots & \vdots \\ \Phi_{1N} & \cdots & \Phi_{NN} \end{bmatrix},$$

$$\Psi_i = \begin{bmatrix} A_{ik_1^i} \otimes A_{ik_1^i} \\ \vdots \\ A_{ik_{\tau^i}^i} \otimes A_{ik_{\tau^i}^i} \end{bmatrix}, \quad \Psi = \mathrm{diag}[\Psi_i], \quad \mathcal{A} = \Phi\Psi. \tag{16}$$

Consider $u(k)$ as in (12). Define as before $Q_i(k) = E(x(k)x(k)^* \mathbf{1}_{\{\theta(k)=i\}})$, $i \in \mathbb{N}$, and $Q(k) = (Q_1(k) \ldots, Q_N(k)) \in \mathbb{H}^n$. The next proposition gives a time evolution for $Q(k)$.

Proposition 2. *We have that* $Q(k+1) = \mathcal{T}(Q(k))$ *and that* $\hat{\varphi}(Q(k+1)) = \mathcal{A}\hat{\varphi}(Q(k))$.

For the proof, see [5].

6.2 Stochastic Stabilizability

In this subsection we present conditions for stochastic stabilizability of system (4). In subsection 6.2 we derive conditions (sufficient and necessary and sufficient) for the general Markov chain setting. In this scenario, the LMI formulation gives a sufficient condition for stochastic stabilizability. Subsection 6.3 deals with some special cases (including the Bernoulli jump case, which corresponds to the situation in which $p_{ij} = p_j > 0$ for all i, j) where we derive necessary and sufficient LMI conditions.

General Markov Chain For $R = \{R_{i\ell} \in \mathbb{B}(\mathbb{R}^n); i \in \mathbb{N},\ \ell \in \mathcal{I}_i\}$, define $\mathcal{D}_j(R) = \sum_{i=1}^{N} \sum_{\ell \in \mathcal{I}_i} p_{ij} \alpha_{i\ell} R_{i\ell}$, for $j \in \mathbb{N}$. The following result presents conditions for stochastic stabilizability of system (4).

Theorem 8. *Consider the following assertions:*

i) System (4) is stochastically stabilizable.
ii) There exists $K_\ell \in \mathbb{B}(\mathbb{R}^n, \mathbb{R}^m)$, $\ell \in \mathcal{I}$ such that for $A_{i\ell}$ as in (13), and $\mathcal{A}$ as in (16), we have that $r_\sigma(\mathcal{A}) < 1$ (or equivalently, $r_\sigma(\mathcal{T}) < 1$ or $r_\sigma(\mathcal{L}) < 1$).
iii) There exists $K_\ell \in \mathbb{B}(\mathbb{R}^n, \mathbb{R}^m)$, $\ell \in \mathcal{I}$ and $V \in \mathbb{H}^n$, $V > 0$, such that for $A_{i\ell}$ as in (13),

$$V - \mathcal{T}(V) > 0. \tag{17}$$

iv) There exists $K_\ell \in \mathbb{B}(\mathbb{R}^n, \mathbb{R}^m)$, $\ell \in \mathcal{I}$ and $P \in \mathbb{H}^n$, $P > 0$, such that for $A_{i\ell}$ as in (13),

$$P - \mathcal{L}(P) > 0. \tag{18}$$

v) There exists $K_\ell \in \mathbb{B}(\mathbb{R}^n, \mathbb{R}^m)$, G_ℓ, $\ell \in \mathcal{I}$, $R_{i\ell}$, $i \in \mathbb{N}$, $\ell \in \mathcal{I}_i$, such that for $A_{i\ell}$ as in (13),

$$\begin{bmatrix} R_{i\ell} & A_{i\ell} G_\ell \\ \star & G_\ell + G_\ell' - \mathcal{D}_i(R) \end{bmatrix} > 0, \quad i \in \mathbb{N},\ \ell \in \mathcal{I}_i. \tag{19}$$

We have that i) $\Leftrightarrow$ ii) $\Leftrightarrow$ iii) $\Leftrightarrow$ iv) and v) $\Longrightarrow$ iii).

For the proof, see [5].

6.3 Some Necessary and Sufficient LMI Conditions

Let us provide next a hypothesis so that all assertions in Theorem 8 are equivalent.

Assumption 2. *Assume that $M \leq N$, $\mathbb{M} \subset \mathbb{N}$, and for each $i, k \in \mathbb{N}$ and $\xi \in \mathcal{I}_k$, we have that $p_{i\xi} = p_{ik}$.*

Notice that the case $M = N$ and $\alpha_{ii} = 1$ for $i \in \mathbb{N}$, so that $\mathcal{I}_i = \{i\}$ (which corresponds to the situation in which $\widehat{\theta}(k) = \theta(k)$, that is, $\theta(k)$ is known) satisfies Assumption 2 since that in this case $\mathcal{I}_k = \{k\}$ and clearly $p_{i\xi} = p_{ik}$ for $\xi \in \mathcal{I}_k$. The case in which $p_{ij} = \frac{1}{N}$ for all $i, j \in \mathbb{N}$ (which corresponds to the situation in which all modes are independent and equally like to occur) also satisfies Assumption 2. In what follows set $\mathbf{I} = (\mathbf{I}_1, \ldots, \mathbf{I}_N) \in \mathbb{H}^{n+}$ as (notice that, from Assumption 1, $\sum_{i=1}^{N} p_{ij} > 0$) $\mathbf{I}_j = \sum_{i=1}^{N} p_{ij} I > 0$. We have the following proposition:

Proposition 3. *If Assumption 2 is satisfied then in Theorem 8 we have that i) $\Longrightarrow$ v).*

For the proof, see [5].

From Theorem 8 and Proposition 3 we have the following corollary.

Corollary 1. *If there exists L_ℓ, G_ℓ, $R_{i\ell}$, $i \in \mathbb{N}$, $\ell \in \mathcal{I}_i$, such that*

$$\begin{bmatrix} R_{i\ell} & A_i G_\ell + B_i L_\ell \\ \star & G_\ell + G'_\ell - \mathcal{D}_i(R) \end{bmatrix} > 0, \quad i \in \mathbb{N}, \ \ell \in \mathcal{I}_i \tag{20}$$

then system (4) is stochastically stabilizable with $K_\ell = L_\ell G_\ell^{-1}$. Moreover if Assumption 2 holds then system (4) is stochastically stabilizable if and only if there exists L_ℓ, G_ℓ, $R_{i\ell}$, $i \in \mathbb{N}$, $\ell \in \mathcal{I}_i$ satisfying (20).

For the proof, see [5].

The Bernoulli Jump Case:

We consider now another particular scenario in which again a condition written as an LMI formulation is necessary and sufficient for stochastic stabilizability of system (4). This is achieved by assuming that the transition probabilities of the Markov chain satisfy:

Assumption 3. *For some $p_1 > 0, \ldots, p_N > 0$ we have that $p_{ij} = p_j$, $\forall i, j \in \mathbb{N}$, which is the so-called Bernoulli jump case.*

Defining now the operator $\bar{\mathcal{D}}(R) = \sum_{i=1}^{N} \sum_{\ell \in \mathcal{I}_i} p_i \alpha_{i\ell} R_{i\ell}$, we have the following result:

Theorem 9. *The following assertions are equivalent.*

a) System (4) is stochastically stabilizable.
b) There exists $K_\ell \in \mathbb{B}(\mathbb{R}^n, \mathbb{R}^m)$, $\ell \in \mathcal{I}$ and $X > 0$, such that for $A_{i\ell}$ as in (13),

$$X - \sum_{i=1}^{N} \sum_{\ell \in \mathcal{I}_i} p_i \alpha_{i\ell} A_{i\ell} X A'_{i\ell} > 0. \tag{21}$$

c) There exists $K_\ell \in \mathbb{B}(\mathbb{R}^n, \mathbb{R}^m)$, G, $R_{i\ell}$, $i \in \mathbb{N}$, $\ell \in \mathcal{I}_i$, such that for $A_{i\ell}$ as in (13),

$$\begin{bmatrix} R_{i\ell} & A_{i\ell} G \\ \star & G + G' - \bar{\mathcal{D}}(R) \end{bmatrix} > 0, \quad i \in \mathbb{N}, \ \ell \in \mathcal{I}_i. \tag{22}$$

For the proof, see [5].

From Theorem 9 the following corollary is immediate.

Corollary 2. *System (4) is stochastically stabilizable if and only if there exists L_ℓ, G, $R_{i\ell}$, $i \in \mathbb{N}$, $\ell \in \mathcal{I}_i$ such that*

$$\begin{bmatrix} R_{i\ell} & A_i G + B_i L_\ell \\ \star & G + G' - \bar{\mathcal{D}}(R) \end{bmatrix} > 0, \quad i \in \mathbb{N}, \ \ell \in \mathcal{I}_i. \tag{23}$$

Moreover if (23) holds then system (4) is stochastically stabilizable with $K_\ell = L_\ell G^{-1}$.

For the proof, see [5].

6.4 H_2-Control

General Markov Chain. We consider now the following controlled discrete-time linear system with Markov jumps:

$$x(k+1) = A_{\theta(k)}x(k) + B_{\theta(k)}u(k) + E_{\theta(k)}w(k), \tag{24}$$
$$z(k) = C_{\theta(k)}x(k) + D_{\theta(k)}u(k), \tag{25}$$
$$x(0) = x_0, \ \ \theta(0) \sim \mu. \tag{26}$$

where the output variable is given by $z(k) \in \mathbb{R}^p$, the input variable by $w(k) \in \mathbb{R}^r$ and the initial probability for θ_0 given by $P(\theta_0 = i) = \mu_i > 0$, $i \in \mathbb{N}$. In what follows we consider system (24) with a stochastically stabilizing control $u(k) = K_{\widehat{\theta}(k)}x(k)$ as in (12). Set for $i \in \mathbb{N}$, $\ell \in \mathcal{I}_i$, $C_{i\ell} = C_i + D_i K_\ell$. The H_2 norm associated to system (24)-(25) with the feedback control gain K_ℓ, $\ell \in \mathcal{I}$, denoted by $\|\mathcal{G}_K\|_2$, is defined as in (5) in Definition 3. As in Theorem 1 we have the following result.

Proposition 4. *Let $V \in \mathbb{H}^{n+}$ and $P \in \mathbb{H}^{n+}$ be the unique solution of the observability and controllability Gramians respectively:*

$$V = \mathcal{T}(V) + \mathbf{E}, \ \textit{(Controllability Gramian)} \tag{27}$$
$$P = \mathcal{L}(P) + \mathbf{C}, \ \textit{(Observability Gramian)} \tag{28}$$

where $\mathbf{E} = (\mathbf{E}_1, \ldots, \mathbf{E}_N) \in \mathbb{H}^n$, $\mathbf{C} = (\mathbf{C}_1, \ldots, \mathbf{C}_N) \in \mathbb{H}^n$ are given by $\mathbf{E}_j = \sum_{i=1}^N \mu_i p_{ij} E_i E_i'$, $\mathbf{C}_i = \sum_{\ell \in \mathcal{I}_i} \alpha_{i\ell} C_{i\ell}' C_{i\ell}$. Then $\|\mathcal{G}_K\|_2^2 = \sum_{i=1}^N \mu_i \operatorname{tr}(E_i' \mathcal{E}_i(P) E_i) = \sum_{i=1}^N \sum_{\ell \in \mathcal{I}_i} \alpha_{i\ell} \operatorname{tr}(C_{i\ell} V_i C_{i\ell}')$.

For the proof, see [5].

Consider now the following LMI optimization problem.

$$\Upsilon = \inf_{W_{i\ell}, R_{i\ell}, G_\ell, L_\ell} \sum_{i=1}^N \sum_{\ell \in \mathcal{I}_i} \alpha_{i\ell} \operatorname{tr}(W_{i\ell})$$

subject to: for $i \in \mathbb{N}$, $\ell \in \mathcal{I}_i$,

$$\begin{bmatrix} R_{i\ell} - \mu_i E_i E_i' & A_i G_\ell + B_i L_\ell \\ \star & G_\ell + G_\ell' - \mathcal{D}_i(R) \end{bmatrix} > 0, \begin{bmatrix} W_{i\ell} & C_i G_\ell + D_i L_\ell \\ \star & G_\ell + G_\ell' - \mathcal{D}_i(R) \end{bmatrix} \geq 0. \tag{29}$$

From Proposition 4 we get the following result.

Theorem 10. *For any feasible solution $W_{i\ell}$, $R_{i\ell}$, G_ℓ, L_ℓ of the LMIs (29), we get that* $K = \{K_\ell = L_\ell G_\ell^{-1}, \ell \in \mathcal{I}\} \in \mathcal{K}$, *and* $\|\mathcal{G}_K\|_2^2 \leq \sum_{i=1}^N \sum_{\ell\in\mathcal{I}_i} \alpha_{i\ell}\,\mathrm{tr}(W_{i\ell})$. *Thus,* $\inf_{K\in\mathcal{K}} \|\mathcal{G}_K\|_2^2 \leq \Upsilon$.

For the proof, see [5].

Some Strengthened Results. As in Proposition 3, for the case in which Assumption 2 is satisfied, Theorem 10 can be strengthened as follows.

Proposition 5. *If Assumption 2 holds then* $\inf_{K\in\mathcal{K}} \|\mathcal{G}_K\|_2^2 = \Upsilon$ *in Theorem 10.*

For the proof, see [5].

The Bernoulli Jump Case. We suppose now that Assumption 3 holds. We have the following result.

Proposition 6. *We have that* $\|\mathcal{G}_K\|_2^2 = \sum_{i=1}^N \sum_{\ell\in\mathcal{I}_i} p_i\alpha_{i\ell}\,\mathrm{tr}\{C_{i\ell}XC_{i\ell}'\}$, *where* $X \in \mathbb{R}^{n\times n}$ *is the unique solution to the following generalized Lyapunov equation:* $X = \sum_{i=1}^N \sum_{\ell\in\mathcal{I}_i} p_i\alpha_{i\ell}A_{i\ell}XA_{i\ell}' + \sum_{i=1}^N \mu_i E_i E_i'$.

For the proof, see [5].

The preceding proposition yields the following result, regarding the *synthesis* of H_2 optimal controllers. Set $\zeta(W) = \sum_{i=1}^N \sum_{\ell\in\mathcal{I}_i} p_i\alpha_{i\ell}\,\mathrm{tr}\{W_{i\ell}\}$. It will be shown next that the design of H_2 optimal controllers may be expressed by the following optimization problem, with decision matrices $W_{i\ell}$, G, $R_{i\ell}$, and L_ℓ of appropriate dimensions for $\ell \in \mathcal{I}$, $i \in \mathbb{N}$:

$$\widehat{\Upsilon} = \inf \zeta(W) \tag{30a}$$

subject to

$$\begin{bmatrix} W_{i\ell} & C_iG + D_iL_\ell \\ \star & G + G' - \bar{\mathcal{D}}(R) \end{bmatrix} > 0, \quad \begin{bmatrix} R_{i\ell} - \frac{\mu_i}{p_i}E_iE_i' & A_iG + B_iL_\ell \\ \star & G + G' - \bar{\mathcal{D}}(R) \end{bmatrix} > 0. \tag{30b}$$

Theorem 11. *For any feasible solution of the LMIs* (30b)*, we get that* $K = \{K_\ell = L_\ell G^{-1}, \ell \in \mathcal{I}\} \in \mathcal{K}$, *and* $\|\mathcal{G}_K\|_2^2 \leq \zeta(W)$. *Moreover* $\inf_{K\in\mathcal{K}} \|\mathcal{G}_K\|_2^2 = \widehat{\Upsilon}$.

For the proof, see [5].

6.5 Numerical Example - An Unmanned Aerial Vehicle

Consider the lateral-directional motion for a small UAV in steady flight with the following 3 operation modes: *no faults*, *fault 1:* aileron control surface is inactive; *fault 2:* aileron is inactive, and rudder gets inverted. A detector is able to distinguish between faulty modes with probability ρ. The following emission probability matrix is therefore adopted: $[\alpha_{ij}] = \begin{bmatrix} 1 & 0 & 0 \\ 0 & \rho & 1-\rho \\ 0 & 1-\rho & \rho \end{bmatrix}$, $\rho \in [0,1]$, where ρ is a parameter which determines the quality of the estimator. The control objective is to maintain the linearized model as close as possible to the origin, in order to prevent the aircraft of drifting away from trim conditions. In figure 1 we present the obtained results. For further details, see [5].

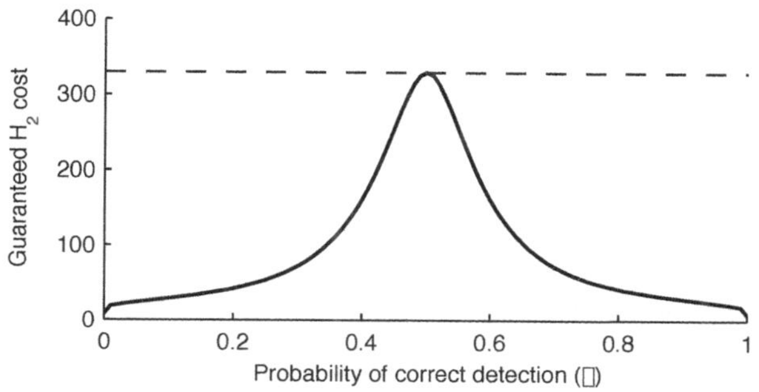

Fig. 1. The obtained H_2 costs span various detection probabilities. The worst scenario coincides with cluster observations, the best case is that of complete observation.

7 Conclusion

We summarize below the main results presented in this paper.
Problem 1: [$\theta(k)$ known, $x(k)$ unknown - H_2 control] - In this case a separation principle for the H_2-control of discrete-time MJLS was derived, considering dynamic MJLS controllers. The result is, mutatis mutandis, in the spirit of that for the classical H_2-control of discrete-time linear systems. An optimal dynamic MJLS controller for the problem was obtained from the mean square stabilizing solution for the CARE associated to the filtering problem and from the mean square stabilizing solution for the CARE associated to the optimal control problem.
Problem 2: [$\theta(k)$ and $x(k)$ unknown - linear filtering] - In this case it was presented the LMMSE for MJLS, based on estimating $x(k)\mathbf{1}_{\{\theta(k)=i\}}$ instead of estimating directly $x(k)$. The LMMSE filter and associated stationary filter produce a recursive scheme suitable for computer implementation which allows some offline computation that alleviates the computational burden. The associated stationary filter for the LMMSE was also obtained. In addition, it contemplates uncertainties in the parameters through, for instance, an LMI approach (not seen in this paper).
Problem 3: [$\widehat{\theta}(k)$ and $x(k)$ known - H_2 control] - It should be noticed that our partial observations setup encompasses cluster observations, full information and mode-independent controls as particular cases. In this case we don't use nonlinear filtering to estimate $\theta(k)$ but, instead, controls are derived directly from the detected symbols $\widehat{\theta}(k)$. Stochastic stabilizability is fully characterized via the spectrum of some matrix lying in the unit circle or Lyapunov-like matrix inequalities. The design of controllers in LMI (linear matrix inequalities) form is amenable to efficient solution via convex programming, and furthermore in particular scenarios, the designs are not conservative. Moreover robustness with respect to polytopic uncertainty can also be considered.

8 Acknowledgement

The authors were partially supported by the Brazilian national research council - CNPq, under grants 302501/2010-0, 304091/2014-6, and 458456/2014-4.

References

1. Brewer, J. W.: Kronecker products and matrix calculus in system theory. *IEEE Trans. Circuits and Systems*, **CAS-25** (1978) 772-781.
2. Caines, P. E. and Zhang, J.: On the adaptive control of jump parameter systems via nonlinear filtering. *SIAM J. Control Optim.*, **33** (1995) 1758-1777.
3. Costa, O. L. V. and Fragoso, M. D.: A separation principle for the H_2-control of continuous-time infinite Markov jump linear systems with partial observations. *J. Math. Analysis and Appl.*, **331** (2007) 97-120.
4. Costa, O. L. V., Fragoso, M. D. and Marques, R. P. *Discrete-Time Markov Jump Linear Systems.* Springer, NY, 2004.
5. Costa, O. L. V., Fragoso, M. D. and Todorov, M. G.: A detector-based approach for the H_2 control of Markov jump linear systems with partial information. *IEEE Trans. Autom. Control*, **60** (2015) 1219-1234.
6. Costa, O. L. V. and Tuesta, E. F.: Finite horizon quadratic optimal control and a separation principle for Markovian jump linear systems. *IEEE Trans. Autom. Control*, **48** (2003) 1836-1842.
7. de Farias, D. P., Geromel, J. C., do Val, J. B. R. and Costa, O. L. V.: Output feedback control of Markov jump linear systems in continuous-time. *IEEE Trans. Autom. Control*, **48** (2000) 944-949.
8. do Val, J. B. R., Geromel, J. C. and Gonçalves, A. P. C.: The H_2-control for jump linear systems: Cluster observations of the Markov state. *Automatica*, **38** (2002) 343-349.
9. Dufour, F. and Elliott, R. J.: Adaptive control of linear systems with Markov perturbations. *IEEE Trans. Autom. Control*, **43** (1998) 351-372.
10. Everdij, M. H. C. and Blom, H. A. P.: Embedding adaptive JLQG into LQ martingale control with a completely observable stochastic control matrix. *IEEE Trans. Autom. Control*, **41** (1996) 424-430.
11. Fragoso, M. D.: On a partially observable LQG problem for systems with Markovian jumping parameters. *Systems and Control Letters*, **10** (1988) 349-356.
12. Fragoso, M. D.: A small random perturbation analysis of a partially observable LQG problem with jumping parameter. *IMA Journal of Math. Control and Inform.*, **7** (1990) 293-305.
13. Fragoso, M. D. and Costa, O. L. V.: A separation principle for the continuous-time LQ-problem with Markovian jump parameters. *IEEE Trans. Autom. Control*, **55** (2010) 2692-2707.

A Functional Analytic Approach to Approximate Iterative Algorithms

Anthony Almudevar

University of Rochester
Dept. of Biostatistics and Computational Biology, Rochester, USA 14642-0630
anthony_almudevar@urmc.rochester.edu

Abstract. In this paper, it is shown how basic ideas of functional analysis may be used to develop a general theory of approximate iterative algorithms. Convergence behavior may be resolved for general classes of algorithms, in contrast to the algorithm-centered approaches often encountered. The theory is demonstrated using a Markov decision process model for the control of tandem queues.
Keywords: Banach spaces, iterative algorithms, approximation theory, Markov decision processes.
AMS 2000 subject classification: Primary 41A65, Secondary 90C40

1 Introduction

Suppose $(\mathcal{V}, \|\cdot\|)$ is a Banach space on which T is a ρ-contractive operator. It is well known that by the Banach fixed point theorem, T possesses unique fixed point $V^* = TV^* \in \mathcal{V}$, which is the limit of the exact iterative algorithm (EIA)

$$V_k = TV_{k-1}, \;\; k = 1, 2, \ldots, \;\; V_0 \in \mathcal{V}. \tag{1}$$

For any number of reasons, we may wish to replace T with an approximate operator $\hat{T}$, or more generally, a sequence of approximate operators $\hat{T}_k$, yielding an *approximate iterative algorithm* (AIA)

$$V_k = \hat{T}_k V_{k-1}, \;\; k = 1, 2, \ldots, \;\; V_0 \in \mathcal{V}. \tag{2}$$

In this paper recent work by the author [1–3] is presented which aims to resolve the following questions:

(Q1) If (1) converges to V^*, under what conditions does (2) also converge to V^*?
(Q2) How does the approximation affect the limiting properties of (2)? How close is the limit of (2) to V^*, and what is the rate of convergence (particularly in comparison to that of (1))?
(Q3) If (2) is subject to design, in the sense that an approximation parameter, such as grid size, can be selected for each $\hat{T}_k$, can an approximation schedule be determined which minimizes approximation error as a function of computation time?

Question (Q1) has been resolved to a great degree of generality, the approximation know to converge to the exact solution with vanishing error [18, 19]. To the best of the author's knowledge, the remaining questions, (Q2) especially, have been solved for specific algorithms, but without a simple general principle having emerged. Here, we will show how the language of functional analysis can be used to unify ideas in control theory, probability, real analysis and algebra to develop a single body of theory from which these questions may be resolved with a high degree of generality. Thus almost all results are expressed in terms of Banach and Hilbert spaces, in contrast with the algorithm-centered approach common in the literature.

We briefly outline the paper. In Chapter 2, the analysis of the relevant theory of algorithm (1) defined on a Banach space is presented. The emphasis is on the development of seminorm iteration. This idea was developed for Markov decision processes (MDP) in [20]. Here we argue that seminorm iteration is quite widely applicable, and in particular can be used generally in place of supremum norm iteration in MDPs, depending on a few basic ideas from spectral analysis. In Chapter 3 a basic theory for the AIA of (2) is introduced, which provides a general resolution of question (Q2). Chapter 4 addresses question (Q3), providing a general principle for the selection of approximation schedules. Of special interest are so-called *coarse-to-fine* approximation methods, in which coarse approximations are used during the initial iterations to save computation time, then gradually refined to yield arbitrarily small error tolerance. In Chapter 5 an MDP model is introduced, and placed in the context of the preceding theory. Chapter 6 introduces a general approximation technique for MDPs. It is noted that the distributions used to construct MDPs often contain large numbers of support points of negligible probability. It would therefore be reasonable to base an approximation method on the truncation of these points, yielding savings in computation time. It is shown how the theory of Chapter 3 can be used to derive convergence rates, and how the theory of Chapter 4 yields a near optimal approximation schedule for coarse-to-fine implementations. In Chapter 7 a tandem queue control model is introduced, and used to demonstrate the proposed theory.

2 Iterative Algorithms on Banach Spaces

The definition of the EIA (1) on a Banach space implies certain assumptions. First, $\mathcal{V}$ must be a vector space and $T : \mathcal{V} \mapsto \mathcal{V}$ an operator on $\mathcal{V}$. In addition, $\|V\| < \infty$ and $\|TV\| < \infty$ for all $V \in \mathcal{V}$, so that (1) is well defined for any initial value $V_0 \in \mathcal{V}$, that is, $V_k \in \mathcal{V}$ for all $k \geq 1$. Furthermore, in a Banach space, convergence of any sequence V_k to limit V' is defined by the condition $\lim_{k\to\infty} \|V_k - V'\| = 0$. Formally, if $\mathcal{V}$ is a vector space, and $\|\cdot\|$ is a norm on $\mathcal{V}$ then the pair $(\mathcal{V}, \|\cdot\|)$ is a normed vector space, and is also a Banach space if it is complete, in the sense that it contains the limit of all of its Cauchy sequences.

2.1 Some Definitions for Banach Spaces

In this section we review a number of important concepts from the theory of Banach spaces. The presentation follows [3], although in that source only those notions most relevant to the proposed theory are summarized. Suitable general introductions to functional analysis and related subjects include [4, 17, 21]. A useful but similarly dedicated summary of functional analysis can be found in [5]. Readers may also find a functional analytic treatment of MDPs in [6, 12, 20]. Suitable sources for probability and measure theory include [7, 9–11], while [15] proves invaluable for an understanding of matrix analysis.

Lipschitz Continuity. The Lipschitz constant L of T is the smallest constant for which

$$\|TV_1 - TV_2\| \leq L\|V_1 - V_2\|, \quad \forall V_1, V_2 \in \mathcal{V}. \tag{3}$$

If $L < \infty$ then T is *Lipschitz continuous.* Then T is nonexpansive or contractive if $L \leq 1$ or $L < 1$ respectively. *Pseudocontraction* is a weaker form of contraction, defined by

$$\|TV - V^*\| \leq \rho\|V - V^*\|, \quad \forall V \in \mathcal{V}, \tag{4}$$

for some $\rho \in (0, 1)$. By the Banach fixed point theorem, if T is contractive then there exists a unique fixed point $V^* = TV^*$ which is the limit of the sequence (1) for any initial point $V_0 \in \mathcal{V}$. This holds also under condition (4), with the additional requirement that V^* is the unique fixed point.

Lipschitz constants of operators are *submultiplicative.* Suppose T_1, T_2 are two operators on normed vector space $(\mathcal{V}, \|\cdot\|)$. Then T_1T_2 is a composite operator defined by the calculation $T_1(T_2V)$ on $V \in \mathcal{V}$. If L_i is the Lipschitz constant of T_i, then the Lipschitz constant of T_1T_2 satisfies

$$L_{1,2} \leq L_1L_2. \tag{5}$$

Linear Operators. The linear operator plays a central role in the theory of Banach spaces. Many important operators, including the dynamic programming operator, while not themselves linear, are constructed from linear operators. An operator $Q : \mathcal{V} \mapsto \mathcal{V}$ on vector space $\mathcal{V}$ is linear if $Q(aV + bW) = aQV + bQW$ for any scalars a, b and vectors V, W. Given normed vector space $(\mathcal{V}, \|\cdot\|)$ the *operator norm* of Q is

$$|||Q||| = \sup\{\|QV\| \mid \|V\| \leq 1\}.$$

If $|||Q||| < \infty$ then Q is a *bounded linear operator*, and is Lipschitz continuous with Lipshitz constant equal to $|||Q|||$. By (5) for any two linear operators Q_1, Q_2 we have the submultiplicative property $|||Q_1Q_2||| \leq |||Q_1|||\ |||Q_2|||$.

Norm Equivalence and Contraction Rates. Two norms $\|\cdot\|_\alpha$, $\|\cdot\|_\beta$ on vector space $\mathcal{V}$ are equivalent, written $\|\cdot\|_\alpha \sim \|\cdot\|_\beta$, if and only if the set of sequences convergent with respect to the two norms are identical. This condition is equivalent to:

$$\sup_{V:\|V\|_\alpha>0} \frac{\|V\|_\beta}{\|V\|_\alpha} < \infty \quad \text{and} \quad \sup_{V:\|V\|_\beta>0} \frac{\|V\|_\alpha}{\|V\|_\beta} < \infty.$$

Any property which holds for equivalent norms is a *norm equivalence property.*

The contraction constant ρ of an operator T is not a norm equivalence property, but a more general notion of a contraction rate is. Consider the Jth iteration T^J of T, which possesses Lipschitz constant ρ_J. By (5) if T is Lipschitz continuous then so if T^J for any $J \geq 1$. In addition, if T is Lipschitz continuous, and $\rho_J < 1$ for some J, then T is J-stage contractive, and the conclusion of the Banach fixed point theorem holds. We may define $\bar{\rho}_J = \rho_J^{1/J}$ as the Jth stage contraction rate.

We have the following theorem (see Theorems 6.3, 7.4 of [3]):

Theorem 1. *If T is a Lipschitz continuous mapping on a Banach space $(\mathcal{V}, \|\cdot\|)$ then there exists a finite constant ρ such that $\lim_{n\to\infty} \bar{\rho}_n = \rho$. Furthermore, ρ is the best possible contraction rate in the sense that $\bar{\rho}_n \geq \rho$, for all $n \geq 1$, and for any $\epsilon > 0$, there exists n_ϵ for which $\bar{\rho}_n \leq \rho + \epsilon$, for all $n \geq n_\epsilon$. Finally, the limit ρ is a norm equivalence property.*

When appropriate we refer to ρ as the asymptotic contraction rate.

Quotient Spaces. Let $\mathcal{N}$ be a subspace of $\mathcal{V}$ (a vector space contained in $\mathcal{V}$). For any $V \in \mathcal{V}$ the *coset* of V modulo $\mathcal{N}$ is the subset $[V] = \{V + n \mid n \in \mathcal{N}\}$. If $W \in [V]$, then $W = V + n_W$ for some $n_W \in \mathcal{N}$. Since $\mathcal{N}$ is a subspace, we have $-n_W \in \mathcal{N}$, so that $V \in [W]$. This means that $[V] = [W]$ for any $W \in [V]$, so that the definition of a coset does not depend on the representative vector. In addition, it may be verified that the space of cosets is a vector space under the linear composition $a[V]+b[W] = [aV + bW]$, with additive identity $[\mathbf{0}] = \mathcal{N}$, where $\mathbf{0}$ is the additive identity of $\mathcal{V}$.

The condition $V - W \in \mathcal{N}$ defines an equivalence relationship $V \sim W$, under which the cosets $[V]$ are equivalence classes. The vector space of equivalence sets is referred to as a *quotient space* $\mathcal{V}/\mathcal{N}$.

Next, associate the following quantity with each coset:

$$\|[V]\|_\alpha = \inf\{\|W\| \mid W \in [V]\} = \inf_{n\in\mathcal{N}} \|V - n\|. \tag{6}$$

Then $\|[V]\|_\alpha$ can be interpreted as the minimum normed distance of V from $\mathcal{N}$. Note that although $\|\cdot\|_\alpha$ of (6) is defined for cosets, it also extends to $\mathcal{V}$ consistently. If we set $\|V\|_\alpha = \inf_{n\in\mathcal{N}} \|V - n\|$ for $V \in \mathcal{V}$ then $\|V\|_\alpha = \|W\|_\alpha$ for any other $W \in [V]$, and becomes a seminorm.

We have Theorem 6.18 from [3]:

Theorem 2. *If $\mathcal{N}$ is a vector subspace of a normed vector space $(\mathcal{V}, \|\cdot\|)$, then the quantity $\|\cdot\|_\alpha$ defined in (6) is a seminorm on the quotient space $\mathcal{V}/\mathcal{N}$ with additive identity $\mathcal{N}$.*

If $\mathcal{N}$ is closed (wrt $\|\cdot\|$) then $\|\cdot\|_\alpha$ is also a norm, and if in addition $(\mathcal{V}, \|\cdot\|)$ is a Banach space, then so is $(\mathcal{V}/\mathcal{N}, \|\cdot\|_\alpha)$.

Function Spaces. Let $\mathcal{F}(\mathcal{X})$ be the vector space of measurable real valued functions on a Borel space $\mathcal{X}$. Then given norm $\|\cdot\|$ we let $\mathcal{F}(\mathcal{X}, \|\cdot\|) \subset \mathcal{F}(\mathcal{X})$ denote the normed vector space of finite normed elements of $\mathcal{F}(\mathcal{X})$. We also define the set of *weight functions* $\mathcal{W}(\mathcal{X}) = \{w \in \mathcal{F}(\mathcal{X}) \mid w(x) > 0\}$. For any weight function w we define the weighted supremum norm $\|\cdot\|_w$ by

$$\|V\|_w = \sup_{x\in\mathcal{X}} |V(x)/w(x)|, \quad V \in \mathcal{F}(\mathcal{X}),$$

denoting the unweighted supremum norm $\|\cdot\|_{sup} = \|\cdot\|_{\mathbf{1}}$. It is easily verified that $\mathcal{F}(\mathcal{X}, \|\cdot\|_{sup})$ is a Banach space, as is any $\mathcal{F}(\mathcal{X}, \|\cdot\|_w)$, being isomorphisms of $\mathcal{F}(\mathcal{X}, \|\cdot\|_{sup})$.

We may define an equivalence class on weight functions by setting $w_1 \sim w_2$ if and only if $\max(\|w_1\|_{w_2}, \|w_2\|_{w_1}) < \infty$. It may be verified that $\|\cdot\|_{w_1} \sim \|\cdot\|_{w_2}$ if and only if $w_1 \sim w_2$ (Theorem 6.23, [3]).

Span Seminorm. Next, suppose we are given Banach space $\mathcal{F}(\mathcal{X}, \|\cdot\|_w)$, and weight function $\nu \in \mathcal{F}(\mathcal{X}, \|\cdot\|_w)$. The space $\mathcal{N}_\nu = \{a\nu \mid a \in \mathbb{R}\}$ is a subspace of $\mathcal{F}(\mathcal{X}, \|\cdot\|_w)$. The conditions hold under which Theorem 2 implies that the quotient space $\mathcal{V}/\mathcal{N}_\nu$ paired with norm (6) is a Banach space. In this case $\|\cdot\|_\alpha$ is equivalent to the *span seminorm*

$$\|V\|_{SP} = \sup_{x\in\mathcal{X}} V(x) - \inf_{x\in\mathcal{X}} V(x) \leq 2\|V\|_{sup}, \quad V \in \mathcal{F}(\mathcal{X}), \tag{7}$$

(that it is a seminorm is easily verified), and the *weighted span seminorm*

$$\|V\|_{SP(\nu)} = \|\nu^{-1}V\|_{SP} = \sup_{x\in\mathcal{X}} \nu(x)^{-1}V(x) - \inf_{x\in\mathcal{X}} \nu(x)^{-1}V(x),$$

for weight function $\nu \in \mathcal{W}$. This is summarized in the following theorem (Theorem 6.26, [3]):

Theorem 3. *If we have Banach space $(\mathcal{V}, \|\cdot\|_w)$ which contains w and identity $\mathbf{0}$, and $\nu \sim w$, then $(\mathcal{V}/\mathcal{N}_\nu, \|\cdot\|_{SP(\nu)})$ is also a Banach space.*

2.2 Iterative Algorithms on Quotient Spaces

Given that a Banach space $(\mathcal{V}/\mathcal{N}_\nu, \|\cdot\|_{SP(\nu)})$ might be induced from a Banach space $(\mathcal{V}, \|\cdot\|_\nu)$ we can then consider whether or not it is possible to define EIA (1) on the quotient space and, if so, whether this yields any advantage. We need to consider the following questions:

1. Can the operator T be defined on a quotient space and, if so, what subspace $\mathcal{N}$ must be used?
2. Convergence in a quotent space is to a subset of $\mathcal{V}$. Given that the objective is a unique fixed point in $\mathcal{V}$, can the problem even be solved in a quotient space?
3. Assuming the answer to the first two questions is yes, how do the contraction rates of the two Banach spaces compare?

Operators on Quotient Spaces. We first consider whether or not T can be meaningfully defined on a quotient space. An operator T (not necessarily linear) on vector space $\mathcal{V}$ possesses an eigenpair $(\beta, \nu) \in \mathbb{R} \times \mathcal{V}$ if for any scalar c and $V \in \mathcal{V}$ we have

$$T(V + c\nu) = TV + \beta c\nu. \tag{8}$$

As a technical matter, although T is constructed as a mapping between single elements of $\mathcal{V}$ we may also regard it as a mapping between sets of vectors, so that if $E \subset \mathcal{V}$, we take $TE \subset \mathcal{V}$ to be a mapping of subsets, evaluated as the image of E under T.

However, under this extension it is only meaningful to describe T as an operator on a quotient space $\mathcal{V}/\mathcal{N}$ if $TE \in \mathcal{V}/\mathcal{N}$ for all $E \in \mathcal{V}/\mathcal{N}$. In terms of the original meaning of the operator, this is equivalent to the claim that for any $E_1 \in \mathcal{V}/\mathcal{N}$ there is another $E_2 \in \mathcal{V}/\mathcal{N}$ such that $T : E_1 \mapsto E_2$ is a bijective mapping. The condition (8) suffices for this to hold for $\mathcal{V}/\mathcal{N}_\nu$.

Implementation. To summarize, if we are given operator T on Banach space $(\mathcal{V}, \|\cdot\|_\nu)$ which possesses eigenpair (β, ν) satisfying (8) then T is also an operator on $(\mathcal{V}/\mathcal{N}_\nu, \|\cdot\|_{SP(\nu)})$.

Of course, as a practical matter we will usually be interested in operations and solutions in the original Banach space $(\mathcal{V}, \|\cdot\|_\nu)$. If T is J-stage contractive then there will exist a fixed point $V^* = TV^* \in \mathcal{V}$ which will be the solution of interest. In addition, by (7) convergence in the supremum norm implies convergence in the span seminorm, but the latter only yields a coset fixed point $[V'] = T[V']$. However, since $\|V^* - TV^*\|_{SP(\nu)} = 0$, V^* and TV^* must be in the same coset. Since $[V']$ is the unique fixed point of T in $\mathcal{V}/\mathcal{N}_\nu$ we must have $V^* \in [V']$, and so may be expressed $V^* = V' + a\nu$ for some scalar a. Applying the fixed point equation, with (8), gives $V' + a\nu = T(V' + a\nu) = TV' + \beta a\nu$. We may then solve for a, giving

$$V^* = HV' = V' + (1 - \beta)^{-1}(TV' - V'), \tag{9}$$

where we refer to H as the *span adjustment.* This means that whenever $\beta < 1$ we may obtain the fixed point $V^* = TV^*$ in $\mathcal{V}$ from any element of the fixed point coset $[V'] = T[V']$.

A similar issue arises in the actual implementation of the iterations. Whether we define the algorithm on $\mathcal{V}$ or $\mathcal{V}/\mathcal{N}_\nu$ the iterates will still take the form $V_{k+1} = TV_k$, where V_k is a specific vector in $\mathcal{V}$. However, if we may only claim convergence in the span seminorm then the iterates $V_{k+1} = TV_k$ need not converge, and a modification is needed to yield a numerically stable algorithm. Fortunately, we need only rely on the fact that a coset may be represented by any of its elements. Therefore, define a canonical member $V_0 \in [V]$, and use evaluations $T[V] = TV_0$. For example, if $1 \in \mathcal{X}$, we may take V_0 to be the unique element of $[V]$ for which $V(1) = 0$. The iterations then assume the form $V_{k+1} = T(V_k - (V_k(1)/\nu(1))\nu)$. Essentially, we have devised a modified operator $T_0 V = T(V - (V(1)/\nu(1))\nu)$, which yields iterations $V_{k+1} = T_0 V_k$.

In general, we may use a stopping rule of the form (Section 7.6, [3])

$$\|V_k - V^*\| \leq \frac{\rho_J}{1-\rho_J} \|V_k - V_{k-J}\|. \tag{10}$$

We may therefore specify a fixed tolerance ϵ, leading to the claim $\|V_N - V^*\| \leq \epsilon$ for the smallest N for which the upper bound (10) does not exceed ϵ. While this suffices for iteration in $\|\cdot\|_\nu$ one more step is needed for iteration in $\|\cdot\|_{SP(\nu)}$. This can be based on the following theorem (Theorem 7.11, [3]):

Theorem 4. *Let T be a monotone operator on a vector space $\mathcal{V}$. Suppose there exists $\nu \in \mathcal{V}$ and constant $\beta < 1$ for which $T(V + a\nu) = TV + \beta a\nu$ for all $V \in \mathcal{V}$ and scalars a. Suppose $V^* = TV^*$ is the fixed point of T. Then*

$$\|H\hat{V} - V^*\|_\nu \leq \frac{1+\beta/2}{(1-\beta)} \|\hat{V} - V^*\|_{SP(\nu)} \tag{11}$$

where H is the span adjustment defined by (9).

Convergence Rates in Quotient Spaces. By Theorem 6.27 of [3] if a linear operator Q possesses eigenpair (β, ν) satisfying (8), then $|||Q|||_{SP(\nu)} \leq |||Q|||_\nu$.

Consider pair $\pi = (R, Q)$, where $R \in \mathcal{V}$ and $Q : \mathcal{V} \mapsto \mathcal{V}$ is a bounded linear operator. Define operator $T_\pi V = R + QV$. Then we have

$$\|T_\pi V_1 - T_\pi V_2\| = \|Q(V_1 - V_2)\| \leq |||Q||| \, \|V_1 - V_2\|,$$

so that $|||Q|||$ is the Lipschitz constant for T_π, and it is easily verified by extending the argument that $|||Q^J|||$ is the Lipschitz constant for T_π^J.

To fix ideas, consider a situation commonly encountered in MDP models. Suppose $\beta \in (0,1)$ is a scalar (a discount factor, typically), and that $Q = \beta Q_0$, where Q_0 is a nonexpansive operator. Then we have $|||Q^J|||_{SP(\nu)} = |||\beta^J Q_0^J|||_{SP(\nu)} = \beta^J |||Q_0^J|||_{SP(\nu)}$, whereas we usually have $|||Q|||_\nu = \beta$ for iteration in the supremum norm. Thus there are two sources of contraction in the span seminorm, the standard discount factor β, and the quantity $|||Q_0^J|||_{SP(\nu)}$.

Suppose Q_0 is an irreducible stochastic matrix, and $\nu = \mathbf{1}$. Then $(1, \mathbf{1})$ is an eigenpair of Q_0. Then $|||Q_0|||_{SP}$ is Dobrushin's ergodic coefficient, and is an upper bound for $|\lambda_{SLEM}| < 1$, where λ_{SLEM} is the second largest eigenvalue in magnitude (see, for example, Theorem 7.2 of [8]). We also have $|||Q_0^J|||_{SP} \approx |\lambda_{SLEM}|^J$ so that $|\lambda_{SLEM}|$ is the asymptotic contraction rate of Q_0, and so $\beta|\lambda_{SLEM}|$ is the asymptotic contraction rate of T_π. We may therefore expect convergence to be considerably faster in the seminorm.

3 Approximate Iterative Algorithms

The discussion so far has been limited to EIAs of the form (1). We now consider AIAs (2), which we express in the form

$$V_k = \hat{T}_k V_{k-1} = TV_{k-1} + U_k \;\; k = 1, 2, \ldots, \;\; V_0 \in \mathcal{V} \tag{12}$$

where U_k is intended to model approximation error. While (2) and (12) are nominally the same, the expression in (12) (uniquely) relies on vector addition, so that while (2) is defined on a metric space, (12) is defined on a vector space. If we accept this additional structure, it becomes possible to characterize the behavior of (12) in very general ways.

To consider approximation error, it will be useful to consider an upper envelope

$$\|U_k\| \le d_k, \;\; k \ge 1.$$

In some of the theory, it will be necessary to require that the sequence d_k varies smoothly. In [3] the *lower and upper convergence rates* of a positive sequence a_k, $k \ge 1$ are defined as

$$\lambda^l\{a_k\} = \underline{\lim}_{k\to\infty} a_{k+1}/a_k \text{ and } \lambda^u\{a_k\} = \overline{\lim}_{k\to\infty} a_{k+1}/a_k.$$

If $\lambda^l\{a_k\} = \lambda^u\{a_k\} = \rho < 1$ then $a_k \to 0$ smoothly with a linear convergence rate of ρ. In contrast $\lambda^l\{a_k\} = \lambda^u\{a_k\} = 1$ for polynomial rates $a_k \propto 1/k^p$, $p > 0$. In addition we let $\hat{\lambda}\{a_k\} = \lim_{k\to\infty} a_k^{1/k}$ when the limit exists (with $\hat{\lambda}^l$,$\hat{\lambda}^u$ similarly defined). Order o_ℓ denotes domination by linear convergence rate.

The strategy is to characterize the envelope d_k in terms of its linear convergence rate compared to ρ (or sublinear convergence rate of 1). Accordingly, for $\rho \in (0, 1)$ define the following families of sequences:

$$\begin{aligned} \mathcal{F}^L_\rho &= \left\{ \{d_k\} : \sum_{k=1}^{\infty} \rho^{-k} d_k < \infty \right\}, \\ \mathcal{F}_\rho &= \left\{ \{d_k\} : \hat{\lambda}\{d_k\} = \rho \right\}, \text{ and} \\ \mathcal{F}^U_\rho &= \left\{ \{d_k\} : \lambda^l\{d_k\} > \rho \right\}. \end{aligned} \tag{13}$$

We present Theorem 10.19 from [3]:

Theorem 5. *Suppose $(\mathcal{V}, \|\cdot\|)$ is a Banach space on which T is a ρ-contractive operator (which therefore possesses fixed point $V^* = TV^* \in \mathcal{V}$). Suppose $\hat{T}_k$ is a sequence of operators on $(\mathcal{V}, \|\cdot\|)$ which defines AIA (2) for which*

$$\|\hat{T}_k V_{k-1} - TV_{k-1}\| \le a_k + b_k \|V_{k-1}\|_0, \;\; k \ge 1,$$

for sequences a_k, b_k, where $\|\cdot\|_0$ is a seminorm for which $\|V\|_0 \le \kappa_0 \|V\|$ for all $V \in \mathcal{V}$ for some finite constant κ_0. Define $d^*_k = a_k + b_k \|V^*\|_0$. Then the following statements hold:

(*i*) *If* $\overline{\lim}_k \kappa_0 b_k = \delta \in [0, 1-\rho)$ *then*

$$\overline{\lim_k} \|V_k - V^*\| \leq (1-\rho-\delta)^{-1} \overline{\lim_k} d_k^*$$

(*ii*) *If* $\{d_k^*\} \in \mathcal{F}_\rho^L$ *then* $\|V_k - V^*\| \leq O(\rho^k)$.
(*iii*) *If* $\{d_k^*\} \in \mathcal{F}_\rho$ *then* $\hat{\lambda}^u \{\|V_k - V^*\|\} \leq \rho$.
(*iv*) *If* $\{d_k^*\} \in \mathcal{F}_\rho^U$ *and* $d_k^* \to 0$ *then*

$$\overline{\lim_k} [d_k^*]^{-1} \|V_k - V^*\| \leq (1 - \rho/\lambda^l\{d_k^*\})^{-1}.$$

More informally, we have $\|V_k - V^*\| \leq O(\max(d_k^*, (\rho+\epsilon)^k))$ for all $\epsilon > 0$ if $d_k^* \to 0$ at a linear rate of at least ρ, and if $d_k^* \to 0$ at a linear rate less than ρ, then $\|V_k - V^*\| \to 0$ at the best possible linear rate ρ. That these rates are tight is quite reasonable to expect, and is verfied in the following theorem (Theorem 10.18, [3]):

Theorem 6. *In algorithm* (12) *let* $d_n = \sup_{n' \geq n} \|U_{n'}\|$ *for* $n \geq 1$. *If* V^* *is a fixed point of* T, *and there exists a finite constant* L *such that* $\|TV - TV^*\| \leq L\|V - V^*\|$ *for all* $V \in \mathcal{V}$, *then* $\overline{\lim}_{n\to\infty} \|V_{n-1} - V^*\|/d_n > 0$.

4 Optimal Approximation Schedules

Questions (Q1) and (Q2) have been resolved with some considerable generality, so we next consider question (Q3) (the reader may be referred to [2] for more detail on this topic). We can write an approximating inequality for an AIA in the form

$$\|V_k - V^*\| \leq \eta_k = B\alpha^k + u_k, \quad k \geq 1. \tag{14}$$

We refer to η_k as the *algorithm tolerance* and u_k is the *approximation tolerance*, that is, the excess tolerance attributable to the use of approximate operators. Without approximation, $B\alpha^k$ remains the tolerance of the EIA. We also refer to any bound $\|U_k\| \leq \epsilon_k$ as the *operator tolerance*. This form is discussed in detail in Chapter 10 of [3].

Next, let g_k be the computation cost for the evaluation of the kth iteration. If g_k is approximately constant, then η_k is an appropriate metric with which to evaluate algorithmic efficiency. On the other hand, suppose this does not hold, that is, the *cumulative cost*

$$\bar{G}_k = \sum_{i=1}^{k} g_i$$

does not satisfy $\bar{G}_k = O(k)$. Then η_k must be transformed to some function which behaves more like $\hat{\eta}_t \approx \eta_k$ when $t = \bar{G}_k$, which we refer to as the *computational algorithm tolerance*, and is interpretable as the algorithm tolerance achieved after a computational effort of t. Accordingly, define

$$k_\eta(v) = \inf\{k : \eta_k \leq v\},$$

so that

$$\bar{G}_\eta(v) = \bar{G}_{k_\eta(v)}$$

is the computation cost required to attain an algorithm tolerance of v.

The situation we anticipate is that there is a family of approximate operators $\hat{T}_\tau$ for T indexed by some parameter τ. For example, suppose T involves integration, but only numerical quadrature methods are feasible. Then τ would define the grid size, and we assume that the algorithm used to evaluate TV may use any value of τ. Clearly, smaller τ would yield smaller tolerance at the cost of greater computation time. Furthermore, we may choose to vary τ within an AIA, yielding approximate operators $\hat{T}_k = \hat{T}_{\tau_k}$ for some sequence τ_k, $k \geq 1$. The intuition here is that initial approximations may be coarse, saving computation time in the early stages, then gradually refined to yield arbitrarily small algorithm tolerances, and, ideally, an AIA proven to converge to the exact solution.

4.1 Tolerance Model

We assume that we have control, through the selection of a sequence of approximate operators, of the sequence u_k in (14). In this context, this sequence is referred to as an *approximation schedule*. We also define a *computation function* G, which gives the cost $g_k = G(u_k)$ of attaining approximation tolerance u_k (by choice of approximate operator $\hat{T}_k$) at the kth iteration. This leads to the definition of the *tolerance model*.

Definition 1. *A set $\mathcal{A}^S$ of AIAs conforms to a* tolerance model $M = (\alpha, G)$ *if a schedule $S = \{u_k\}$ may be associated with each algorithm for which (i) (14) holds for some finite B and for which (ii) the computation cost of the kth iteration is given by $g_k = G(u_k)$.*

This definition may seem problematic, since $G(u_k)$ models the cost of iteration k, while u_k represents the cumulative effect of the entire approximation history. However, we may just as easily use rates given by Theorem 5. When operator tolerance ϵ_k converges at a slower rate than α^k we have $u_k = O(\epsilon_k)$, and when ϵ_k converges at a faster rate, (14) will also hold for some $u_k = O(\epsilon_k)$.

4.2 Main Theorem

Suppose we are given computation function G and two schedules $S = \{u_k\}$ and $S' = \{u'_k\}$. We make use of the following regularity conditions:

(A) The function $G : [0, \infty) \to [0, \infty]$ is finite on $(0, \infty)$, and satisfies the following properties for any two schedules $\{u_k\}, \{u'_k\}$:

(*i*) if $u'_k = o(u_k)$ then $\underline{\lim}_{k\to\infty}(G(u'_k) - G(u_k)) = d_G$ for some $d_G > 0$ (possibly, $d_G = \infty$),
(*ii*) if $\lim_{k\to\infty} G(u'_k)/G(u_k) = \infty$ then $u'_k = o(u_k)$.

We will also employ the following growth conditions on G and schedule S:

(B1) There exists a finite constant r_G for which the pair (G, S) satisfies

$$\overline{\lim_{k\to\infty}}\, G(u_{k+1})/G(u_k) = r_G.$$

(B2) The pair (G, S) satisfies $\lim_{k\to\infty} G(u_k)/\sum_{j=1}^{k} G(u_j) = 0$.

We will sometimes need to strengthen an ordering $u'_k = o(u_k)$ of two schedules $S = \{u_k\}$ and $S' = \{u'_k\}$. It will suffice to impose the following condition:

(C) Given two schedules S, S' we have $u'_k = o(u_{k+d})$ for all integers $d \geq 0$.

It is easily verified that (C) holds when $u'_k = o(u_k)$ and $\lambda^l\{u_k\} > 0$.

Theorems 7, 8 and 9 together establish a general principle. See [2] (Theorems 2-4) for the proofs. Suppose we are given two schedules $\{u_k\}$ and $\{u'_k\}$, with respective computational algorithm tolerances $\hat{\eta}_t$ and $\hat{\eta}'_t$. If schedule $\{u_k\}$ is closer than $\{u'_k\}$ to the linear rate of convergence α^k, then $\hat{\eta}_t$ is not worse, and may be strictly better than, $\hat{\eta}'_t$.

The first theorem deals with the lower bound case. Suppose $u_k = O(\alpha^k)$. If $u'_k = o(u_k)$ then $\hat{\eta}_t$ cannot be improved by $\hat{\eta}'_t$, and if in addition condition (C) holds, $\hat{\eta}_t$ will be strictly better.

Theorem 7. *Given tolerance model $M = (\alpha, G)$, with G satisfying condition (A), suppose we are given two schedules S and S', with constants B, B'. Suppose $u_k \leq O(\alpha^k)$. Then (i) if $u'_k = o(u_k)$ then $\hat{\eta}_t \leq O(\hat{\eta}'_t)$, (ii) if S, S' satisfy condition (C) then $\hat{\eta}_t = o(\hat{\eta}'_t)$.*

For the upper bound case it is assumed that $u_k \geq O(\alpha^k)$ and that $u_k = o_\ell(u'_k)$. Under the weaker condition (B1) $\hat{\eta}_t$ cannot be improved by $\hat{\eta}'_t$, and if u'_k converges sublinearly then $\hat{\eta}_t$ is strictly better (Theorem 8). Under condition (B2) $\hat{\eta}_t$ is strictly better in either case (Theorem 9).

Theorem 8. *Suppose a tolerance model $M = (\alpha, G)$ and two schedules S, S' satisfy $(B1)$ with $u_k = o_\ell(u'_k)$. Suppose also that schedule S satisfies*

$$\overline{\lim_{k\to\infty}}\, B\alpha^k/u_k = \kappa < \infty. \tag{15}$$

Then for any positive constant $r < 1$ we have (i) $\underline{\lim}_{v\to 0}\, \bar{G}'_\eta\left(r(1+\kappa)^{-1}v\right)/\bar{G}_\eta(v) \geq r_G^{-1}$, (ii) for any finite constant γ there exists $\beta_\gamma > 0$ such that if $-\log(\lambda^l\{u'_k\}) \leq \beta_\gamma$ then $\underline{\lim}_{v\to 0}\, \bar{G}'_\eta\left(r(1+\kappa)^{-1}v\right)/\bar{G}_\eta(v) \geq \gamma$, (iii) furthermore, if S' converges sublinearly then $\lim_{v\to 0} \bar{G}'_\eta\left(r(1+\kappa)^{-1}v\right)/\bar{G}_\eta(v) = \infty$.

Theorem 9. *Suppose a tolerance model $M = (\alpha, G)$ and two schedules S, S' satisfy $(B2)$ with $u_k = o_\ell(u'_k)$. Suppose, in addition, that $\underline{\lim}_{i\to\infty} u_i/\alpha^i > 0$. Then $\hat{\eta}_t = o(\hat{\eta}'_t)$.*

5 Markov Decision Processes

A Markov decision process (MDP) will be made of the following elements (see, for example, [13, 14], or Section 12.1 of [3]):

(M1) A Borel space $\mathcal{X}$. We refer to $\mathcal{X}$ as the *state space.*
(M2) A Borel space $\mathcal{A}$. We refer to $\mathcal{A}$ as the *action space.*
(M3) With each $x \in \mathcal{X}$ associate $\mathcal{K}_x \subset \mathcal{A}$, with $\mathcal{K}_x \neq \emptyset$. The *state/action space* $\mathcal{K} = \{(x,a) \in \mathcal{X} \times \mathcal{A} : a \in \mathcal{K}_x\}$ is assumed to be a measurable subset of $\mathcal{X} \times \mathcal{A}$.
(M4) A *measurable stochastic kernel* $Q : \mathcal{K} \to \mathcal{M}(\mathcal{X})$, the space of probability measures on $\mathcal{X}$.
(M5) A measurable mapping $R : \mathcal{K} \to \mathbb{R}$, referred to as the *cost function.*
(M6) A *discount factor* $\beta \geq 0$.

The process occurs on discrete time points, or stages, $k = 1, 2, \ldots$. Stage k is associated with state/action pair $z_k = (x_k, a_k) \in \mathcal{K}$, at which a cost of $R(x_k, a_k)$ is assumed, the process then transitioning to state x_{k+1} according to probability distribution $Q(\cdot \mid x_k, a_k)$. A control policy Φ is a method of selecting action a_k given history $H_k = (x_1, \ldots, x_k)$. A *Markovian deterministic policy* is a mapping $\phi : \mathcal{X} \mapsto \mathcal{A}$ which forces $a_k = \phi(x_k)$. It will be convenient to refer to model $\pi = (R, Q)$.

Given π, Φ, β, and initial state $X_1 = x$, the MDP generates state/action pairs (X_k, A_k) for stage k, assuming expected discounted cost

$$V_\pi^\Phi(x) = E_x^\Phi \left[\sum_{n=1}^{\infty} \beta^{n-1} R(X_n, A_n) \right], \quad x \in \mathcal{X}. \tag{16}$$

The problem is to determine the policy Φ^* which minimizes $V_\pi^\Phi(x)$ for any x.

The theory of MDPs is well known, and under general conditions the optimal policy is a Markovian deterministic policy $\Phi^* = \phi^*$. Of central importance is the *dynamic programming operator* (DPO), defined elementwise by the computation

$$\bar{T}_\pi V(x) = \inf_{a \in \mathcal{K}_x} R(x,a) + \beta \int_{y \in \mathcal{X}} V(y) dQ(y \mid x, a). \tag{17}$$

If we define *value function* $\bar{V}_\pi(x) = \inf_\Phi V_\pi^\Phi(x)$ for each $x \in \mathcal{X}$, that is, the minimum attainable cost given $X_1 = x$, then under general conditions it satisfies fixed point equation $\bar{V}_\pi = \bar{T}_\pi \bar{V}_\pi$, and is the limit of the iterative algorithm $V_{k+1} = \bar{T}_\pi V_k$, given suitable initial solution V_0, generally referred to as *value iteration* (VI).

If we select Markovian deterministic policy ϕ, we have $R^\phi(x) = R(x, \phi(x))$ and $Q^\phi(\cdot \mid x) = Q(\cdot \mid x, \phi(x))$, defining operator $T_\pi^\phi V = R^\phi + Q^\phi V$. This operator is simpler than the DPO, but similiar in that the fixed point solution $V_\pi^\phi = T_\pi^\phi V_\pi^\phi$ equals the policy value function (16) for $\Phi = \phi$, which is the limit of the iterative algorithm $V_{k+1} = T_\pi^\phi V_k$. It will sometimes be convenient to consider T_π^ϕ as a special case of a DPO by restricting the state/action space to $\mathcal{K}^\phi = \{(x, \phi(x)) \mid x \in \mathcal{X}\}$.

5.1 DPO Norms

At this point we introduce two quantities which will be central to our approximation theory. First, we introduce the total variation norm on the space of signed measures on Borel space $\mathcal{X}$:

$$\|\mu\|_{TV} = \sup_E \mu(E) - \inf_E \mu(E) = \int |f| d\nu,$$

where the supremum and infimum are taken over all measurable subsets, and f is a density of μ with respect to measure ν (the definition obviously does not depend on the choice of measure ν, as long as the density exists). The second formulation makes possible the weighted total variation norm

$$\|\mu\|_{TV(w)} = \int w|f| d\nu,$$

for positive weight function w. Two measures can be compared with $\|\cdot\|_{TV(w)}$, but the quantity $\|\mu_1 - \mu_2\|_{TV(w)}$ can be made arbitrarily close to zero only if the two measures possess densities with respect to a common measure.

For a model π we may define

$$\eta^w_Q = \sup_{z\in\mathcal{K}} w(x)^{-1}\|Q(\cdot \mid z)\|_{TV(w)} \text{ and } \eta^w_R = \sup_{z\in\mathcal{K}} w(x)^{-1}|R(z)|. \tag{18}$$

The definition extends to R^ϕ and Q^ϕ by restricting the state/action space to $\mathcal{K}^\phi$, in which case we have

$$\eta^w_{Q^\phi} = \left\|\left|Q^\phi\right|\right\|_w \leq \eta^w_Q,$$

and η^w_Q is in fact the Lipschitz constant for the DPO $\bar{T}_\pi$ (Theorem 12.14, [3]).

However, the quantities η^w_R and η^w_Q also play a role in approximation, defining a model distance between $\pi = (R, Q)$ and $\hat{\pi} = (\hat{R}, \hat{Q})$, given that these quantities are well defined for differences $R - \hat{R}$ and signed measure kernel $Q - \hat{Q}$. We summarize Theorem 14.4 of [3]:

Theorem 10. *The operator tolerance $\|\bar{T}_{\hat{\pi}}V - \bar{T}_\pi V\|_w$ is bounded by*

$$\|\bar{T}_{\hat{\pi}}V - \bar{T}_\pi V\|_w \leq \eta^w_{R-\hat{R}} + \eta^w_{Q-\hat{Q}}\|V\|_w. \tag{19}$$

Furthermore, if (β, ν) is a positive eigenpair for any Q^ϕ then

$$\|\bar{T}_{\hat{\pi}}V - \bar{T}_\pi V\|_\nu \leq \eta^\nu_{R-\hat{R}} + \frac{1}{2}\eta^\nu_{Q-\hat{Q}}\|V\|_{SP(\nu)}. \tag{20}$$

6 Functional Approximation by Truncation

Theorem 10 provides a way of quantifying directly the effect of model approximation on algorithm tolerance for MDPs. One simple example is a truncation scheme for approximating Q. This is typically constructed from well known probability distributions, for example, the Poisson distribution for a queueing model. Suppose $\mathcal{K}$ is finite. An evaluation of $\bar{T}_\pi$ requires the integration $\int V(y)dQ(y \mid z)$ to be performed for each state /action pair in $\mathcal{K}$, which is an order $O(|\mathcal{X}|^2|\mathcal{A}|)$ computation. Possibly, the distribution $Q(\cdot \mid z)$ contains large numbers of negligible probabilities. If these were truncated from the distribution, we may have an accurate approximation while yielding considerable savings of computation time. We show how this question may be resolved using the preceding approximation theory.

Let $f(y \mid z)$ be the density function of $Q(y \mid z)$ and let $E \subset \mathcal{X}$. We define approximation

$$\bar{f}_E(y \mid z) = \frac{f(y \mid z)I\{y \in E\}}{\int_{y \in E} f(y \mid z)d\mu}.$$

The total variation norm between f and $\bar{f}_E$ is then

$$\|f - \bar{f}_E\|_{TV(w)} = \int_{y \in E} w(P(E)^{-1} - 1)f(y \mid z)d\mu + \int_{y \in E^c} wf(y \mid z)d\mu.$$

If weight function $w \equiv \mathbf{1}$ we have

$$\|f - \bar{f}_E\|_{TV(w)} = 2P(E^c). \tag{21}$$

If we associate a subset $E_z \subset \mathcal{X}$ with each state/action pair and accept $\bar{f}_{E_z}(y \mid z)$ as an approximate stochastic kernel, then we have

$$\eta^w_{Q-\hat{Q}} = \max_{z \in \mathcal{K}} 2P(E^c_z) \tag{22}$$

and Theorem 10 may be used directly, assuming we are using the exact cost function R.

We may then define a parametrized family of approximate operators $\hat{T}_\tau$. For any τ, for each density $f(y \mid z)$ select $E_{z,\tau}$ to be the smallest set for which $2P(E^c_{z,\tau}) \leq \tau$ under distribution $Q(\cdot \mid z)$ (this would be facilitated by an ordering of the probabilities). In addition, set $n_{max}(z, \tau) = |E_{z,\tau}|$. We may then claim $\eta^w_{Q-\hat{Q}} \leq \tau$.

Following Section 4, an approximation schedule τ_k would then be selected for which $\tau_k = O(\beta^k)$, where β is the linear convergence rate of the EIA. Regularity conditions for this approximation method hold, and are discussed in Section 16.2 of [3].

7 Example: Tandem Queue Models

Suppose we are given a control problem involving N tandem $M/G/m/K$ queues. Each queue has Poisson arrivals with rates λ_i, $i = 1, \ldots, N$. We assume that the control problem involves allocation of service to the queues. Much of the discussion will be independent of the particular control structure, however, we will illustrate the proposed methodology using the *switchover queue*, in which a single server is assigned to a specific queue in response to current queue sizes. Decision epochs occur at the end of a service, or at the next arrival when the system is empty. The decision a is the queue to which the server will instantaneously switch (if the server does not switch then a is the current queue). The state is $x = (y, w_1, \ldots, w_N)$ where y is the current server queue, and $w_1, \ldots, w_N$ are the current queue sizes. We assume there is a unit cost to switching. In addition a cost of $C(w_1, \ldots, w_N)$ is assumed at each decision epoch, giving cost function

$$R(x, a) = I\{y \neq a\} + C(w_1, \ldots, w_N).$$

The purpose of C is to control excessive queue length, so we use for our example $C(w_1, \ldots, w_N) = (\max_i w_i)^{1.5}$.

To fix ideas assume that the service time is deterministic of unit time length. The queueing model is constructed using Kendall's embedded Markov chain method [16]. This means that the distributions defining transition kernel Q are based on the product density of N independent Poisson distributions with mean vector $(\lambda_1, \ldots, \lambda_N)$.

7.1 Supremum and Seminorm Convergence

To iterate in the supremum norm, we may used bound (10), the simplest rule obtained by setting $J = 1$,

$$\|V_k - V^*\|_{sup} \leq \frac{\beta}{1-\beta}\|V_k - V_{k-1}\|_{sup}. \tag{23}$$

We require two inequalities to implement iteration in the span seminorm. The objective is to bound $\|\hat{V}_k - V^*\|_{sup}$, where $\hat{V}_k$ is the current approximation. However, as discussed in Section 2.2 the approximate solution is not the current iterate V_k but the span adjustment HV_k defined in (9). Combining Theorem 4 with (10) gives

$$\|HV_k - V^*\|_{SP(\nu)} \leq \frac{1+\beta/2}{1-\beta} \times \frac{\rho_J}{1-\rho_J}\|V_k - V_{k-J}\|_{SP(\nu)}, \tag{24}$$

for $J \geq 1$. Note that an approximately sharp bound for $\rho_J/(1-\rho_J)$ which is independent of β (that is, does not involve the potentially very large factor $(1-\beta)^{-1}$) will exist, as discussed in Section 2.2.

As a demonstration VI in both the supremum norm and the span seminorm was applied to the switchover queue problem with $N = 2$, $K = 50$, $\lambda_i \equiv 1/8$, $\beta = 0.999$ and tolerance $\epsilon = 0.001 \times (1-\beta)^{-1}$. Stopping rules (23) - (24) were used. In our example we find that convergence in the span seminorm is so rapid, that little is lost with the conservative (but still quite correct) choice of $J = 1$ and $\rho_1 = \beta$ for stopping rule (24). Convergence was obtained in 15,208 and 199 iterations (4853.545 and 71.706 seconds) for stopping rules (23) and (24) respectively, with the solutions being identical within the given tolerance.

7.2 Analysis of Truncation Approximation

The queue occupancy is a vector in $\mathcal{S}^N$ where $\mathcal{S} = (0, 1, \ldots, K)$. The state-space will then be $\mathcal{S}^M \times \mathcal{E}$, where $\mathcal{E}$ defines service states associated with the control problem. If we assume $\mathcal{E}$ is not too complicated (server allocation, for example) then the complexity of the state space can be taken to be of order $O\left((K+1)^N\right)$.

Let W be a vector of N independent Poisson random variables truncated at K, with distribution F_W. The transition distribution from $z \in \mathcal{K}$ can be expressed $F(w, z) = F_W(t(w, z))$ for a suitable transformation $t(w, z)$. The cardinality of the image $t(\mathcal{S}^N, z)$ may vary considerably with z, but can be expected to be a significant fraction of $(K+1)^N$ over a large subset of $\mathcal{K}$, (ie. when current queue occupancy is low).

For unmodified VI, each iteration is assumed to be of order $O(|\mathcal{X}|^2|\mathcal{A}|)$, given our assumptions regarding the transition kernel. If we start with initial solution $V_0 \equiv 0$ we have directly

$$\|V_k - V^*\|_{sup} \leq \beta^k \|V^*\|_{sup}$$

so if we set tolerance ϵ relative to $\|V^*\|_{sup}$ we will require $k_\epsilon = \log(\epsilon)/\log(\beta) \approx -(1-\beta)^{-1}\log(\epsilon)$ iterations.

Because the complexity varies by state, and the contributions per state are additive, it is most natural to analyze cumulative complexity on a per state-action pair basis. For unmodifed VI, the complexity per iteration for z is $|t(\mathcal{S}^N, z)| = O((K+1)^N)$ for a significant fraction of states. The worst case cumulative cost required for a specific $z \in \mathcal{K}$ is therefore

$$\hat{\eta}_t^{vi}(z) = \beta^{t/(K+1)^N}.$$

For a first comparison, we consider an AIA with a fixed truncation level. Using Theorem 5, (22) and Theorem 10 we have asymptotic approximation tolerance

$$u_k \approx \frac{\tau}{1-\beta}\|V_{max}\|_{sup}, \quad \text{where } \tau = \max_{z \in \mathcal{K}} 2P(E^c_{z,\tau}).$$

The reliance here on $\|V_{max}\|_{sup}$ can be relaxed, but will serve our purpose. A more refined analysis could use instead $\|V^*\|_{sup}$, so we will assume the two quantities are comparable. It is important to note that while knowing the asymptotic tolerance is crucial, the contribution from the exact tolerance is needed for a meaningful comparison:

$$\hat{\eta}_t^{tr}(z) = \beta^{t/n_{max}(z,\tau)} + \tau(1-\beta)^{-1}.$$

Comparison of VI with Truncation Approximation Algorithms. The following strategy will be used to compare $\hat{\eta}_t^{vi}(z)$ and $\hat{\eta}_t^{tr}(z)$. We first note that $\hat{\eta}_t^{tr}(z)$ decreases to asymptotic bound $\tau(1-\beta)^{-1}$. Therefore, the analyst would select the truncation parameter according to a design tolerance. Of course, to

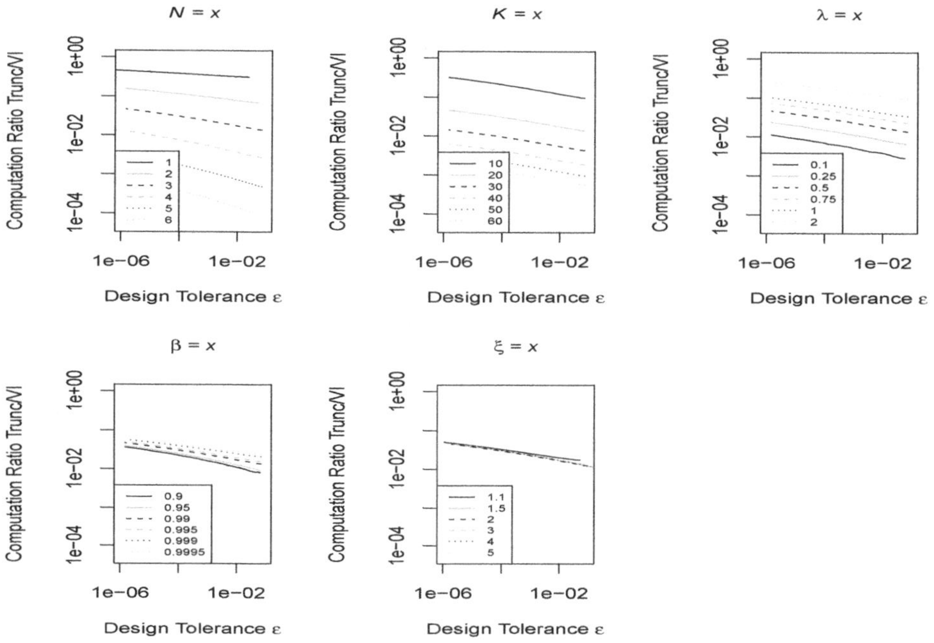

Fig. 1. Plots for analysis of Section 7.2 (fixed truncation algorithm). Horizontal axis gives algorithm tolerance $\epsilon = v$.

determine $\bar{G}_\eta^{tr}(v)$, the computational effort needed for algorithm tolerance v, we must set the asymptotic bound strictly smaller than v:

$$\bar{G}_\eta^{vi}(v) = (K+1)^N \frac{\log(v)}{\log(\beta)}, \quad v > 0$$

$$\bar{G}_\eta^{tr}(v) = n_{max}(z,\tau) \frac{\log\left(v - \tau(1-\beta)^{-1}\right)}{\log(\beta)}, \quad v > \tau(1-\beta)^{-1}.$$

An important conclusion can be readily made. Suppose interest is in the queueing model $M/G/1/\infty$, which is to be approximated using a large truncation parameter K. The total computation order will be $O(\log(\epsilon)(1-\beta)^{-1}K^2|\mathcal{A}|)$. If we employ a fixed truncation scheme, we would have a computation of order $O(\log(\epsilon)(1-\beta)^{-1}K n_{max}(z,\tau)|\mathcal{A}|)$. What is of interest here is that the parameter $n_{max}(z,\tau)$ does not depend on K. Therefore, the complexity scales linearly with K for any fixed ϵ. Thus, any relative advantage of the fixed truncation method over VI can be expected to become more significant as K increases.

For a numerical demonstration, we will assume that the tolerance $v = \xi\tau(1-\beta)^{-1}$ for some fixed factor ξ. We then vary τ, which results in a sequence of pairs $(\tau, n_{max}(z,\tau))$. For fixed ξ this yields target tolerance v, following which $\bar{G}_\eta^{vi}(v)$ and $\bar{G}_\eta^{tr}(v)$ are evaluated for those specific parameters. We then plot the ratio

$\bar{G}^{tr}_{\eta}(v)/\bar{G}^{vi}_{\eta}(v)$ against v, representing the reduction of per state computational effort required to achieve algorithm tolerance v.

We start with the baseline parameter set $N = 3$, $K = 20$, $\lambda = 0.5$, $\beta = 0.99$, $\xi = 1.5$. Figure 1 shows examples of these functions obtained by varying each parameter separately from the baseline set. The observed factors range from approximately 0.5 to 10^{-4}. As can be expected, the factors are smallest for more complex models, both in terms of the number of queues N and the truncation parameter K. We also find that the increase in efficiency is greater for smaller values of λ, which is to be expected given the higher concentration of probability mass on fewer support points. The discount rate β, while a significant factor in the overall computation time for either algorithm, has less effect on the relative efficiency than the model parameters. The value ξ was observed to affect the relative efficiency, and can be appropriately tuned, but in our demonstration the effect is small relative to the model parameters. A value of 1.5 was use on this basis. Finally, the efficiency gains of the truncated method were greater for larger values of $\epsilon = v$, but this variation was also relatively small relative to that induced by varying the model parameters.

For the baseline model, the number of occupancy states is $(K+1)^N = 9,261$, so that the total number of states for a model with a simple control structure will be a fixed multiple of this number. The more complex models have 226,981 ($N = 3$, $K = 60$) and 85,766,121 ($N = 6$, $K = 20$).

If we then apply the truncation algorithm with a tolerance schedule, we set schedule $\tau_k = r^k$, and the cumulative cost per state/action pair at iteration k becomes

$$G_k(z) = \sum_{i=1}^{k} n_{max}(z, r^k).$$

Then the tolerance is

$$\eta_k^{vi} = \beta^k \left(1 + \sum_{i=1}^{k} \tau_k/\beta^k\right),$$

following Theorem 5.

We next present a numerical example. First, take $N = 1$, $K = 200$, $\beta = 0.99$ and $\epsilon = 0.001$. Figure 2 (Plot A) shows the cumulative computation required for tolerance ϵ for varying values of r in the tolerance schedule. Clearly, the choice is very consequential, and the optimal is near, if not exactly equal to, the discount rate β.

Next, for two models we plot algorithm tolerance as a function of cumulative computational effort per state for unmodified VI, and the truncation algorithm with tolerance schedule, and for several fixed truncation schemes of varying tolerance ϵ. Here, we use $N = 2, 4$, $K = 10$, $\beta = 0.99$, $r = \beta$ and $\epsilon = 0.0001, 0.001, 0.01$ for the fixed truncation schemes. Clearly, truncation outperforms unmodified VI, while able to achieve arbitrarily small tolerance. Interestingly, the fixed truncation schemes are comparable to the schedule truncation algorithm, until the target tolerance is reached (Figure 2, Plots B, C).

Finally, the scheduled truncation was compared to unmodified VI for a switchover queue model with parameters $N = 2$, $K = 5, 10, \ldots, 50$, $\lambda_i \equiv 3/8$, $\beta = 0.999$ and tolerance $\epsilon = 0.0001 \times (1 - \beta)^{-1}$. Convergence was tested using the span seminorm, so the approximation schedule τ_k was set by estimating the convergence rate using the methods proposed in Section 16.3 of [3]. The total computational effort was captured. The ratio of effort ranged from 1.03 to 7.08 as K varied from 5 to 50, for unmodifed VI relative to scheduled truncation.

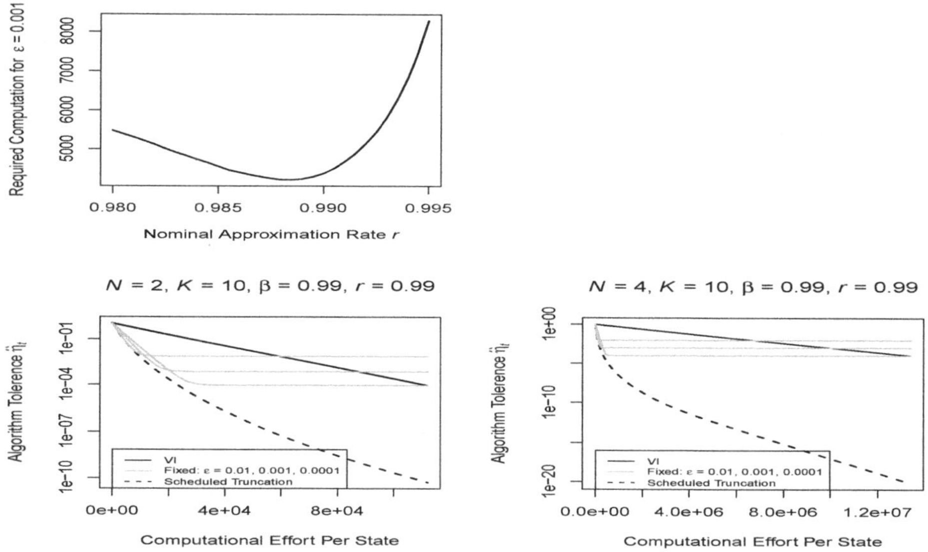

Fig. 2. Plots for analysis of Section 7.2 (scheduled truncation algorithm).

References

1. Almudevar, A.: Approximate fixed point iteration with an application to infinite horizon Markov decision processes. *SIAM J. Control Optim.*, **47** (2008) 2303-2347.
2. Almudevar, A. and de Arruda, E.F: Optimal approximation schedules for a class of iterative algorithms, with an application to multigrid value iteration. *IEEE Trans Automat. Control*, **57** (2012) 3132-3146.
3. Almudevar, A. *Approximate Iterative Algorithms.* CRC Press, 2014.
4. Ash, R.B. and Dolacutuseans-Dade, C.A. *Real Analysis and Probability.* Academic Press, San Diego, Second Edition, 2000.
5. Atkinson, K.A. and Han, W. *Theoretical Numerical Analysis: A Functional Analysis Framework.* Springer-Verlag, New York, NY, 2001.
6. Bertsekas, D.P. and Shreve, S.E. *Stochastic Optimal Control: The Discrete-Time Case.* Academic Press, New York, NY, 1978.

7. Billingsley, P. *Probability and Measure.* John Wiley and Sons, New York, NY, Third Edition, 1995.
8. Bremaud, P. *Markov Chains: Gibbs Fields, Monte Carlo Simulation and Queues.* Springer, New York, NY, 1999.
9. Durrett, R. *Probability: Theory and Examples.* Cambridge University Press, New York, NY, Fourth Edition, 2010.
10. Feller, W. *Probability Theory and Its Applications, Volume 1.* John Wiley and Sons, New York, NY, Third Edition, 1968.
11. Feller, W. *Probability Theory and Its Applications, Volume 2.* John Wiley and Sons, New York, NY, Second Edition, 1971.
12. Hernandez-Lerma, O. *Adaptive Markov Control Processes.* Springer-Verlag, New York, NY, 1989.
13. Hernandez-Lerma, O. and Lasserre, J.B. *Discrete-Time Markov Control Processes: Basic Optimality.* Springer, New York, NY, 1996.
14. Hernandez-Lerma, O. and Lasserre, J.B. *Further Topics on Discrete-Time Markov Control Processes.* Springer, New York, NY, 1999.
15. Horn, R.A and Johnson, C.R. *Matrix Analysis.* Cambridge University Press, Cambridge, UK, 1985.
16. Kendall, D.G.: Stochastic processes occurring in the theory of queues and their analysis by the method of the imbedded Markov chain. *Ann. Math. Statist.*, **24** (1953) 338–354.
17. Kolmogorov, A.N. and Fomin, S.V. *Introductory Real Analysis.* Dover, Mineola, New York, 1970.
18. Ortega, J.M. and Rheinboldt, W.C.: On a class of approximate iterative processes. *Archive for Rational Mechanics and Analysis*, (1967) 352–365.
19. Ostrowski, A. *The Rounding-off Stability of Iterations.* Basel Math. Notes, BMN-12, 1964.
20. Puterman, M.L. *Markov Decision Processes: Discrete Stochastic Dynamic Programming.* John Wiley and Sons, Inc., New York, NY, 1994.
21. Royden, H.L. *Real Analysis.* MacMillan Publishing, New York, NY, Second Edition, 1968.

Finite Approximations to Markov Decision Processes with Borel Spaces

Naci Saldi, Serdar Yüksel and Tamás Linder

Queen's University
Dept. of Mathematics and Statistics, Kingston, ON, Canada
nsaldi@mast.queensu.ca
yuksel@mast.queensu.ca
linder@mast.queensu.ca

Abstract. The purpose of this paper is to study finite model approximations for stochastic control problems with Borel state and action spaces, for both discounted and average cost problems. We study both the finite-action and finite-state approximations of a discrete time Markov decision process (MDP). Under mild continuity conditions, we establish that optimal policies for finite models obtained through quantization of the state and action spaces are approximately optimal for the original MDP. For both problems, we also obtain explicit bounds on the approximation error in terms of the number of representation points in the quantizer, under further conditions. This paper is a review of the authors' recent results in a number of publications.
Keywords: Markov decision processes, stochastic control, finite state approximation, quantization.
AMS 2000 subject classification: 93E20, 90C40, 90C39.

1 Introduction

In this chapter, we provide a condensed review of some of our recent results on the finite action and state approximation of optimal control policies for a discrete time Markov decision processes (MDPs) with Borel state and action spaces, under discounted and average cost criteria.

Various methods have been developed to compute approximately optimal control policies in the literature. For rather complete surveys of these techniques, we refer the reader to [2, 3, 6, 7, 16, 17, 29] and references therein. With the exception of [7, 16], these works in general study either a finite horizon cost or a discounted infinite horizon cost. As well, the majority of these results are for MDPs with discrete (i.e., finite or countable) state and action spaces, or a bounded one-stage cost function (see, e.g., [2, 3, 16–18]). Those that consider general state and action spaces (see, e.g., [2, 4–7]) assume in general Lipschitz type continuity conditions on the components of the control model, in order to provide a rate of convergence analysis for the approximation error.

Our contributions in this chapter allow for more general settings in one or more of the following directions: (i) we consider a general setup, where the state and action spaces are Borel (with the action space being compact), and the one-stage cost function is possibly unbounded, (ii) the continuity assumptions we impose on the components of the control model are weaker than the conditions imposed in prior works that considered general state and action spaces, (iii) we also consider the challenging average cost criterion under reasonable assumptions. (iv) Moreover, under additional continuity conditions on the transition probability and the one stage cost function we also obtain bounds for a rate of approximation in terms of the number of points used to discretize the state space, thereby providing a tradeoff between the computation cost and the performance loss in the system. We refer to [19–22] for a detailed literature review, as well as proofs of the results presented in this chapter.

A discrete-time Markov decision process (MDP) can be described by a five-tuple $\big(\mathsf{X}, \mathsf{A}, \{\mathsf{A}(x) : x \in \mathsf{X}\}, p, c\big)$ where Borel spaces X and A denote the *state* and *action* spaces, respectively. The collection $\{\mathsf{A}(x) : x \in \mathsf{X}\}$ is a family of nonempty subsets $\mathsf{A}(x)$ of A, which gives the admissible actions for the state $x \in \mathsf{X}$. The *stochastic kernel* $p(\,\cdot\,|x, a)$ denotes the *transition probability* of the next state given that previous state-action pair is (x, a) [11]. Hence, it satisfies: (i) $p(\,\cdot\,|x, a)$ is an element of $\mathcal{P}(\mathsf{X})$ (the set of probability measures on X) for all (x, a), and (ii) $p(D|\,\cdot\,, \cdot\,)$ is a measurable function from $\mathsf{X} \times \mathsf{A}$ to $[0, 1]$ for each $D \in \mathcal{B}(\mathsf{X})$, where $\mathcal{B}(\mathsf{X})$ is the Borel σ-field over X. The *one-stage cost* function c is a measurable function from $\mathsf{X} \times \mathsf{A}$ to $[0, \infty)$. In the remainder of this chapter, it is assumed that $\mathsf{A}(x) = \mathsf{A}$ for all $x \in \mathsf{X}$. Define the history spaces $\mathsf{H}_0 = \mathsf{X}$ and $\mathsf{H}_t = (\mathsf{X} \times \mathsf{A})^t \times \mathsf{X}$, $t = 1, 2, \ldots$ endowed with their product Borel σ-algebras generated by $\mathcal{B}(\mathsf{X})$ and $\mathcal{B}(\mathsf{A})$. A *policy* is a sequence $\pi = \{\pi_t\}$ of stochastic kernels on A given H_t. The set of all policies is denoted by Π. Let Φ denote the set of stochastic kernels φ on A given X, and let $\mathbb{F}$ denote the set of all measurable functions f from X to A. A *randomized stationary* policy is a constant sequence $\pi = \{\pi_t\}$ of stochastic kernels on A given X such that $\pi_t(\,\cdot\,|x) = \varphi(\,\cdot\,|x)$ for all t for some $\varphi \in \Phi$. A *deterministic stationary* policy is a constant sequence of stochastic kernels $\pi = \{\pi_t\}$ on A given X such that $\pi_t(\,\cdot\,|x) = \delta_{f(x)}(\,\cdot\,)$ for all t for some $f \in \mathbb{F}$. The set of randomized and deterministic stationary policies are identified with the sets Φ and $\mathbb{F}$, respectively. According to the Ionescu Tulcea theorem [11], an initial distribution μ on X and a policy π define a unique probability measure P_μ^π on $\mathsf{H}_\infty = (\mathsf{X} \times \mathsf{A})^\infty$, which is called a *strategic measure* [8]. The expectation with respect to P_μ^π is denoted by $\mathbb{E}_\mu^\pi$. If $\mu = \delta_x$, we write P_x^π and $\mathbb{E}_x^\pi$ instead of $P_{\delta_x}^\pi$ and $\mathbb{E}_{\delta_x}^\pi$. The cost functions to be minimized in this paper are the β-discounted cost and the average cost, respectively given by

$$J(\pi, x) = \mathbb{E}_x^\pi \biggl[\sum_{t=0}^{\infty} \beta^t c(X_t, A_t) \biggr],$$

$$V(\pi, x) = \varlimsup_{T \to \infty} \frac{1}{T} \mathbb{E}_x^\pi \biggl[\sum_{t=0}^{T-1} c(X_t, A_t) \biggr].$$

With this notation, the discounted and average value functions of the control problem are defined as

$$J^*(x) := \inf_{\pi \in \Pi} J(\pi, x), \qquad V^*(x) := \inf_{\pi \in \Pi} V(\pi, x).$$

A policy π^* is said to be optimal if $J(\pi^*, x) = J^*(x)$ (or $V(\pi^*, x) = V^*(x)$ for the average cost) for all $x \in \mathsf{X}$. Under fairly mild conditions, the set $\mathbb{F}$ of deterministic stationary policies contains an optimal policy for discounted cost (see, e.g., [11, 9]) and average cost optimal control problems (under somewhat stronger continuity/recurrence conditions, see, e.g., [9]).

Summary of the chapter. In Section 2, we study the quantization of the action space, under strong continuity and weak continuity conditions. In Section 3, we study finite model approximations for the state space obtained through quantization. This section also studies non-compact state spaces as well as rates of convergence as the size of the finite model increases, under stronger continuity conditions. The chapter ends with some concluding remarks.

Notation and Conventions. For a metric space E, the Borel σ-algebra is denoted by $\mathcal{B}(\mathsf{E})$. We let $B(\mathsf{E})$ and $C_b(\mathsf{E})$ denote the set of all bounded Borel measurable and continuous real functions on E, respectively. For any $u \in C_b(\mathsf{E})$ or $u \in B(\mathsf{E})$, let $\|u\| := \sup_{e \in \mathsf{E}} |u(e)|$ which turns $C_b(\mathsf{E})$ and $B(\mathsf{E})$ into Banach spaces. Given any Borel measurable function $w : \mathsf{E} \to [1, \infty)$ and any real valued Borel measurable function u on E, we define the w-norm of u as $\|u\|_w := \sup_{e \in \mathsf{E}} \frac{|u(e)|}{w(e)}$, and let $B_w(\mathsf{E})$ and $C_w(\mathsf{E})$ denote the Banach space of all real valued measurable and continuous functions u on E with finite w-norm [12], respectively. Let $\mathcal{P}(\mathsf{E})$ denote the set of all probability measures on E. A sequence $\{\mu_n\}$ of measures on E is said to converge weakly (resp., setwise) to a measure μ if $\int_{\mathsf{E}} g(e)\mu_n(de) \to \int_{\mathsf{E}} g(e)\mu(de)$ for all $g \in C_b(\mathsf{E})$ (resp., for all $g \in B(\mathsf{E})$). For any $\mu, \nu \in \mathcal{P}(\mathsf{E})$, the total variation distance between μ and ν, denoted as $\|\mu - \nu\|_{TV}$, is defined as $\|\mu - \nu\|_{TV} := 2\sup_{D \in \mathcal{B}(\mathsf{E})} |\mu(D) - \nu(D)|$. For any $\nu \in \mathcal{P}(\mathsf{E})$ and measurable real function g on E, we define $\nu(g) := \int g d\nu$. Unless otherwise specified, the term 'measurable' will refer to Borel measurability in the rest of the chapter.

2 Finite Action Approximation to MDPs

To give a precise definition of the problem we study in this section, we first give the definition of a quantizer from the state to the action space.

Definition 1. *A measurable function* $q : \mathsf{X} \to \mathsf{A}$ *is called a* quantizer *from* X *to* A *if the range of* q, *i.e.,* $q(\mathsf{X}) = \{q(x) \in \mathsf{A} : x \in \mathsf{X}\}$, *is finite.*

The elements of $q(\mathsf{X})$ (the possible values of q) are called the *levels* of q. The rate $R = \log_2 |q(\mathsf{X})|$ of a quantizer q (approximately) represents the number of bits needed to losslessly encode the output levels of q using binary codewords of equal length. Let $\mathcal{Q}$ denote the set of all quantizers from X to A. A *deterministic stationary quantizer* policy is a constant sequence $\pi = \{\pi_t\}$ of stochastic kernels on A given X such that $\pi_t(\,\cdot\,|x) = \delta_{q(x)}(\,\cdot\,)$ for all t for some $q \in \mathcal{Q}$. For any finite set $\Lambda \subset \mathsf{A}$, let $\mathcal{Q}(\Lambda)$ denote the set of all elements in $\mathcal{Q}$ having range Λ. Analogous with $\mathbb{F}$, the set of all deterministic stationary quantizer policies induced by $\mathcal{Q}(\Lambda)$ will be identified with the set $\mathcal{Q}(\Lambda)$.

Our main objective in this section is to find conditions on the components of the MDP under which there exists a sequence of finite subsets $\{\Lambda_n\}_{n\geq 1}$ of A for which the following holds:

(P) For any initial point x, we have $\lim_{n\to\infty} \inf_{q\in\mathcal{Q}(\Lambda_n)} J(q,x) = \inf_{f\in\mathbb{F}} J(f,x)$ (or $\lim_{n\to\infty} \inf_{q\in\mathcal{Q}(\Lambda_n)} V(q,x) = \inf_{f\in\mathbb{F}} V(f,x)$ for the average cost), provided that the set $\mathbb{F}$ of deterministic stationary policies is an optimal class for the MDP.

In other words, if for each n, MDP_n is defined as the Markov decision process having the components $\{\mathsf{X}, \Lambda_n, p, c\}$, then **(P)** is equivalent to stating that value function of MDP_n converges to the value function of the original MDP.

2.1 Near Optimality of Quantized Policies Under Strong Continuity

In this section we consider the problem **(P)** for the MDPs with strongly continuous transition probability. We impose the assumptions below on the components of the Markov decision process.

Assumption 1.

(a) The one stage cost function c is nonnegative and bounded satisfying $c(x,\,\cdot\,) \in C_b(\mathsf{A})$ for all $x \in \mathsf{X}$.
(b) The stochastic kernel $p(\,\cdot\,|x,a)$ is setwise continuous in $a \in \mathsf{A}$.
(c) A is compact.

Let d_{A} denote the metric on A. Since the action space A is compact and thus totally bounded, one can find a sequence of finite sets $\Lambda = \{a_{n,1},\dots,a_{n,k_n}\} \subset \mathsf{A}$ such that for all n,

$$\min_{i\in\{1,\dots,k_n\}} d_{\mathsf{A}}(a,a_{n,i}) < 1/n \text{ for all } a \in \mathsf{A}.$$

In other words, Λ_n is a $1/n$-net in A. In the rest of Section 2, we assume that the sequence $\{\Lambda_n\}_{n\geq 1}$ is fixed. To ease the notation in the sequel, let us define the mapping $\Upsilon_n : \mathbb{F} \to \mathcal{Q}(\Lambda_n)$ as

$$\Upsilon_n(f)(x) = \arg\min_{a\in\Lambda_n} d_{\mathsf{A}}(f(x),a), \tag{1}$$

where ties are broken so that $\Upsilon_n(f)(x)$ is measurable.

Discounted Cost We consider here problem **(P)** for the discounted cost with a discount factor $\beta \in (0,1)$ under the Assumption 1. The following theorem is the main result of this section which states that for any $f \in \mathbb{F}$, the discounted cost function of $\Upsilon_n(f) \in \mathcal{Q}(\Lambda_n)$ converges to the discounted cost function of f as $n \to \infty$. Therefore, it implies that the discounted value function of the MDP_n converges to the discounted value function of the original MDP.

Theorem 1. *[20] Let $f \in \mathbb{F}$ and $\{\Upsilon_n(f)\}$ be the quantized approximations of f. Then, $J(\Upsilon_n(f), x) \to J(f, x)$ as $n \to \infty$, for all $x \in \mathsf{X}$.*

The proof of Theorem 1 is a consequence of the proposition below. Before stating the proposition, we first define the ws^∞ topology on $\mathcal{P}(\mathsf{H}_\infty)$ which was first introduced by Schäl in [26]. Let $\mathcal{C}(\mathsf{H}_0) = B(\mathsf{X})$ and let $\mathcal{C}(\mathsf{H}_t)$ $(t \geq 1)$ be the set of real valued functions g on H_t such that $g \in B(\mathsf{H}_t)$ and $g(x_0, \cdot, x_1, \cdot, \ldots, x_{t-1}, \cdot, x_t) \in C_b(\mathsf{A}^t)$ for all $(x_0, \ldots, x_t) \in \mathsf{X}^{t+1}$. The ws^∞ topology on $\mathcal{P}(\mathsf{H}_\infty)$ is defined as the smallest topology which renders all mappings $P \mapsto P(g)$, $g \in \bigcup_{t=0}^{\infty} \mathcal{C}(\mathsf{H}_t)$, continuous.

Proposition 1. *[20] Suppose Assumption 1 holds. Then for any $f \in \mathbb{F}$, the strategic measures $\{P_x^{\Upsilon_n(f)}\}$ induced by the quantized approximations $\{\Upsilon_n(f)\}_{n\geq 1}$ of f converge to the strategic measure P_x^f of f in the ws^∞ topology, for all $x \in \mathsf{X}$.*

Average Cost In contrast to the discounted cost criterion, the expected average cost is in general not sequentially continuous with respect to strategic measures for the ws^∞ topology under practical assumptions. Hence, we develop an approach based on the convergence of the sequence of invariant probability measures under quantized stationary policies to solve **(P)** for the average cost criterion. First, observe that any deterministic stationary policy f defines a stochastic kernel on X given X via

$$Q_f(\,\cdot\,|x) := p(\,\cdot\,|x, f(x)). \tag{2}$$

Let Q_f^t denote the t-step transition probability of this Markov chain. If Q_f admits an ergodic invariant probability measure ν_f, then by [13, Theorem 2.3.4 and Proposition 2.4.2], there exists an invariant set $\mathsf{M}_f \in \mathcal{B}(\mathsf{X})$ with full ν_f measure such that for all x in that set we have

$$V(f, x) = \int_{\mathsf{X}} c(x, f(x))\nu_f(dx). \tag{3}$$

The following assumptions will be imposed for the average cost case.

Assumption 2. *Suppose Assumption 1 holds. In addition, we have*

(e) For any $f \in \mathbb{F}$, Q_f has a unique invariant probability measure ν_f.

(f1) The set of invariant probability measures $\Gamma_{\mathbb{F}} := \{\nu \in \mathcal{P}(\mathsf{X}) : \nu Q_f = \nu$ for some $f \in \mathbb{F}\}$ is relatively sequentially compact in the setwise topology.

(f2) There exists $x \in \mathsf{X}$ such that for all $B \in \mathcal{B}(\mathsf{X})$, $Q_f^t(B|x) \to \nu_f(B)$ uniformly in $f \in \mathbb{F}$.

(g) $\mathsf{M} := \bigcap_{f \in \mathbb{F}} \mathsf{M}_f \neq \emptyset$.

The following theorem is the main result of this section. It states that for any $f \in \mathbb{F}$, the average cost function of $\Upsilon_n(f) \in \mathcal{Q}(\Lambda_n)$ converges to the average cost function of f as $n \to \infty$. In other words, the average value function of MDP$_n$ converges to the average value function of the original MDP.

Theorem 2. *[20] Let $x \in \mathsf{M}$ and $f \in \mathbb{F}$. Then, we have $V(\Upsilon_n(f), x) \to V(f, x)$ as $n \to \infty$, under Assumption 2 with (f1) or (f2).*

Note that Assumption 2-(e),(f2),(g) are satisfied under any of the conditions Ri, $i \in \{0, 1, 1(a), 1(b), 2, \ldots, 6\}$ in [14]. Moreover, $\mathsf{M} = \mathsf{X}$ in Assumption 2-(g) if at least one of the above conditions holds. Moreover, the condition

(e1) $p(\,\cdot\,|x, a) \leq \zeta(\,\cdot\,)$ for all $x \in \mathsf{X}$, $a \in \mathsf{A}$ for some finite measure ζ on X,

implies Assumption 2-(f1).

2.2 Near Optimality of Quantized Policies Under Weak Continuity

In this section, we consider **(P)** for MDPs with weakly continuous transition probability. Specifically, we will show that the value function of MDP_n converges to the value function of the original MDP, which is equivalent to **(P)**.

Discounted Cost Here, we consider the discounted cost case with a discount factor $\beta \in (0, 1)$. The following assumptions will be imposed for both the discounted cost and the average cost. These assumptions are used in the literature for studying discounted Markov decision processes with unbounded one-stage cost and weakly continuous transition probability.

Assumption 3.
(a) The one stage cost function c is nonnegative and continuous.
(b) The stochastic kernel $p(\,\cdot\,|x, a)$ is weakly continuous in $(x, a) \in \mathsf{X} \times \mathsf{A}$.
(c) A is compact.
(d) There exist nonnegative real numbers M and $\alpha \in [1, \frac{1}{\beta})$, and a continuous weight function $w : \mathsf{X} \to [1, \infty)$ such that for each $x \in \mathsf{X}$, we have

$$\sup_{a \in \mathsf{A}} c(x, a) \leq M w(x), \tag{4}$$

$$\sup_{a \in \mathsf{A}} \int_{\mathsf{X}} w(y) p(dy|x, a) \leq \alpha w(x), \tag{5}$$

and $\int_{\mathsf{X}} w(y) p(dy|x, a)$ is continuous in (x, a).

For any real-valued measurable function u on X, let $Tu : \mathsf{X} \to \mathbb{R}$ is given by

$$Tu(x) := \min_{a \in \mathsf{A}} \biggl[c(x, a) + \beta \int_{\mathsf{X}} u(y) p(dy|x, a) \biggr]. \tag{6}$$

In the literature T is called the *Bellman optimality operator* for the MDP. Analogously, let us define the Belmann optimality operator T_n of MDP_n as

$$T_n u(x) := \min_{a \in \Lambda_n} \biggl[c(x, a) + \beta \int_{\mathsf{X}} u(y) p(dy|x, a) \biggr]. \tag{7}$$

It can be shown that both T and T_n are contraction operators with modulus $\sigma = \alpha\beta$ mapping $C_w(\mathsf{X})$ into itself. Furthermore, value functions of MDP and MDP$_n$ are fixed points of these operators; that is, $TJ^* = J^*$ and $T_n J_n^* = J_n^*$. Let us define $v^0 = v_n^0 = 0$, and $v^{t+1} = Tv^t$ and $v_n^{t+1} = T_n v_n^t$ for $t \geq 1$; that is, $\{v^t\}_{t\geq 1}$ and $\{v_n^t\}_{t\geq 1}$ are successive approximations to the discounted value functions of the MDP and MDP$_n$, respectively.

Lemma 1. *[21] For any compact set $K \subset \mathsf{X}$ and $t \geq 1$, we have*

$$\lim_{n\to\infty} \sup_{x\in K} |v_n^t(x) - v^t(x)| = 0. \tag{8}$$

The theorem below is the main result of this section which states that the discounted value function of MDP$_n$ converges to the discounted value function of the original MDP. It can be proved by using Lemma 1 and taking into account that $\{v^t\}_{t\geq 1}$ and $\{v_n^t\}_{t\geq 1}$ are successive approximations to J^* and J_n^*, respectively.

Theorem 3. *[21] For any compact set $K \subset \mathsf{X}$ we have*

$$\lim_{n\to\infty} \sup_{x\in K} |J_n^*(x) - J^*(x)| = 0. \tag{9}$$

Average Cost Here, we consider problem **(P)** for the average cost criterion. We prove an approximation result analogous to Theorem 3. To do this, some new assumptions are needed on the components of the original MDP in addition to the conditions in Assumption 3. A version of these assumptions were used in [10] and [27] to study the existence of the solution to the Average Cost Optimality Equality (ACOE) and Inequality (ACOI).

Assumption 4. *Suppose Assumption 3 holds with (5) replaced by condition (e) below. Moreover, suppose there exist a probability measure λ on X and a continuous function $\phi : \mathsf{X} \times \mathsf{A} \to [0, \infty)$ such that*

(e) $\int_{\mathsf{X}} w(y)p(dy|x,a) \leq \alpha w(x) + \lambda(w)\phi(x,a)$ *for all* $(x,a) \in \mathsf{X} \times \mathsf{A}$, *where* $\alpha \in (0,1)$.
(f) $p(D|x,a) \geq \lambda(D)\phi(x,a)$ *for all* $(x,a) \in \mathsf{X} \times \mathsf{A}$ *and* $D \in \mathcal{B}(\mathsf{X})$.
(g) The weight function w is μ-integrable.
(h) $\int_{\mathsf{X}} \phi(x, f(x))\lambda(dx) > 0$ *for all* $f \in \mathbb{F}$.

The following theorem is a consequence of [27, Theorems 3.3 and 3.6].

Theorem 4. *[21] Under Assumption 4 the following holds. For each $f \in \mathbb{F}$, the stochastic kernel $Q_f(\,\cdot\,|x)$ has an unique invariant probability measure ν_f. Furthermore, w is ν_f-integrable, and therefore, $\rho_f := \int_{\mathsf{X}} c(x, f(x))\nu_f(dx) < \infty$. There exist $f^* \in \mathbb{F}$ and $h^* \in C_w(\mathsf{X})$ such that the triplet (h^*, f^*, ρ_{f^*}) satisfies the average cost optimality equality (ACOE) and therefore, for all $x \in \mathsf{X}$*

$$\inf_{\pi\in\Pi} V(\pi, x) := V^*(x) = \rho_{f^*}.$$

By [27, Theorem 3.5], h^* is the unique fixed point of the following contraction operator with modulus α mapping $C_w(\mathsf{X})$ into itself

$$Fu(x) := \min_{a\in\mathsf{A}} \left[c(x,a) + \int_{\mathsf{X}} u(y)p(dy|x,a) - \lambda(u)\phi(x,a) \right].$$

Note that all the statements in Theorem 4 are also valid for MDP_n with an optimal policy f_n^* and a canonical triplet $(h_n^*, f_n^*, \rho_{f_n^*})$. Analogous with F, define the contraction operator F_n (with modulus α) corresponding to MDP_n as

$$F_n u(x) := \min_{a\in\Lambda_n} \left[c(x,a) + \int_{\mathsf{X}} u(y)p(dy|x,a) - \lambda(u)\phi(x,a) \right],$$

and therefore, $h_n^* \in C_w(\mathsf{X})$ is its fixed point. The following lemma can be proved similar to Theorem 3.

Lemma 2. *For any compact set $K \subset \mathsf{X}$, we have*

$$\lim_{n\to\infty} \sup_{x\in K} |h_n^*(x) - h^*(x)| = 0.$$

The next theorem is the main result of this section which states that the average cost value function, denoted as V_n^*, of MDP_n converges to the average cost value function V^* of the original MDP. It can be proved by taking the limit in ACOE of MDP_n and using Lemma 2.

Theorem 5. *[21] We have $\lim_{n\to\infty} |V_n^* - V^*| = 0$, where V_n^* and V^* are both constants.*

Remark 1. When one considers partially observed MDPs (POMDPs), it is known that any POMDP can be reduced to a (completely observable) MDP [30] whose states are the posterior state distributions or beliefs of the observer. One can show that setwise continuity of the reduced MDP is not possible even under very strict conditions, whereas weak continuity can be satisfied under reasonable conditions on the transition kernel and the continuity of the measurement channel. Thus the results in this section are applicable to POMDPs; see [21].

2.3 Rates of Convergence

In Sections 2.1 and 2.2 we considered the convergence of the finite-action models MDP_n to the original model. In this section we obtain performance bounds on the approximation errors due to quantization of the action space in terms of the number of points used to discretize action space. Namely, we study the following problem.

(Pr) For any $f \in \mathbb{F}$ and initial point x, the approximating sequence $\{\Upsilon_n(f)\}$ in **(P)** is such that $|W(f,x) - W(\Upsilon_n(f),x)|$ can be explicitly upper bounded by a term depending on the cardinality of Λ_n, where $W \in \{J, V\}$.

Thus **(Pr)** implies that the approximation error in **(P)** can be explicitly controlled by the number of points used to discretize the action space. We will impose a new set of assumptions in this section.

Assumption 5.

(h) A *is infinite compact subset of* $\mathbb{R}^d$ *for some* $d \geq 1$.
(j) c *is bounded and* $|c(x,\tilde{a}) - c(x,a)| \leq K_1 d_{\mathsf{A}}(\tilde{a},a)$ *for all* x, *and some* $K_1 \geq 0$.
(k) $\|p(\cdot\,|x,\tilde{a}) - p(\cdot\,|x,a)\|_{TV} \leq K_2 d_{\mathsf{A}}(\tilde{a},a)$ *for all* x, *and some* $K_2 \geq 0$.
(l) There exists positive constants C *and* $\beta \in (0,1)$ *such that for all* $f \in \mathbb{F}$, *there is a (necessarily unique) probability measure* $\nu_f \in \mathcal{P}(\mathsf{X})$ *satisfying* $\|Q_f^t(\cdot\,|x) - \nu_f\|_{TV} \leq C\kappa^t$ *for all* $x \in \mathsf{X}$ *and* $t \geq 1$.

Assumption 5-(l) implies that for any policy $f \in \mathbb{F}$, the stochastic kernel Q_f, defined in (2), has a unique invariant probability measure ν_f and satisfies *geometric ergodicity* [13]. Note that Assumption 5-(l) holds under any of the conditions Ri, $i \in \{0,1,1(a),1(b),2,\ldots,5\}$ in [14]. For further conditions that imply Assumption 5-(l) we refer the reader to [13, 14].

The following result is a consequence of the fact that if A is a compact subset of $\mathbb{R}^d$ then there exist a constant $\alpha > 0$ and finite subsets $\Lambda_n \subset \mathsf{A}$ with cardinality $|\Lambda_n| = n$ such that $\max_{x\in\mathsf{A}} \min_{y\in\Lambda_n} d_{\mathsf{A}}(x,y) \leq \alpha(1/n)^{1/d}$ for all n, where d_{A} is the Euclidean distance on A inherited from $\mathbb{R}^d$.

Lemma 3. *Let* $\mathsf{A} \subset \mathbb{R}^d$ *be compact. Then for any* $f \in \mathbb{F}$ *the sequence* $\{\Upsilon_n(f)\}$ *satisfies* $\sup_{x\in\mathsf{X}} d_{\mathsf{A}}(\Upsilon_n(f)(x), f(x)) \leq \alpha(1/n)^{1/d}$ *for some constant* α.

The following proposition is the key result in this section.

Proposition 2. *[20] Let* $f \in \mathbb{F}$ *and* $\{q_n\}$ *be the quantized approximations of* f, *i.e.,* $\Upsilon_n(f) = q_n$. *For any initial point* x, *we have*

$$\|Q_f^t(\cdot\,|x) - Q_{q_n}^t(\cdot\,|x)\|_{TV} \leq \alpha K_2(2t-1)(1/n)^{1/d} \tag{10}$$

for all $t \geq 1$ *under Assumption 5-(h),(j),(k).*

Discounted Cost The following result solves **(Pr)** for the discounted cost criterion. The proof of it essentially follows from Proposition 2.

Theorem 6. *[20] Let* $f \in \mathbb{F}$ *and* $\{q_n\}$ *be the quantized approximations of* f, *i.e.,* $\Upsilon_n(f) = q_n$. *For any initial point* x, *we have*

$$|J(f,x) - J(q_n,x)| \leq K(1/n)^{1/d}, \tag{11}$$

where $K = \frac{\alpha}{1-\beta}(K_1 - \beta K_2 M + \frac{2\beta M K_2}{1-\beta})$ *with* $M := \|c\|$, *under Assumption 5-(h),(j),(k).*

Average Cost Note that for any $f \in \mathbb{F}$, Assumption 5-(l) implies that ν_f is an unique invariant probability measure for Q_f and that $V(f,x) = \int_{\mathsf{X}} c(x,f(x))\nu_f(dx)$ for all x. The following theorem basically follows from Proposition 2 and the Assumption 5-(l). It solves **(Pr)** for the average cost criterion.

Theorem 7. *[20] Let $f \in \mathbb{F}$ and $\{q_n\}$ be the quantized approximations of f, i.e., $\Upsilon_n(f) = q_n$. Then, under Assumption 5, for any $x \in \mathsf{X}$ we have*

$$|V(f,x) - V(q_n,x)| \leq 2MC\kappa^t + K_t(1/n)^{1/d} \tag{12}$$

for all $t \geq 0$, where $K_t = \big((2t-1)K_2\alpha M + K_1\alpha\big)$ and $M := \|c\|$.

Observe that depending on the values of C and κ, we can first make the first term in the upper bound small enough by choosing sufficiently large t, and then for this t we can choose n large enough such that the second term in the upper bound is small.

3 Finite State Approximation to MDPs

In this section our aim is to study the finite-state approximation problem for discrete time Markov decision processes, by reducing it to a finite state MDP obtained through quantization of the state space on a finite grid. In particular, we study the following two problems.

(Q1) Under what conditions on the components of the MDP do the true cost functions of the policies obtained from finite models converge to the optimal value function as the number of grid points goes to infinity?

(Q2) Can we obtain bounds on the performance loss due to discretization in terms of the number of grid points if we strengthen the conditions sufficient in **(Q1)**?

Our approach to solve problem **(Q1)** can be summarized as follows: (i) first, we obtain approximation results for the compact-state case, (ii) we find conditions under which a compact representation leads to near optimality for non-compact state MDPs, (iii) we obtain the convergence of the finite-state models to non-compact models. A by-product of this analysis, we obtain *compact-state-space approximations* for an MDP with non-compact Borel state space. In particular, our findings directly lead to finite models if the state space is countable.

3.1 Finite State Approximation of Compact State MDPs

In this section we consider **(Q1)** for the MDPs with compact state space. To distinguish compact-state MDPs from non-compact ones, the state space of the compact-state MDPs will be denoted by Z instead of X. We impose the assumptions below on the components of the Markov decision process.

Assumption 6.

(a) The one-stage cost function c is in $C_b(\mathsf{Z} \times \mathsf{A})$.

(b) The stochastic kernel $p(\,\cdot\,|z,a)$ is weakly continuous in (z,a) and setwise continuous in a.

(c) Z and A are compact.

Let d_{Z} denote the metric on Z. Since the state space Z is assumed to be compact and thus totally bounded, one can find a sequence $\big(\{z_{n,i}\}_{i=1}^{k_n}\big)_{n\geq 1}$ of finite grids in Z such that for all n,

$$\min_{i\in\{1,\ldots,k_n\}} d_{\mathsf{Z}}(z, z_{n,i}) < 1/n \text{ for all } z \in \mathsf{Z}.$$

Let $\mathsf{Z}_n := \{z_{n,1}, \ldots, z_{n,k_n}\}$ and define function Q_n mapping Z to Z_n by

$$Q_n(z) := \operatorname*{arg\,min}_{z_{n,i}\in\mathsf{Z}_n} d_{\mathsf{Z}}(z, z_{n,i}),$$

where ties are broken so that Q_n is measurable. For each n, Q_n induces a partition $\{\mathcal{S}_{n,i}\}_{i=1}^{k_n}$ of the state space Z given by

$$\mathcal{S}_{n,i} = \{z \in \mathsf{Z} : Q_n(z) = z_{n,i}\}.$$

Let $\{\nu_n\}$ be a sequence of probability measures on Z satisfying $\nu_n(\mathcal{S}_{n,i}) > 0$ for al i, n. We let $\nu_{n,i}$ be the restriction of ν_n to $\mathcal{S}_{n,i}$ defined by $\nu_{n,i}(\,\cdot\,) := \frac{\nu_n(\,\cdot\,)}{\nu_n(\mathcal{S}_{n,i})}$. The measures $\nu_{n,i}$ will be used to define a sequence of finite-state MDPs, denoted as MDP$_n$ ($n \geq 1$), to approximate the original model. To this end, for each n define the one-stage cost function $c_n : \mathsf{Z}_n \times \mathsf{A} \to [0, \infty)$ and the transition probability p_n on Z_n given $\mathsf{Z}_n \times \mathsf{A}$ by

$$c_n(z_{n,i}, a) := \int_{\mathcal{S}_{n,i}} c(z, a)\nu_{n,i}(dz),$$

$$p_n(\,\cdot\,|z_{n,i}, a) := \int_{\mathcal{S}_{n,i}} Q_n * p(\,\cdot\,|z, a)\nu_{n,i}(dz),$$

where $Q_n * p(\,\cdot\,|z, a) \in \mathcal{P}(\mathsf{Z}_n)$ is the pushforward of the measure $p(\,\cdot\,|z, a)$ with respect to Q_n; that is,

$$Q_n * p(z_{n,j}|z, a) = p\big(\{z \in \mathsf{Z} : Q_n(z) = z_{n,j}\}|z, a\big),$$

for all $z_{n,j} \in \mathsf{Z}_n$. For each n, we define MDP$_n$ as a Markov decision process with the following components: Z_n is the state space, A is the action space, p_n is the transition probability and c_n is the one-stage cost function. History spaces, policies and cost functions are defined in a similar way as in the original model.

Discounted Cost Here we consider **(Q1)** for the discounted cost criterion with a discount factor $\beta \in (0, 1)$. Recall the Bellman optimality operator T defined in (6). Define also the operator T_n, which is the Bellman optimality operator for MDP$_n$, by

$$T_n u(z_{n,i}) = \min_{a\in\mathsf{A}} \int_{\mathcal{S}_{n,i}} \left[c(z, a) + \beta \int_{\mathsf{Z}} \hat{u}(y)p(dy|z, a)\right] \nu_{n,i}(dz),$$

where $u : \mathsf{Z}_n \to \mathbb{R}$ and $\hat{u}$ is the piecewise constant extension of u to Z given by $\hat{u}(z) = u\circ Q_n(z)$. Under Assumption 6, the fixed point of T_n is the value function

J_n^* of MDP_n and there exists an optimal stationary policy f_n^* for MDP_n. Hence, we have $J_n^* = T_n J_n^* = T_n J_n(f_n^*, \cdot) = J_n(f_n^*, \cdot)$, where J_n denotes the discounted cost for MDP_n. Let us extend $\hat{T}_n$ to the set of all bounded measurable functions on Z as follows:

$$\hat{T}_n u(z) := \min_{a \in \mathsf{A}} \int_{\mathcal{S}_{n,i_n(z)}} \left[c(x,a) + \beta \int_{\mathsf{Z}} u(y) p(dy|x,a) \right] \nu_{n,i_n(z)}(dx), \tag{13}$$

where $i_n : \mathsf{Z} \to \{1, \ldots, k_n\}$ maps z to the index of the partition $\{\mathcal{S}_{n,i}\}$ it belongs to. Since $\hat{T}_n(u \circ Q_n) = (T_n u) \circ Q_n$ for all $u \in B(\mathsf{Z}_n)$, we have

$$\hat{T}_n(J_n^* \circ Q_n) = (T_n J_n^*) \circ Q_n = J_n^* \circ Q_n.$$

Hence, the fixed point of $\hat{T}_n$ is the piecewise constant extension of the fixed point of T_n.

Remark 2. In the rest of this chapter, when we take the integral of any function with respect to $\nu_{n,i_n(z)}$, it is tacitly assumed that the integral is taken over all set $\mathcal{S}_{n,i_n(z)}$. Hence, we can drop $\mathcal{S}_{n,i_n(z)}$ in the integral for the ease of notation.

We now define another operator F_n on $B(\mathsf{Z})$ by simply interchanging the order of the minimum and the integral in (13), i.e.,

$$F_n u(z) := \int \min_{a \in \mathsf{A}} \left[c(x,a) + \beta \int_{\mathsf{Z}} u(y) p(dy|x,a) \right] \nu_{n,i_n(z)}(dx).$$

We note that F_n is the extension (to infinite state spaces) of the operator defined in [18, p. 236] for the proposed approximate value iteration algorithm. However, unlike in [18], F_n will serve here as an intermediate point between T and $\hat{T}_n$ (or T_n) to solve **(Q1)** for the discounted cost. The following theorem states that the fixed point, say u_n^*, of F_n converges to the fixed point J^* (i.e., the value function) of T as n goes to infinity.

Theorem 8. *If u_n^* is the unique fixed point of F_n, then $\lim_{n\to\infty} \|u_n^* - J^*\| = 0$.*

The next step is to show that the fixed point $\hat{J}_n^*$ of $\hat{T}_n$ converges to the fixed point J^* of T. This follows from Theorem 8 and the following result: for any $u \in C_b(\mathsf{Z})$, $\|\hat{T}_n u - F_n u\| \to 0$ as $n \to \infty$.

Theorem 9. *[22, 25] The fixed point $\hat{J}_n^*$ of $\hat{T}_n$ converges to the fixed point J^* of T. Therefore, the (extended) value function $J_n^* \circ Q_n$ of MDP_n converges to the value function of MDP.*

Recall the optimal stationary policy f_n^* for MDP_n and extend it to Z by letting $\hat{f}_n(z) = f_n^* \circ Q_n(z)$. Since $\hat{J}_n^* = J_n^* \circ Q_n$, one can show that $\hat{f}_n$ is the optimal selector of $\hat{T}_n \hat{J}_n^*$; that is, $\hat{T}_n \hat{J}_n^* = \hat{J}_n^* = \hat{T}_{\hat{f}_n} \hat{J}_n^*$, where $\hat{T}_{\hat{f}_n}$ is defined as

$$\hat{T}_{\hat{f}_n} u(z) := \int \left[c(x, \hat{f}_n(x)) + \beta \int_{\mathsf{Z}} u(y) p(dy|x, \hat{f}_n(x)) \right] \nu_{n,i_n(z)}(dx).$$

Define analogously

$$T_{\hat{f}_n} u(z) := c(z, \hat{f}_n(z)) + \beta \int_{\mathsf{Z}} u(y) p(dy|z, \hat{f}_n(z)).$$

It is known that the fixed point of $T_{\hat{f}_n}$ is the true cost function of the stationary policy $\hat{f}_n$ (i.e., $J(\hat{f}_n, z)$).

The following theorem states that the cost function under the policy $\hat{f}_n$ converges to the value function J^* as $n \to \infty$. It follows from Theorem 9 and the following result: $\|\hat{T}_{\hat{f}_n} u - T_{\hat{f}_n} u\| \to 0$ as $n \to \infty$, for any $u \in C_b(\mathsf{Z})$.

Theorem 10. *[22, 25] The discounted cost of the policy $\hat{f}_n$, obtained by extending the optimal policy f_n^* of MDP$_n$ to Z, converges to the optimal value function J^* of the original MDP* $\lim_{n\to\infty} \|J(\hat{f}_n, \cdot\,) - J^*\| = 0$.

Therefore, to find a near optimal policy for the original MDP, it is sufficient to compute the optimal policy of MDP$_n$ for sufficiently large n, and then extend this policy to the original state space.

Average Cost In this section we impose some new conditions on the components of the original MDP in addition to Assumption 6 to solve **(Q1)** for the average cost. A version of first two conditions were imposed in [27] to show the existence of the solution to the Average Cost Optimality Equation (ACOE) and the optimal stationary policy by using the fixed point approach.

Assumption 7. *Suppose Assumption 6 holds with item (b) replaced by condition (f) below. In addition, there exist a non-trivial finite measure ζ on Z, a nonnegative measurable function θ on $\mathsf{Z} \times \mathsf{A}$, and a constant $\lambda \in (0,1)$ such that for all $(z,a) \in \mathsf{Z} \times \mathsf{A}$*

(d) $p(B|z,a) \geq \zeta(B)\theta(z,a)$ *for all* $B \in \mathcal{B}(\mathsf{Z})$,
(e) $\frac{1-\lambda}{\zeta(\mathsf{Z})} \leq \theta(z,a)$,
(f) The stochastic kernel $p(\,\cdot\,|z,a)$ is continuous in (z,a) with respect to the total variation distance.

The following theorem is a consequence of [10, Lemma 3.4 and Theorem 2.6] and [27, Theorems 3.3].

Theorem 11. *[22, 25] Under Assumptions 7 the following holds. For each $f \in \mathbb{F}$, the stochastic kernel $Q_f(\,\cdot\,|x)$ has an unique invariant probability measure μ_f. Therefore, we have $V(f,z) = \int_{\mathsf{Z}} c(z, f(z))\mu_f(dz) =: \rho_f$. There exist positive real numbers R and $\kappa < 1$ such that for every $z \in \mathsf{Z}$*

$$\sup_{f \in \mathbb{F}} \|Q_f^t(\,\cdot\,|z) - \mu_f\|_{TV} \leq R\kappa^t,$$

where R and κ continuously depend on $\zeta(\mathsf{Z})$ and λ. Furthermore, there exist $f^* \in \mathbb{F}$ and $h^* \in B(\mathsf{Z})$ such that the triplet (h^*, f^*, ρ_{f^*}) satisfies the average cost optimality inequality (ACOI) and therefore,

$$\inf_{\pi \in \Pi} V(\pi, z) =: V^*(z) = \rho_{f^*}.$$

For each n, define the one-stage cost function $b_n : \mathsf{Z} \times \mathsf{A} \to [0, \infty)$ and the stochastic kernel q_n on Z given $\mathsf{Z} \times \mathsf{A}$ as

$$b_n(z, a) := \int c(x, a) \nu_{n,i_n(z)}(dx),$$
$$q_n(\,\cdot\,|z, a) := \int p(\,\cdot\,|x, a) \nu_{n,i_n(z)}(dx).$$

Observe that c_n (i.e., the one stage cost function of MDP_n) is the restriction of b_n to Z_n, and p_n (i.e., the stochastic kernel of MDP_n) is the pushforward of the measure q_n with respect to Q_n; that is, $c_n(z_{n,i}, a) = b_n(z_{n,i}, a)$ for all $i = 1, \ldots, k_n$ and $p_n(\,\cdot\,|z_{n,i}, a) = Q_n * q_n(\,\cdot\,|z_{n,i}, a)$.

For each n, let $\widetilde{\mathrm{MDP}}_n$ be defined as a Markov decision process with the following components: Z is the state space, A is the action space, q_n is the transition probability, and b_n is the one-stage cost function. History spaces, policies and cost functions are defined in a similar way as before. We note that a careful analysis of $\widetilde{\mathrm{MDP}}_n$ reveals that its Bellman optimality operator is essentially the operator $\hat{T}_n$. Hence, the value function of $\widetilde{\mathrm{MDP}}_n$ is the piecewise constant extension of the value function of MDP_n for the discounted cost. A similar conclusion will be made for the average cost in Lemma 4.

First, notice that if we define

$$\theta_n(z, a) := \int \theta(y, a) \nu_{n,i_n(z)}(dy),$$
$$\zeta_n := Q_n * \zeta \text{ (i.e., pushforward of } \zeta \text{ with respect to } Q_n),$$

then it is straightforward to prove that for all n, $\widetilde{\mathrm{MDP}}_n$ satisfies Assumption 7-(d),(e) when θ is replaced by θ_n, and Assumption 7-(d),(e) is true for MDP_n when θ and ζ are replaced by the restriction of θ_n to Z_n and ζ_n, respectively. Hence, Theorem 11 holds (with the same R and κ) for $\widetilde{\mathrm{MDP}}_n$ and MDP_n for all n. Therefore, we denote by $\tilde{f}_n^*$ and f_n^* the optimal stationary policies of $\widetilde{\mathrm{MDP}}_n$ and MDP_n with the corresponding average costs $\tilde{\rho}^n_{\tilde{f}_n^*}$ and $\rho^n_{f_n^*}$, respectively. Furthermore, we also write $\tilde{\rho}^n_f$ and ρ^n_f to denote the average cost of any stationary policy f for $\widetilde{\mathrm{MDP}}_n$ and MDP_n, respectively. The corresponding invariant probability measures are also denoted in a same manner, with μ replacing ρ.

The following lemma essentially says that MDP_n and $\widetilde{\mathrm{MDP}}_n$ are not very different.

Lemma 4. *The stationary policy given by the piecewise constant extension of the optimal policy f_n^* of MDP_n to Z (i.e., $f_n^* \circ Q_n$) is optimal for $\widetilde{MDP}_n$ with the same cost function $\rho^n_{f_n^*}$. Hence, $\tilde{f}_n^* = f_n^* \circ Q_n$ and $\tilde{\rho}^n_{\tilde{f}_n^*} = \rho^n_{f_n^*}$.*

By Lemma 4, we can consider $\widetilde{\mathrm{MDP}}_n$ in place of MDP_n. The following theorem states that the value function of $\widetilde{\mathrm{MDP}}_n$ converges to the value function of MDP as $n \to \infty$.

Theorem 12. *[22, 25] We have* $\sup_{f\in\mathbb{F}}|\tilde{\rho}_f^n - \rho_f| \to 0$ *as* $n \to \infty$*. Therefore,* $|\tilde{\rho}_{\tilde{f}_n^*}^n - \rho_{f^*}| \to 0$ *as* $n \to \infty$

The following theorem states that if one applies the piecewise constant extension of the optimal stationary policy of MDP_n to the original MDP, the resulting cost function will converge to the value function of the original MDP. It follows from Lemma 4 and Theorem 12.

Theorem 13. *[22, 25] The average cost of the optimal policy* $\tilde{f}_n^*$ *for* $\widetilde{MDP}_n$*, obtained by extending the optimal policy* f_n^* *of* MDP_n *to* Z*, converges to the optimal value function* $J^* = \rho_{f^*}$ *of the original MDP, i.e.,*

$$\lim_{n\to\infty} |\rho_{\tilde{f}_n^*} - \rho_{f^*}| = 0.$$

Therefore, to find a near optimal policy for the original MDP, it is sufficient to compute the optimal policy of MDP_n for sufficiently large n, and then extend this policy to the original state space.

3.2 Finite State Approximation of Non-Compact State MDPs

In this section we consider **(Q1)** for noncompact state MDPs with unbounded one-stage cost. We impose the assumptions below on the components of the Markov decision process; additional assumptions will be imposed for the average cost problem.

Assumption 8.

(a) The one-stage cost function c *is nonnegative and continuous.*
(b) The stochastic kernel $p(\,\cdot\,|x,a)$ *is weakly continuous in* (x,a) *and setwise continuous in* a*.*
(c) X *is locally compact and* A *is compact.*
(d) There exist nonnegative real numbers M *and* $\alpha \in [1, \frac{1}{\beta})$*, and a continuous weight function* $w : \mathsf{X} \to [1,\infty)$ *such that for each* $x \in \mathsf{X}$*, we have*

$$\sup_{a\in\mathsf{A}} c(x,a) \leq Mw(x), \tag{14}$$

$$\sup_{a\in\mathsf{A}} \int_{\mathsf{X}} w(y)p(dy|x,a) \leq \alpha w(x), \tag{15}$$

and $\int_{\mathsf{X}} w(y)p(dy|x,a)$ *is continuous in* a*.*

Since X is locally compact separable metric space, there exists a nested sequence of compact sets $\{K_n\}$ such that $K_n \subset \operatorname{int} K_{n+1}$, where $\operatorname{int} D$ denotes the interior of D, and $\mathsf{X} = \bigcup_{n=1}^{\infty} K_n$ [1, Lemma 2.76]. Let $\{\nu_n\}$ be a sequence of probability measures such that for each $n \geq 1$, $\nu_n \in \mathcal{P}(K_n^c)$ and

$$\gamma_n := \int_{K_n^c} w(x)\nu_n(dx) < \infty, \tag{16}$$

$$\gamma = \sup_n \tau_n := \sup_n \Big\{ 0 \vee \sup_{(x,a)\in\mathsf{X}\times\mathsf{A}} \int_{K_n^c} \big(\gamma_n - w(y)\big)p(dy|x,a) \Big\} < \infty, \tag{17}$$

where D^c denotes the complement of the set D and $x \vee y = \max(x, y)$.

Similar to the finite-state MDP construction, we define a sequence of compact-state MDPs, denoted as c-MDP$_n$, to approximate the original model. To this end, for each n let $\mathsf{X}_n = K_n \cup \{\Delta_n\}$, where $\Delta_n \in K_n^c$ is a so-called pseudo-state. To ease the notation, for any $\vartheta \in \mathcal{P}(\mathsf{X})$ we define $\Lambda_n \vartheta(\,\cdot\,) := \vartheta(\,\cdot\, \cap K_n) + \vartheta(K_n^c)\delta_{\Delta_n}(\,\cdot\,)$, and for a measurable real function u on X we let $\Gamma_n[u(x)] := \int_{K_n^c} u(x)\nu_n(dx)$. If u is a function of (x, a), we denote $\int_{\mathsf{X}} u(x, a)\nu_n(dx)$ by $\Gamma_n[u(x, a)]$. We define the transition probability p_n on X_n given $\mathsf{X}_n \times \mathsf{A}$ and the one-stage cost function $c_n : \mathsf{X}_n \times \mathsf{A} \to [0, \infty)$ by

$$p_n(\,\cdot\,|x, a) = \begin{cases} \Lambda_n p(\,\cdot\,|x, a), & \text{if } x \in K_n \\ \Gamma_n\big[\Lambda_n p(\,\cdot\,|z, a)\big], & \text{if } x = \Delta_n, \end{cases}$$

$$c_n(x, a) = \begin{cases} c(x, a), & \text{if } x \in K_n \\ \Gamma_n[c(z, a)], & \text{if } x = \Delta_n. \end{cases}$$

With these definitions, c-MDP$_n$ is defined as a Markov decision process with the components $(\mathsf{X}_n, \mathsf{A}, p_n, c_n)$. History spaces and costs are defined in a similar way as in the original model. Let Π_n, Φ_n, and $\mathbb{F}_n$ denote the set of all, randomized stationary and deterministic stationary policies of c-MDP$_n$, respectively. For each policy $\pi \in \Pi_n$ and initial distribution $\mu \in \mathcal{P}(\mathsf{X}_n)$, we denote the cost functions for c-MDP$_n$ by $J_n(\pi, \mu)$ and $V_n(\pi, \mu)$.

To obtain the main result of this section, we introduce, for each n, another MDP, denoted by $\overline{\text{MDP}}_n$, with the components $(\mathsf{X}, \mathsf{A}, q_n, b_n)$ where

$$q_n(\,\cdot\,|x, a) = \begin{cases} p(\,\cdot\,|x, a), & \text{if } x \in K_n \\ \Gamma_n\big[p(\,\cdot\,|z, a)\big], & \text{if } x \in K_n^c, \end{cases}$$

$$b_n(x, a) = \begin{cases} c(x, a), & \text{if } x \in K_n \\ \Gamma_n[c(z, a)], & \text{if } x \in K_n^c. \end{cases}$$

For each policy $\pi \in \Pi$ and initial distribution $\mu \in \mathcal{P}(\mathsf{X})$, we denote the cost functions for $\overline{\text{MDP}}_n$ by $\bar{J}_n(\pi, \mu)$ and $\bar{V}_n(\pi, \mu)$.

Discounted Cost In this section we consider **(Q1)** for the discounted cost criterion with a discount factor $\beta \in (0, 1)$. The following result states that c-MDP$_n$ and $\overline{\text{MDP}}_n$ are equivalent for the discounted cost.

Lemma 5. *[22, 23] We have*

$$\bar{J}_n^*(x) = \begin{cases} J_n^*(x), & \text{if } x \in K_n \\ J_n^*(\Delta_n), & \text{if } x \in K_n^c, \end{cases}$$

where $\bar{J}_n^*$ is the discounted value function of $\overline{\text{MDP}}_n$ and J_n^* is the discounted value function of c-MDP$_n$. Furthermore, if, for any deterministic stationary *policy* $f \in \mathbb{F}_n$, *we define* $\bar{f}(x) = f(x)$ *on* K_n *and* $\bar{f}(x) = f(\Delta_n)$ *on* K_n^c, *then*

$$\bar{J}_n(\bar{f}, x) = \begin{cases} J_n(f, x), & \text{if } x \in K_n \\ J_n(f, \Delta_n), & \text{if } x \in K_n^c. \end{cases}$$

In particular, if the deterministic stationary policy $f_n^* \in \mathbb{F}_n$ *is optimal for c-MDP*$_n$, *then its extension* $\bar{f}_n^*$ *to* X *is also optimal for* $\overline{MDP}_n$.

For each n, let us define w_n by letting $w_n(x) = w(x)$ on K_n and $w_n(x) = \Gamma_n[w(z)] =: \gamma_n$ on K_n^c. Hence, $w_n \in B(\mathsf{X})$ by (16). Then, the components of $\overline{\text{MDP}}_n$ satisfy the following for any $x \in \mathsf{X}$:

$$\sup_{a \in \mathsf{A}} b_n(x, a) \leq M w_n(x) \tag{18}$$

$$\sup_{a \in \mathsf{A}} \int_{\mathsf{X}} w_n(y) q_n(dy|x, a) \leq \alpha w_n(x) + \gamma \text{ (see (17))}. \tag{19}$$

Using same arguments in [12, Remark 8.3.5, p. 46], we can use (18) and (19) to prove that the components of $\overline{\text{MDP}}_n$ satisfy similar conditions as in Assumption 8.

Lemma 6. *[22, 23] Let* $\alpha_0 := \frac{1}{\beta_0} \in (\alpha, \frac{1}{\beta})$. *Then, for each* n *there exists* $C_n : \mathsf{X} \to [1, \infty)$ *which satisfies*

$$\sup_{a \in \mathsf{A}} b_n(x, a) \leq C_n(x) \tag{20}$$

$$\sup_{a \in \mathsf{A}} \int_{\mathsf{X}} C_n(y) q_n(dy|x, a) \leq \alpha_0 C_n(x), \tag{21}$$

for all $x \in \mathsf{X}$. *Letting* $L_1 = \frac{L}{1-\beta_0\alpha}$ *and* $L_2 = \frac{L\beta_0}{(1-\beta_0)(1-\beta_0\alpha)}\gamma$, *we also have that* C_n *is upper bounded by* $L_1 w_n + L_2$.

Since (i) $b_n(x, a)$ is continuous in a for all $x \in \mathsf{X}$, (ii) $q_n(\cdot\,|x, a)$ is setwise continuous in a for all $x \in \mathsf{X}$, (iii) $C_n \in B(\mathsf{X})$, and (iv) $\alpha_0 < \frac{1}{\beta}$, $\overline{\text{MDP}}_n$ satisfies the assumptions in [12, Theorem 8.3.6, p. 47]. Let us define the Bellman optimality operator $\overline{T}_n : B(\mathsf{X}) \to B(\mathsf{X})$ of $\overline{\text{MDP}}_n$ by

$$\overline{T}_n u(x) = \begin{cases} \min_{a \in \mathsf{A}} \big[c(x, a) + \beta \int_{\mathsf{X}} u(y) p(dy|x, a)\big], & \text{if } x \in K_n \\ \min_{a \in \mathsf{A}} \Gamma_n \big[c(z, a) + \beta \int_{\mathsf{X}} u(y) p(dy|z, a)\big], & \text{if } x \in K_n^c. \end{cases}$$

Then successive approximations to the discounted value function of $\overline{\text{MDP}}_n$ are given by $v_n^0 = 0$ and $v_n^{t+1} = \overline{T}_n v_n^t$ $(t \geq 1)$. Similar to v_n^t, let us define $v^0 = 0$ and $v^{t+1} = T v^t$, where T is the Bellman optimality operator for the original MDP.

Using [12, Theorem 8.3.6, p. 47] one can prove the following lemma.

Lemma 7. *[22, 23] Suppose Assumption 8 holds. Then, for all* t, *and for any compact set* $K \subset \mathsf{X}$, *we have* $\lim_{n \to \infty} \sup_{x \in K} |v_n^t(x) - v^t(x)| = 0$.

The following theorem states that the discounted value function of $\overline{\text{MDP}}_n$ converges to the discounted value function of the original MDP uniformly on each compact set $K \subset \mathsf{X}$.

Theorem 14. *[22, 23] Suppose Assumption 8 holds. Then, for any compact set* $K \subset \mathsf{X}$ *we have* $\lim_{n\to\infty} \sup_{x\in K} |\bar{J}_n^*(x) - J^*(x)| = 0$.

In the remainder of this section, we use the above results and Theorem 10 to compute a near optimal policy for the original MDP. It is straightforward to check that for each n, c-MDP$_n$ satisfies the assumptions in Theorem 10. Let $\{\varepsilon_n\}$ be a sequence of positive real numbers such that $\lim_{n\to\infty} \varepsilon_n = 0$.

By Theorem 10, for each $n \geq 1$, there exists a deterministic stationary policy $f_n \in \mathbb{F}_n$, obtained from the finite state approximations of c-MDP$_n$, such that

$$\sup_{x\in\mathsf{X}_n} |J_n(f_n, x) - J_n^*(x)| \leq \varepsilon_n,$$

where for each n, finite-state models are constructed replacing $(\mathsf{Z}, \mathsf{A}, p, c)$ with the components $(\mathsf{X}_n, \mathsf{A}, p_n, c_n)$ of c-MDP$_n$ in Section 3.1. By Lemma 5, for each $n \geq 1$ we also have

$$\sup_{x\in\mathsf{X}} |\bar{J}_n(f_n, x) - \bar{J}_n^*(x)| \leq \varepsilon_n, \tag{22}$$

where, with an abuse of notation, we also denote the extended policy by f_n.

Analogous with Theorem 14, we can prove the following lemma.

Lemma 8. *[22, 23] Suppose Assumption 8 holds. Then, for any compact set* $K \subset \mathsf{X}$, *we have* $\lim_{n\to\infty} \sup_{x\in K} |\bar{J}_n(f_n, x) - J(f_n, x)| = 0$. *This holds for all sequences of policies* $\{f_n\} \subset \mathbb{F}$.

The following theorem states that the true cost functions of the policies obtained from finite state models converge to the value function of the original MDP.

Theorem 15. *[22, 23] Suppose Assumption 8 holds. Then, for any compact set* $K \subset \mathsf{X}$, *we have* $\lim_{n\to\infty} \sup_{x\in K} |J(f_n, x) - J^*(x)| = 0$.

Average Cost In this section we obtain approximation results, analogous to Theorems 14 and 15, for the average cost criterion. To do this, we impose some new assumptions on the components of the original MDP in addition to Assumption 8. These assumptions are the unbounded counterpart of Assumption 7. With the exception of Assumption 9-(j), they are very similar to Assumption 4. In what follows, for any finite signed measure ϑ, we let

$$\|\vartheta\|_w := \sup_{\|g\|_w \leq 1} \left| \int_{\mathsf{X}} g(x)\vartheta(dx) \right|.$$

Here $\|\vartheta\|_w$ is called the w-norm of ϑ.

Assumption 9. *Suppose Assumption 8 holds with item (b) and (15) replaced by conditions (j) and (e) below, respectively. In addition, there exist a probability measure* η *on* X *and a positive measurable function* $\phi : \mathsf{X} \times \mathsf{A} \to (0, \infty)$ *such that for all* $(x, a) \in \mathsf{X} \times \mathsf{A}$

(e) $\int_{\mathsf{X}} w(y) p(dy|x, a) \leq \alpha w(x) + \eta(w)\phi(x, a)$, *where* $\alpha \in (0, 1)$.
(f) $p(D|x, a) \geq \eta(D)\phi(x, a)$ *for all* $D \in \mathcal{B}(\mathsf{X})$.
(g) The weight function w *is* η*-integrable, i.e.,* $\eta(w) < \infty$.
(h) For each $n \geq 1$, $\inf_{(x,a) \in K_n \times \mathsf{A}} \phi(x, a) > 0$.
(j) The stochastic kernel $p(\,\cdot\,|x, a)$ *is continuous in* (x, a) *with respect to the* w*-norm.*

Analogous with Theorems 4 and 11, the following theorem is a consequence of [27, Theorems 3.3] and [10, Lemma 3.4 and Theorem 2.6] (see also [12, Proposition 10.2.5])

Theorem 16. *Under Assumption 9 the following hold. For each* $f \in \mathbb{F}$, *the stochastic kernel* $Q_f(\,\cdot\,|x)$ *has an unique invariant probability measure* μ_f. *Furthermore,* w *is* μ_f*-integrable, and therefore,* $\rho_f := \int_{\mathsf{X}} c(x, f(x))\mu_f(dx) < \infty$. *There exist positive real numbers* R *and* $\kappa < 1$ *such that*

$$\sup_{f \in \mathbb{F}} \|Q_f^t(\,\cdot\,|x) - \mu_f\|_w \leq R w(x) \kappa^t \tag{23}$$

for all $x \in \mathsf{X}$, *where* R *and* κ *continuously depend on* α, $\eta(w)$, *and* $\inf_{f \in \mathbb{F}} \eta(\phi(y, f(y)))$. *Furthermore, there exist* $f^* \in \mathbb{F}$ *and* $h^* \in B_w(\mathsf{X})$ *such that the triplet* (h^*, f^*, ρ_{f^*}) *satisfies the average cost optimality inequality (ACOI), and therefore, for all* $x \in \mathsf{X}$

$$\inf_{\pi \in \Pi} V(\pi, x) := V^*(x) = \rho_{f^*},$$

Note that this theorem implies that for each $f \in \mathbb{F}$, the average cost is given by $V(f, x) = \int_{\mathsf{X}} c(y, f(y))\mu_f(dy)$ for all $x \in \mathsf{X}$ (instead of μ_f-a.e.).

Let V_n and $\bar{V}_n$ denote the average costs of c-MDP$_n$ and $\overline{\text{MDP}}_n$, respectively. The value functions for average cost are denoted similar to discounted cost case. Analogous to Lemma 5, the following result states that c-MDP$_n$ and $\overline{\text{MDP}}_n$ are not too different for the average cost.

Lemma 9. *We have*

$$\bar{V}_n^*(x) = \begin{cases} V_n^*(x), & \text{if } x \in K_n \\ V_n^*(\Delta_n), & \text{if } x \in K_n^c. \end{cases}$$

Furthermore, if, for any deterministic stationary policy $f \in \mathbb{F}_n$, *we define* $\bar{f}(x) = f(x)$ *on* K_n *and* $\bar{f}(x) = f(\Delta_n)$ *on* K_n^c, *then*

$$\bar{V}_n(\bar{f}, x) = \begin{cases} V_n(f, x), & \text{if } x \in K_n \\ V_n(f, \Delta_n), & \text{if } x \in K_n^c. \end{cases}$$

In particular, if the deterministic stationary policy $f_n^* \in \mathbb{F}_n$ *is optimal for c-MDP*$_n$, *then its extension* $\bar{f}_n^*$ *to* X *is also optimal for* $\overline{MDP}_n$.

By Lemma 9, in the remainder of this section we consider $\overline{\text{MDP}}_n$ in place of c-MDP$_n$. Recall the definition of constants γ_n and τ_n from (16) and (17). For each $n \geq 1$, we define $\phi_n : \mathsf{X} \times \mathsf{A} \to (0, \infty)$ and $\varsigma_n \in \mathbb{R}$ as

$$\phi_n(x,a) := \begin{cases} \phi(x,a), & \text{if } x \in K_n \\ \Gamma_n[\phi(y,a)], & \text{if } x \in K_n^c, \end{cases}$$

and $\varsigma_n := \int_{K_n^c} w(y)\eta(dy)$. Since $\eta(w) < \infty$ and τ_n can be made arbitrarily small by properly choosing ν_n, we assume, without loss of generality, that

$$\lim_{n\to\infty} (\tau_n + \varsigma_n) = 0. \tag{24}$$

Let $\alpha_n := \alpha + \varsigma_n + \tau_n$. Then one can show that the components of $\overline{\text{MDP}}_n$ satisfy the following conditions similar with Assumption 9: for any $(x,a) \in \mathsf{X} \times \mathsf{A}$,

$$\sup_{a\in\mathsf{A}} b_n(x,a) \leq M w_n(x)$$

$$\int_{\mathsf{X}} w_n(y) q_n(dy|x,a) \leq \alpha_n w_n(x) + \eta(w_n)\phi_n(x,a),$$

$$\int_{\mathsf{X}} w_n(y)\eta(dy) < \infty,$$

$$q_n(D|x,a) \geq \eta(D)\phi_n(x,a) \quad \text{for all } D \in \mathcal{B}(\mathsf{X}).$$

We note that by (24), there exists $n_0 \geq 1$ such that $\alpha_n < 1$ for $n \geq n_0$. Hence, for $n \geq n_0$, Theorem 16 holds for $\overline{\text{MDP}}_n$ with some $R_n \in \mathbb{R}_+$ and $\kappa_n \in (0,1)$, and we have $R_{\max} := \sup_{n\geq n_0} R_n < \infty$ and $\kappa_{\max} := \sup_{n\geq n_0} \kappa_n < 1$.

In the remainder of this section, it is assumed that $n \geq n_0$. Let $\{\varepsilon_n\}$ be a sequence of positive real numbers converging to zero. For each $f \in \mathbb{F}$, let μ_f^n denote the unique invariant probability measure of the transition kernel $q_n(\,\cdot\,|x,f(x))$ and let ρ_f^n denote the average cost of f; that is, $\rho_f^n := \bar{V}_n(f,x) = \int_{\mathsf{X}} b_n(y,f(y))\mu_f^n(dy)$ for all $x \in \mathsf{X}$. Therefore, the value function of $\overline{\text{MDP}}_n$, denoted as $\bar{V}_n^*$, is given by $V_n^*(x) = \inf_{f\in\mathbb{F}} \rho_f^n$, i.e., it is constant on X.

Before making the connection with Theorem 13, we state the following result.

Lemma 10. *Suppose Assumption 9 holds. Then the transition probability p_n of c-MDP$_n$ is continuous in (x,a) with respect to the total variation distance.*

Thus we obtain that for each $n \geq 1$, c-MDP$_n$ satisfies the assumption in Theorem 13 for

$$\zeta(\,\cdot\,) = \Lambda_n \eta(\,\cdot\,),$$

$$\theta(x,a) = \begin{cases} \phi(x,a), & \text{if } x \in K_n \\ \Gamma_n[\phi(y,a)], & \text{if } x = \Delta_n, \end{cases}$$

and some $\lambda \in (0,1)$, where the existence of λ follows from Assumption 9-(h) and the fact $\phi > 0$. Therefore, for each $n \geq 1$, there exists a deterministic stationary policy $f_n \in \mathbb{F}_n$, obtained from the finite state approximations of c-MDP$_n$, such that $\sup_{x \in \mathsf{X}_n} |V_n(f_n, x) - V_n^*(x)| \leq \varepsilon_n$, where finite-state models are constructed replacing $(\mathsf{Z}, \mathsf{A}, p, c)$ with the components $(\mathsf{X}_n, \mathsf{A}, p_n, c_n)$ of c-MDP$_n$ in Section 3.1. By Lemma 9, we also have $|\rho_{f_n}^n - \bar{V}_n^*| \leq \varepsilon_n$, where by abuse of notation we also denote the extended policy by f_n. The following is the main result of this section which states that the true average cost of the policies f_n obtained from finite state approximations of c-MDP$_n$ converges to the average value function V^* of the original MDP.

Lemma 11. *[22, 23] Suppose Assumption 9 holds. Then we have*

$$\lim_{n \to \infty} \sup_{f \in \mathbb{F}} |\rho_f^n - \rho_f| = 0, \qquad \lim_{n \to \infty} |\rho_{f_n} - V^*| = 0. \tag{25}$$

3.3 Discretization of the Action Space

For computing near optimal policies using well known algorithms, such as value iteration, policy iteration, and Q-learning, the action space must be finite. In this section, using results from Section 3.1 we show that, as a pre-processing step, the action space can taken to be finite if it has sufficiently large number of points for accurate approximation.

It was shown in Theorems 1 and 2 that any MDP with (infinite) compact action space and with bounded one-stage cost function can be well approximated by an MDP with finite action space under assumptions that are satisfied by c-MDP$_n$ for each n, for both the discounted cost and the average cost cases. Recall the sequence of finite subsets $\{\Lambda_k\}$ of A from Section 2.1. We define c-MDP$_{n,k}$ as the Markov decision process having the components $\{\mathsf{X}_n, \Lambda_k, p_n, c_n\}$ and we let $\mathbb{F}_n(\Lambda_k)$ denote the set of all deterministic stationary policies for c-MDP$_{n,k}$. Note that $\mathbb{F}_n(\Lambda_k)$ is the set of policies in $\mathbb{F}_n$ taking values only in Λ_k. Therefore, in a sense, c-MDP$_{n,k}$ and c-MDP$_n$ can be viewed as the same MDP, where the former has constraints on the set of policies. For each n and k, by an abuse of notation, let f_n^* and $f_{n,k}^*$ denote the optimal stationary policies of c-MDP$_n$ and c-MDP$_{n,k}$, respectively, for both the discounted and average costs. Then Theorems 1 and 2 show that for all n, we have

$$\lim_{k \to \infty} J_n(f_{n,k}^*, x) = J_n(f_n^*, x) := J_n^*(x)$$
$$\lim_{k \to \infty} V_n(f_{n,k}^*, x) = V_n(f_n^*, x), := V_n^*(x)$$

for all $x \in \mathsf{X}_n$. In other words, the discounted and average value functions of c-MDP$_{n,k}$ converge to the discounted and average value functions of c-MDP$_n$ as $k \to \infty$.

Let us fix $x \in \mathsf{X}$. For n sufficiently large (so $x \in K_n$), we choose k_n such that $|J_n(f_{n,k_n}^*, x) - J_n(f_n^*, x)| < 1/n$ (or $|V_n(f_{n,k_n}^*, x) - V_n(f_n^*, x)| < 1/n$ for the average cost). We note that if A is a compact subset of a finite dimensional Euclidean space, then by using Theorems 6 and 7 one can obtain an explicit expression for k_n in terms of n under further continuity conditions on c and p.

By Lemmas 8 and 11, we have $|\bar{J}_n(f^*_{n,k_n},x) - J(f^*_{n,k_n},x)| \to 0$ and $|\bar{V}_n(f^*_{n,k_n},x) - V(f^*_{n,k_n},x)| \to 0$ as $n \to \infty$, where again by an abuse of notation, the policies extended to X are also denoted by f^*_{n,k_n}. Since $\bar{J}_n(f^*_{n,k_n},x) = J_n(f^*_{n,k_n},x)$ and $\bar{V}_n(f^*_{n,k_n},x) = V_n(f^*_{n,k_n},x)$, it follows that

$$\lim_{n\to\infty} J(f^*_{n,k_n},x) = J^*(x) \qquad \lim_{n\to\infty} V(f^*_{n,k_n},x) = V^*(x).$$

Therefore, before discretizing the state space to compute the near optimal policies, one can discretize, without loss of generality, the action space A in advance on a finite grid using sufficiently large number of grid points.

3.4 Rates of Convergence for Compact-State MDPs

In this section we consider **(Q2)** for MDPs with compact state space; that is, we derive an upper bound on the performance loss due to discretization in terms of the cardinality of the set Z_n (i.e., number of grid points) . To do this, we will impose some new assumptions on the components of the MDP in addition to Assumptions 6 and 7. First, we present some definitions that are needed in the development. For each $g \in C_b(\mathsf{Z})$, let

$$\|g\|_{\mathrm{Lip}} := \sup_{(z,y)\in \mathsf{Z}\times\mathsf{Z}} \frac{|g(z)-g(y)|}{d_{\mathsf{Z}}(z,y)}.$$

If $\|g\|_{\mathrm{Lip}}$ is finite, then g is called Lipschitz continuous with Lipschitz constant $\|g\|_{\mathrm{Lip}}$. $\mathrm{Lip}(\mathsf{Z})$ denotes the set of all Lipschitz continuous functions on Z, i.e.,

$$\mathrm{Lip}(\mathsf{Z}) := \{g \in C_b(\mathsf{Z}) : \|g\|_{\mathrm{Lip}} < \infty\}$$

and $\mathrm{Lip}(\mathsf{Z},K)$ denotes the set of all $g \in \mathrm{Lip}(\mathsf{Z})$ with $\|g\|_{\mathrm{Lip}} \le K$. The *Wasserstein distance of order* 1 [28, p. 95] between two probability measures ζ and ξ over Z is defined as

$$W_1(\zeta,\xi) := \sup\biggl\{\biggl|\int_{\mathsf{Z}} g d\zeta - \int_{\mathsf{Z}} g d\xi\biggr| : g \in \mathrm{Lip}(\mathsf{Z},1)\biggr\}.$$

W_1 is also called the *Kantorovich-Rubinstein distance.* It is known that if Z is compact, then $W_1(\zeta,\xi) \le \mathrm{diam}(\mathsf{Z})\|\zeta-\xi\|_{TV}$ [28, Theorem 6.13]. For compact Z, the Wasserstein distance of order 1 is weaker than total variation distance. Furthermore, for compact Z, the Wasserstein distance of order 1 metrizes the weak topology on the set of probability measures $\mathcal{P}(\mathsf{Z})$ [28, Corollary 6.11] which also implies that convergence in this sense is weaker than setwise convergence.

In this section we impose the following supplementary assumptions in addition to Assumption 6 and Assumption 7.

Assumption 10.

(g) The one-stage cost function c satisfies $c(\,\cdot\,,a) \in \mathrm{Lip}(\mathsf{Z},K_1)$ for all $a \in \mathsf{A}$ for some K_1.

(h) *The stochastic kernel* p *satisfies* $W_1\big(p(\cdot\,|z,a), p(\cdot\,|y,a)\big) \leq K_2 d_{\mathsf{Z}}(z,y)$ *for all* $a \in \mathsf{A}$ *for some* K_2.
(h') *The stochastic kernel* p *satisfies:* $\|p(\cdot\,|z,a) - p(\cdot\,|y,a)\|_{TV} \leq K_2 d_{\mathsf{Z}}(z,y)$ *for all* $a \in \mathsf{A}$ *and for some* K_2.
(j) Z *is an infinite compact subset of* $\mathbb{R}^d$ *for some* $d \geq 1$, *equipped with the Euclidean norm.*

We note that Assumption 10-(j) implies the existence of a constant $\alpha > 0$ and finite subsets $\mathsf{Z}_n \subset \mathsf{Z}$ with cardinality n such that

$$\max_{z \in \mathsf{Z}} \min_{y \in \mathsf{Z}_n} d_{\mathsf{Z}}(z,y) \leq \alpha (1/n)^{1/d} \tag{26}$$

for all n, where d_{Z} is the Euclidean distance on Z. In the following, we replace Z_n defined in Section 3.1 with Z_n satisfying (26) in order to derive *explicit* bounds on the approximation error in terms of the cardinality of Z_n.

Discounted Cost Assumptions 6 and 10 (without Assumption 10-(h')) are imposed throughout this section. Additionally, we assume that $K_2\beta < 1$. The last assumption is the key to prove the next result which states that the value function J^* of the original MDP for the discounted cost is in $\mathrm{Lip}(\mathsf{Z})$ [15].

Theorem 17. *[22] Suppose Assumptions 6, 10 (without Assumption 10-(h')) and* $K_2\beta < 1$ *hold. Then the value function* J^* *for the discounted cost is in* $\mathrm{Lip}(\mathsf{Z}, K)$, *where* $K = K_1 \frac{1}{1-\beta K_2}$.

The following theorem can be proved by a further analysis of the results in Section 3.1 and Theorem 17. Recall that the policy $\hat{f}_n \in \mathbb{F}$ is obtained by extending the optimal policy f_n^* of MDP_n to Z.

Theorem 18. *[22] We have*

$$\|J(\hat{f}_n, \cdot) - J^*\| \leq \frac{\tau(\beta, K_2) K_1 \frac{1}{1-\beta K_2} + \frac{2K_1}{1-\beta}}{1-\beta} 2\alpha (1/n)^{1/d},$$

where $\tau(\beta, K_2) = (2+\beta)\beta K_2 + \frac{\beta^2 + 4\beta + 2}{(1-\beta)^2}$ *and* α *is the coefficient in (26).*

Remark 3. It is important to point out that if we replace Assumption 10-(h) with (h'), then Theorem 18 remains valid (with possibly different constants in front of the term $(1/n)^{1/d}$). However, in this case, we do not need the assumptions $K_2\beta < 1$.

Average Cost In this section, we suppose that Assumptions 7 and 10 (without Assumption 10-(h)) hold. The following theorem is the main result of this section which can be proved by a further analysis of the results in Section 3.1. Recall that the policy $\tilde{f}_n^*$, the optimal policy of $\widetilde{\mathrm{MDP}}_n$, is obtained by extending the optimal policy f_n^* of MDP_n to Z.

Theorem 19. *[22] For all $t \geq 1$, we have*

$$|\rho_{\tilde{f}_n^*} - \rho_{f^*}| \leq 4\|c\|R\kappa^t + 4K_1\alpha(1/n)^{1/d} + 2\|c\|K_2\alpha(1/n)^{1/d}(2^{t+1} - 2).$$

In particular, with $\varrho_1 := -\frac{\ln(\kappa)}{\ln(2/\kappa)}$, $I_1 := 4\|c\|R$, $I_2 := 4K_1\alpha$, *and* $I_3 := 2\|c\|K_2\alpha$ *and* $I_4 := \left(\frac{I_1}{2I_3 \ln(\frac{1}{\kappa})\ln(2)}\right)^{-1}$,

$$|\rho_{\tilde{f}_n^*} - \rho_{f^*}| \leq \left(I_1 I_4^{\varrho_1} + 2I_3 I_4^{\varrho_1 - 1}\right)(1/n)^{\varrho_1/d} + \left(I_2 - 2I_3\right)(1/n)^{1/d}.$$

4 Concluding remarks

In this chapter we established conditions so that an MDP with Borel state and action spaces can be approximated with arbitrary precision by finite state and action models, where finite state models are obtained via quantization of the state space of the original MDP. We considered both discounted and average cost formulations. The results are established for MDPs with both compact state space and non-compact state space under different assumptions. For MDPs with compact state space, we also obtained explicit rate of convergence bounds on the approximation error. We finally note that we have recently started to extend the results reviewed in this chapter to setups with constraints [24], as well as team decision problems.

References

1. Aliprantis, C., and Border, K. *Infinite Dimensional Analysis.* Springer, 2006.
2. Bertsekas, D.: Convergence of discretization procedures in dynamic programming. *IEEE Trans. Autom. Control,* **20** (1975) 415-419.
3. Bertsekas, D. and Tsitsiklis, J. *Neuro-Dynammic Programming.* Athena Scientific, 1996.
4. Chow, C.-S. and Tsitsiklis, J. N.: An optimal one-way multigrid algorithm for discrete-time stochastic control. *IEEE Trans. on Autom. Control,* **36** (1991) 898-914.
5. Dufour, F. and Prieto-Rumeau, T.: Approximation of Markov decision processes with general state space. *J. Math. Anal. Appl.* **388** (2012) 1254-1267.
6. Dufour, F. and Prieto-Rumeau, T.: Finite linear programming approximations of constrained discounted Markov decision processes. *SIAM J. Control Optim.,* **51** (2013) 1298–1324.
7. Dufour, F. and Prieto-Rumeau, T.: Approximation of average cost Markov decision processes using empirical distributions and concentration inequalities. *Stochastics* (2014) 1-35.
8. Feinberg, E.: On measurability and representation of strategic measures in Markov decision processes. *Statistics, Probability and Game Theory,* **30** (1996) 29-43.
9. Feinberg, E., Kasyanov, P. and Zadioanchuk, N.: Average cost Markov decision processes with weakly continuous transition probabilities. *Math. Oper. Res.,* **37** (2012) 591-607.

10. Gordienko, E. and Hernandez-Lerma, O.: Average cost Markov control processes with weighted norms: Existence of canonical policies. *Appl. Math.*, **23** (1995), 199-218.
11. Hernández-Lerma, O. and Lasserre, J. *Discrete-Time Markov Control Processes: Basic Optimality Criteria.* Springer, 1996.
12. Hernández-Lerma, O. and Lasserre, J. *Further Topics on Discrete-Time Markov Control Processes.* Springer, 1999.
13. Hernández-Lerma, O. and Lasserre, J. *Markov Chains and Invariant Probabilities.* Birkhauser, 2003.
14. Hernández-Lerma, O., Montes-De-Oca, R. and Cavazos-Cadena, R.: Recurrence conditions for Markov decision processes with Borel state space: a survey. *Ann. Oper. Res.*, **28** (1991) 29-46.
15. Hinderer, K.: Lipshitz continuity of value functions in Markovian desision processes. *Math. Meth. Oper. Res.*, **62** (2005) 3-22.
16. Ortner, R.: Pseudometrics for state aggregation in average reward Markov decision processes. In *Algorithmic Learning Theory.* Springer-Verlag, 2007, 373-387.
17. Ren, Z. and Krogh, B.: State aggregation in Markov decision processes. In *CDC 2002*, Las Vegas, December 2002, 3819-3824.
18. Roy, B.: Performance loss bounds for approximate value iteration with state aggregation. *Math. Oper. Res.*, **31** (2006) 234-244.
19. Saldi, N. *Optimal Quantization and Approximation in Source Coding and Stochastic Control.* PhD thesis, Queen's Univ., Kingston, ON, Canada, 2015.
20. Saldi, N., Linder, T. and Yüksel, S.: Asymtotic optimality and rates of convergence of quantized stationary policies in stochastic control. *IEEE Trans. Autom. Control*, **60** (2015) 553-558.
21. Saldi, N., Yüksel, S. and Linder, T.: Near optimality of quantized policies in stochastic control under weak continuity conditions. *Journal of Mathematical Analysis and Applications*, to appear, also on arXiv:1410.6985.
22. Saldi, N., Yüksel, S. and Linder, T.: *Asymptotic Optimality of Finite Approximations to Markov Decision Processes with Borel Spaces.* arXiv:1503.02244, 2015.
23. Saldi, N., Yüksel, S. and Linder, T.: Finite-state approximation of Markov decision processes with unbounded costs and Borel spaces. In *IEEE Conf. Decision Control*, Osaka, Japan, December 2015.
24. Saldi, N., Yüksel, S. and Linder, T.: Finite-state approximations to constrained Markov decision processes with Borel spaces. In *Proc. Allerton Conference* Illinois, USA, October 2015.
25. Saldi, N., Yüksel, S. and Linder, T.: Finite state approximations of Markov decision processes with general state and action spaces. In *American Control Conference (ACC)*, 2015, 3589-3594.
26. Schäl, M.: On dynamic programming: compactness of the space of policies. *Stochastic Process. Appl.*, **3** (1975) 345-364.
27. Vega-Amaya, O.: The average cost optimality equation: a fixed point approach. *Bol. Soc. Mat. Mexicana*, **9** (2003) 185-195.
28. Villani, C. *Optimal Transport: Old and New.* Springer, 2009.
29. White, D.: Finite-state approximations for denumerable state infinite horizon discounted Markov decision processes. *J. Math. Anal. Appl.*, **74** (1980) 292-295.
30. Yushkevich, A.: Reduction of a controlled Markov model with incomplete data to a problem with complete information in the case of Borel state and control spaces. *Theory Prob. Appl.*, **21** (1976) 153-158.

Markov Decision Processes with State-Dependent Discount Factors: Stability with Respect to the Prokhorov Metric

Evgueni Gordienko[1] and J. Adolfo Minjárez-Sosa[2]

[1] Departamento de Matemáticas, Universidad Autónoma Metropolitana, Unidad Iztapalapa, México, D.F.
gord@xanum.uam.mx

[2] Departamento de Matemáticas, Universidad de Sonora, Rosales s/n, Col. Centro, 83000, Hermosillo, Sonora, México.
aminjare@gauss.mat.uson.mx

Abstract. In this paper we obtain upper bounds for stability index in discrete-time Markov decision processes under a discounted optimality criterion with state-dependent discount factors. Under Lipschitz conditions the main inequality is expressed in terms of the Prokhorov metric. We illustrate our results with examples on approximation by means of empirical distributions, a water reservoir model, and a radioactive material stock control model.
Keywords: Markov decision processes, state-dependent discount factors, approximating process, stability inequality, Prokhorov metric.
AMS 2010 subject classification: Primary 93E10, Secondary 90C40

1 Introduction

In the last decade there have been several works dealing with Markov decision processes (MDPs) with non constant discount factors under different settings: random discount factors (see, e.g., [4–6]), discount factors as a function of states/actions of controlled processes (see, e.g., [22]), as well as a combination of the previous cases [17]. These works introduce, in certain sense, new approaches and extensions from the basic model proposed many years ago in [13, 19] where optimality equations and existence of optimal policies were investigated for MDPs with state-action- dependent discount factors.

A common motivation for considering such control models are their economic and financial applications, where frequently discounting is uncertain and may depend on the current state of a system. This situation could also occur in MDPs modelling some supply systems (see, e.g., [22]).

In this paper we consider MDPs evolving according to the equation

$$x_t = F(x_{t-1}, a_t, \xi_t), \quad t = 1, 2, ..., \tag{1}$$

under a discounted optimality criterion with state-dependent discount factors of the form

$$\alpha(x_t), \quad t = 1, 2, ..., \tag{2}$$

where x_t and a_t are the state and the action (or control) at time t. The disturbance process $\{\xi_t\}$ is a sequence of independent and identically distributed (i.i.d.) random vectors with distribution μ. In this case, the discount factors (2) play the following role during the evolution of the system. At initial time, when the system is in state x_0, the controller chooses an action a_1 and a cost $c(x_0, a_1)$ is incurred. Then the system moves to a new state x_1 according to a transition law determined by the equation (1). Once the system is in state x_1 the controller selects an action a_2 and incurs a discounted cost $\alpha(x_1)c(x_1, a_2)$. Next the system moves to a state x_2 and the process is repeated. In general, for the stage $t \geq 1$, the controller incurs the discounted cost

$$\prod_{i=0}^{t-1} \alpha(x_{i-1})c(x_{t-1}, a_t), \quad \alpha(x_{-1}) := 1. \tag{3}$$

Throughout the evolution of the system, the actions are selected by means of rules π known as control policies, and the costs (3) are accumulated in an infinite horizon. This scheme defines the total expected discounted cost with state-dependent discount factor which is denoted by $V(x, \pi)$, where x is the initial state. Hence the optimal control problem is to find a policy π_* such that

$$V(x, \pi_*) = \inf_{\pi} V(x, \pi) =: V_*(x).$$

The most of works studying this class of performance indices or some of their variants (see, e.g., [4–6, 13, 17, 19, 22]) are focused on showing existence of optimal policies, or furthermore on the construction of adaptive/minimax control policies under uncertainty of distributions of random vectors governing the controlled processes. In our case we are interested in to study the *stability problem* of MDPs (1)-(2). Specifically, we assume that the distribution μ of the random vectors $\{\xi_t\}$ is at least partly unknown, and it can be approximated by a certain known distribution $\tilde{\mu}$. For instance, $\tilde{\mu}$ can be obtained from some theoretical consideration or by statistical estimations in the case when the process $\{\xi_t\}$ is observable. Under this situation we can consider the "approximating MDP"

$$\tilde{x}_t = F(\tilde{x}_{t-1}, \tilde{a}_t, \tilde{\xi}_t), \quad t = 1, 2, ..., \tag{4}$$

where $\left\{\tilde{\xi}_t\right\}$ are i.i.d. random vectors with the distribution $\tilde{\mu}$. Then, by analyzing the corresponding optimal control problem, it is possible, at least theorically, to get an optimal policy $\tilde{\pi}_*$ for (4).

Hence, the idea behind the "stability estimation" (or "robustness estimation") is to consider the policy $\tilde{\pi}_*$ as a reasonable approximation to the unavailable optimal policy π_*. Thus, if the policy $\tilde{\pi}_*$ is applied to control the original MDP (1), then the corresponding increase of the discounted cost is represented by the so-called *stability index* defined as (see, e.g., [8–10])

$$\Delta(x) := V(x, \tilde{\pi}_*) - \inf_{\pi} V(x, \pi) = V(x, \tilde{\pi}_*) - V_*(x) \geq 0, \quad x \in X.$$

It is worth remarking that, even for a constant discount factor, $\Delta(x)$ can not approach zero when $\tilde{\mu}$ converges weakly to μ (see, e.g., [9] and Example 1 in Section 3). Therefore, under suitable Lipschitz conditions on the components of the control model, our main objective is to estimate $\Delta(x)$ by means of *stability inequalities* defined in terms of probabilistic metrics. Specifically we prove

$$\sup_{x \in X} \Delta(x) \leq K \pi_r(\mu, \tilde{\mu}), \tag{5}$$

where π_r is the Prokhorov metric (metrizing the weak convergence) and K is a constant explicitly calculated .

On the other hand, as is showed in Example 1, inequality (5) can be false if the discount factor function $\alpha(x)$ is not continuous. Taking into account this situation, additionally we prove that without any continuity (or Lipschitz) conditions on $\alpha(x)$ (neither on the function F in (1)), the following stability inequality holds:

$$\sup_{x \in X} \Delta(x) \leq \bar{K} \mathcal{V}_T(\mu, \tilde{\mu}),$$

where $\mathcal{V}_T$ is the total variation distance.

In order to illustrate our results, we present applications of inequality (5) in a simple water release model, as well as in a class of storage models of decayed material (see, e.g., [21]). In addition, assuming observability of the random vectors $\{\xi_t\}$, we also discuss briefly the approximation of the unknown distribution μ by means of the empirical distribution. Analyzing the stability problem from this statistical point of view is important in the field of adaptive control for systems with unknown transition probabilities (see, e.g., [7, 12, 15, 16]. Specifically in our case, by using the stability inequality (5) it is possible to measure the performance of policies constructed by applying empirical estimation and control schemes in MDPs with unknown disturbance distribution.

The remainder of the paper is organized as follows. Section 2 contains the description of the Markov decision model and the stability problem. In Section 3 we introduce the assumptions as well as our main results, whereas in Section 4 some application examples are presented. Finally, the proofs are given in Section 5.

Notation. Throughout the paper we shall use the following notation. Given a *Borel space* Y—that is, a Borel subset of a complete separable metric space—$\mathcal{B}(Y)$ denotes the Borel σ–algebra and "measurability" always means measurability with respect to $\mathcal{B}(Y)$. The class of all probability measures on Y is denoted by $\mathbb{P}(Y)$. Given two Borel spaces Y and Y', a *stochastic kernel* $\varphi(\cdot|\cdot)$ on Y given Y' is a function such that $\varphi(\cdot|y')$ is in $\mathbb{P}(Y)$ for each $y' \in Y'$, and $\varphi(B|\cdot)$ is a measurable function on Y' for each $B \in \mathcal{B}(Y)$. In addition, $\mathbb{B}(Y)$ stands for the space of real-valued bounded measurable functions on Y with the norm $\|v\| := \sup_{x \in Y} |v(y)|$. Finally, $\mathbb{N}$ ($\mathbb{N}_0$) denotes the positive (nonnegative) integers numbers, and $\Re$ is the set of real numbers.

2 The Control Model and Stability Index

2.1 Markov Control Model

Let $\mathcal{M} := (X, A, S, F, \mu, c, \alpha)$ be a discrete-time Markov control model with state-dependent discount factors, associated to the system (1)-(2), satisfying the following conditions. The state space X, the action or control space A, and the disturbance space S are Borel spaces with metrics d_X, d_A, and d_S, respectively. For each state $x \in X$, $A(x)$ is a nonempty Borel subset of A representing the set of admissible controls when the system is in state $x \in X$. The set $\mathbb{K} = \{(x, a) : x \in X, a \in A(x)\}$ is assumed to be a Borel subset of $X \times A$. The dynamic of the system is modeled by a measurable function $F : X \times A \times S \to X$ as in (1), that is

$$x_t = F(x_{t-1}, a_t, \xi_t), \quad t = 1, 2, ..., \tag{6}$$

where $x_t \in X$, $a_t \in A(x_t)$, and $\{\xi_t\}$ is a sequence of $S-$valued, i.i.d. random vectors with distribution $\mu \in \mathbb{P}(S)$. Finally,

$$c : \mathbb{K} \to [-b, b], \quad b \in \Re_+ \tag{7}$$

and

$$\alpha : X \to [0, \bar{\alpha}], \quad \bar{\alpha} < 1, \tag{8}$$

are measurable functions representing a bounded one-stage cost and the discount factor function as in (2), respectively.

Let $\mathbb{H}_0 := X$ and $\mathbb{H}_t := \mathbb{K}^t \times X$, $t \geq 1$ be the spaces of admissible histories up to time t. A control policy (randomized, history-dependent) is a sequence $\pi = \{\pi_t\}$ of stochastic kernels π_t on A given $\mathbb{H}_t$ such that $\pi_t(A(x_t) \mid h_t) = 1$, for all $h_t \in \mathbb{H}_t$, $t \in \mathbb{N}_0$. We denote by Π the set of all control policies and by $\mathbb{F} \subset \Pi$ the subset of stationary policies (see, e.g., [3, 11] for a precise definition). Each stationary policy $\pi \in \mathbb{F}$ is identified with a measurable function $f : X \to A$ such that $\pi_t(\cdot \mid h_t)$ is concentrated at $f(x_t) \in A(x_t)$ for all $h_t \in \mathbb{H}_t$, $t \in \mathbb{N}_0$, so that π is of the form $\pi = \{f, f, f, ...\}$. In this case we denote π by f, and for each $f \in \mathbb{F}$, we write $c(x, f) := c(x, f(x))$, $x \in X$.

According to (3), for a policy $\pi \in \Pi$, given the initial state $x_0 = x$, we define the total expected discounted cost with state-dependent discount factor as

$$V(x, \pi) := E_x^\pi \left[\sum_{t=1}^{\infty} \prod_{i=0}^{t-1} \alpha(x_{i-1}) c(x_{t-1}, a_t) \right], \tag{9}$$

where E_x^π denotes the expectation operator with respect to the probability measure P_x^π induced by the policy π, given $x_0 = x$ (see, e.g., [3] for the construction of P_x^π). Furthermore, the optimal value function is defined as

$$V_*(x) := \inf_{\pi \in \Pi} V(x, \pi), \quad x \in X.$$

2.2 Stability Index

In our stability estimation settings, we assume that $\mu \in \mathbb{P}(S)$ is at least partly unknown, and it can be approximated by a certain known distribution $\tilde{\mu} \in \mathbb{P}(S)$. The typical situation is when $\tilde{\mu}$ is obtained by a suitable statistical estimation procedure of μ (or of some their parameters), or in the case when μ can be known, but it is replaced by some "simpler" distribution. Hence, let $\tilde{\mathcal{M}} := (X, A, S, F, \tilde{\mu}, c, \alpha)$ be the control model associated to the process

$$\tilde{x}_t = F(\tilde{x}_{t-1}, \tilde{a}_t, \tilde{\xi}_t), \quad t = 1, 2, ..., \tag{10}$$

where $\left\{\tilde{\xi}_t\right\}$ is a sequence of i.i.d. random vectors in S with distribution $\tilde{\mu} \in \mathbb{P}(S)$. Both processes, (6) and (10), have the common initial state $x_0 = x \in X$. We denote by $\tilde{E}_x^\pi$ the expectation operator corresponding to the probability measure on the trajectories of the process (10), induced by the $\pi \in \Pi$ and $x \in X$. Then the performance index $\tilde{V}(x, \pi)$ and the value function $\tilde{V}_*(x)$ are defined accordingly as follows:

$$\tilde{V}(x, \pi) := \tilde{E}_x^\pi \left[\sum_{t=1}^{\infty} \prod_{i=0}^{t-1} \alpha(\tilde{x}_{i-1}) c(\tilde{x}_{t-1}, \tilde{a}_t) \right],$$

$$\tilde{V}_*(x) := \inf_{\pi \in \Pi} \tilde{V}(x, \pi).$$

Observe that V and $\tilde{V}$ are well defined due to (7) and (8). In fact, for all $\pi \in \Pi$ and $x \in X$,

$$|V(x, \pi)| \leq \frac{b}{1 - \bar{\alpha}} \quad \text{and} \quad \left|\tilde{V}(x, \pi)\right| \leq \frac{b}{1 - \bar{\alpha}},$$

which implies

$$\|V_*\| \leq \frac{b}{1 - \bar{\alpha}} \quad \text{and} \quad \left\|\tilde{V}_*\right\| \leq \frac{b}{1 - \bar{\alpha}}, \tag{11}$$

that is $V_*, \tilde{V}_* \in \mathbb{B}(X)$.

Assumptions introduced in next section guarantee the existence of stationary optimal policies $f_*, \tilde{f}_* \in \mathbb{F}$:

$$V_*(x) = V(x, f_*), \quad x \in X, \tag{12}$$

$$\tilde{V}_*(x) = \tilde{V}(x, \tilde{f}_*), \quad x \in X. \tag{13}$$

However, the solution of the optimal control problem given in (12) is not possible due to the lack of knowledge of the distribution μ. Hence, we suppose that the controller can find (approximate) the optimal policy $\tilde{f}_*$ in (13), and that it is used to control the original process (6). In this sense, $\tilde{f}_*$ is considered as an available approximation to an unknown optimal policy f_*. Using this procedure, the increase of expected incurred cost is expressed by the following stability index:

$$\Delta(x) := V(x, \tilde{f}_*) - V_*(x) \geq 0, \quad x \in X. \tag{14}$$

Finding upper bounds for Δ, given in terms of certain distances between μ and $\tilde{\mu}$, is the subject of the paper.

3 Assumptions and Stability Inequalities

We will use the metric $d_{\mathbb{K}} := \max(d_X, d_A)$ on the spaces $X \times A$ and $\mathbb{K}$. Let π_r be the Prokhorov (or Lévy-Prokhorov) metric on the space of probability distributions $\mathbb{P}(S)$ defined as (see,e.g., [2]):

$$\pi_r(\mu, \eta) := \inf \{\varepsilon > 0 : \mu(B) \leq \eta(B^\varepsilon) + \varepsilon, \ \eta(B) \leq \mu(B^\varepsilon) + \varepsilon, \ \text{for all } B \in \mathcal{B}(S)\},$$

where $B^\varepsilon := \{s \in S : d_S(s, B) < \varepsilon\}$. The convergence in π_r is equivalent to the weak convergence.

In order to obtain the first stability inequality under the metric π_r, we need the following set of Lipschitz conditions. Let ξ and $\tilde{\xi}$ be generic random vectors for ξ_t and $\tilde{\xi}_t$, respectively.

Assumption 1. *There exist finite constants L_c, L_A, L_α, L_*, and L_F such that for all $k, k' \in \mathbb{K}$ and $x, x' \in X$,*

(a) $|c(k) - c(k')| \leq L_c d_{\mathbb{K}}(k, k')$;

(b) $H(A(x), A(x')) \leq L_A d_X(x, x')$, *where H is the Hausdorff metric;*

(c) $|\alpha(x) - \alpha(x')| \leq L_\alpha d_X(x, x')$;

(d) $\mathcal{V}_T(F(k, \xi), F(k', \xi)) \leq L_* d_{\mathbb{K}}(k, k')$, *where $\mathcal{V}_T$ is the total variation distance between probability distributions (or random vectors) on $(S, \mathcal{B}(S))$;*

(e) $\sup_{k \in \mathbb{K}} d_X(F(k, s), F(k, s')) \leq L_F d_S(s, s')$, $s, s' \in S$;

(f) for every bounded continuous function $u : X \to \Re$ the map

$$k \to E\left(u\left[F\left(k, \tilde{\xi}\right)\right]\right)$$

is continuous on $\mathbb{K}$.

Remark 1. Using Assumption 1 and following similar arguments as the proof of Theorem 3.2 in [22], one can easily establish the following facts:

i) Operators $T, \tilde{T} : (\mathbb{B}, \|\cdot\|) \to (\mathbb{B}, \|\cdot\|)$ defined as

$$Tu(x) := \inf_{a \in A(x)} \{c(x, a) + \alpha(x) E\left(u\left[F\left(x, a, \xi\right)\right]\right)\}, \quad x \in X, \tag{15}$$

and

$$\tilde{T}u(x) := \inf_{a \in A(x)} \left\{c(x, a) + \alpha(x) E\left(u\left[F\left(x, a, \tilde{\xi}\right)\right]\right)\right\}, \quad x \in X, \tag{16}$$

are contraction with modulus $\bar{\alpha}$.

ii) The value functions V_* and $\tilde{V}_*$ are the fixed point of T and $\tilde{T}$ respectively:

$$V_* = TV_*, \quad \tilde{V}_* = \tilde{T}\tilde{V}_*. \tag{17}$$

iii) There exist $f_*, \tilde{f}_* \in \mathbb{F}$ such that

$$TV_*(x) = c(x, f_*(x)) + \alpha(x)E\left(V_*\left[F\left(x, f_*(x), \xi\right)\right]\right), \quad x \in X, \tag{18}$$

and

$$T\tilde{V}_*(x) = c(x, \tilde{f}_*(x)) + \alpha(x)E\left(u\left[F\left(x, \tilde{f}_*(x), \tilde{\xi}\right)\right]\right), \quad x \in X. \tag{19}$$

iv) The policies f_* and $\tilde{f}_*$ are optimal for the MDPs (6) and (10), respectively.

Theorem 1. *Under Assumption 1*

$$\sup_{x\in X} \Delta(x) \le K\pi_r(\mu, \tilde{\mu}), \tag{20}$$

where

$$K = \frac{4\bar{\alpha}}{(1-\bar{\alpha})^2}\left[\frac{b}{1-\bar{\alpha}} + (1+L_A)L_F\left[L_c + (L_\alpha + \bar{\alpha}L_*)\frac{b}{1-\bar{\alpha}}\right]\right]. \tag{21}$$

The most restrictive condition used to prove (20) is the Assumption 1(d). In fact, when X and A are unbounded subsets of Euclidian spaces, such condition basically demands the existence of a density of the random vector ξ with bounded partial derivative, vanishing fast enough in "infinity". However, for MDPs where the set of admissible actions is constant, that is $A(x) = A$, $x \in X$, we can significantly relax Assumption 1 (d) (as well as Assumption 1 (a)), as is stated in the following set of conditions.

Assumption 2. *For all $x \in X$, $A(x) = A$ where A is a compact set. In addition, there exist finite constants $\bar{L}_c$, L_α, L_F, and a measurable function $\bar{L}_* : S \to \Re$ such that for all $x, x' \in X$ we have:*

(a) c is continuous on $\mathbb{K}$ and $\sup_{a\in A} |c(x,a) - c(x',a)| \le \bar{L}_c d_X(x,x')$;

(b) $|\alpha(x) - \alpha(x')| \le L_\alpha d_X(x,x')$;

(c) $\sup_{a\in A} d_X(F(x,a,\xi), F(x',a,\xi)) \le \bar{L}_*(\xi) d_X(x,x')$ and $\bar{\alpha}E\left[\bar{L}_*(\xi)\right] =: \beta < 1$;

(d) $\sup_{k\in\mathbb{K}} d_X(F(k,s), F(k,s')) \le L_F d_S(s,s')$, $s, s' \in S$;

(e) for every bounded continuous function $u : X \to \Re$ the maps

$$k \to E\left(u\left[F\left(k,\xi\right)\right]\right) \quad \text{and} \quad k \to E\left(u\left[F\left(k,\tilde{\xi}\right)\right]\right)$$

are continuous on $\mathbb{K}$.

It is not difficult to verify that under Assumption 2 the equations (17)-(19) are satisfied, and in addition, the existence of stationary optimal policies f_* and $\tilde{f}_*$ holds true.

Theorem 2. *Under Assumption 2*

$$\sup_{x\in X} \Delta(x) \le \bar{K}\pi_r(\mu, \tilde{\mu}), \tag{22}$$

where

$$\bar{K} = \frac{4\bar{\alpha}}{(1-\bar{\alpha})^2}\left[\frac{b}{1-\bar{\alpha}} + \frac{L_F}{1-\beta}\left(\bar{L}_c + L_\alpha\frac{b}{1-\bar{\alpha}}\right)\right]. \tag{23}$$

Remark 2. As it follows from the proof of Theorem 1 in Section 5, under Assumption 1

$$\sup_{x\in X} \Delta(x) \le \frac{K}{2}\mathrm{Dud}(\mu,\tilde{\mu}); \tag{24}$$

and under Assumption 2

$$\sup_{x\in X} \Delta(x) \le \frac{\bar{K}}{2}\mathrm{Dud}(\mu,\tilde{\mu}), \tag{25}$$

where Dud$(\cdot,\cdot)$ is the Dudley metric defined as (see [2])

$$\mathrm{Dud}(\mu,\tilde{\mu}) := \sup_{\varphi\in Lip_{1,\infty}} \left| \int_S \varphi d\mu - \int_S \varphi d\tilde{\mu} \right|, \tag{26}$$

and

$$Lip_{1,\infty} := \left\{ \varphi : S \to \Re : \sup_{s\in S} |\varphi(s)| + \sup_{s\neq s'} \frac{|\varphi(s)-\varphi(s')|}{d_S(s,s')} \le 1 \right\}. \tag{27}$$

The following counterexample shows that the Lipschitz condition on the discount function $\alpha(x)$, $x\in X$, is essential to get stability inequalities as in (22) and (25).

Example 1. Let us consider the following pair of deterministic controlled processes defined on $X=[0,\infty)$ and $A=\{0,1\}$:

$$x_t = a_t\xi_t, \qquad \tilde{x}_t = \tilde{a}_t\tilde{\xi}_t, \quad t\in\mathbb{N},$$

where, for $\varepsilon>0$, $\xi_t=1$, $\tilde{\xi}_t=1+\varepsilon$, and the common initial state is $x=x_0=1$. Let

$$c(x,1)=3, \quad x\ge 0;$$

$$c(x,0)=\begin{cases} 1+2x, & 0\le x\le 1; \\ 3, & x>1; \end{cases}$$

$$\alpha(x)=\begin{cases} \frac{1}{2}, & 0\le x\le 1; \\ \frac{1}{10}, & x>1. \end{cases}$$

It is clear that the optimal policy for the process $\{\tilde{x}_t\}$ is $\tilde{f}_*=\{1,1,...\}$. Applying this policy to control the process $\{x_t\}$ we have

$$V(1,\tilde{f}_*) = 3\sum_{t=0}^{\infty}\left(\frac{1}{2}\right)^t = 6.$$

On the other hand, for $f=\{0,0,...\}$

$$V(1,f) = 3+\sum_{t=1}^{\infty}\left(\frac{1}{2}\right)^t = 4.$$

Thus, for every $\varepsilon > 0$

$$\Delta(1) \geq V(1, \tilde{f}_*) - V(1, f) = 2.$$

Now we note that $\pi_r(\mu, \mu_\varepsilon) \to 0$ as $\varepsilon \searrow 0$ ($\mu = \delta_1$, $\mu_\varepsilon = \delta_{1+\varepsilon}$), and that Assumption 2, except part (b), is satisfied.

Regarding the issue on last example, it is worth noting that if we replace the Prokhorov metric π_r in the right-hand side of (22) by the total variation metric $\mathcal{V}_T$, the continuity of $\alpha(x)$ is not required, neither the conditions (a), (c) and (d) in Assumption 2.

Going back to the case in which $A(x)$ can depend on $x \in X$, we introduce the last set of conditions. These conditions are used basically to guarantee the existence of stationary optimal policies f_* and $\tilde{f}_*$ for the processes (6) and (10), respectively.

Assumption 3. *(a) The cost function c is lower semicontinuous on* $\mathbb{K}$.

(b) The set-valued mapping $x \to A(x)$ *is upper semicontinuous with respect to the Hausdorff metric.*

(c) For every bounded continuous function $u : X \to \Re$ *the maps*

$$k \to E\left(u\left[F\left(k, \xi\right)\right]\right) \quad \textit{and} \quad k \to E\left(u\left[F\left(k, \tilde{\xi}\right)\right]\right)$$

are continuous on $\mathbb{K}$.

Theorem 3. *Under Assumption 3*

$$\sup_{x \in X} \Delta(x) \leq K_* \mathcal{V}_T(\mu, \tilde{\mu}), \tag{28}$$

where

$$K_* = \frac{2\bar{\alpha}b}{(1-\bar{\alpha})^3}.$$

Inequality (28) is consequence of (11), the definition of the total variation metric $\mathcal{V}_T$, and inequality (46) in the proof of Theorem 1.

The requirement of bounded cost (7) might be restrictive. However, within the context of the Assumptions 1 and 2, such a condition is necessary; otherwise the process could be unstable, as shown in the following example.

Example 2. We consider the following controlled processes with state and action spaces $X = [0, \infty)$ and $A = A(x) = \{0, 1\}$, given by $x_0 = \tilde{x}_0 = 0$, and for $t \in \mathbb{N}$,

$$x_t = a_t \left(x_{t-1} + \xi_t\right), \tag{29}$$

$$\tilde{x}_t = a_t \left(\tilde{x}_{t-1} + \tilde{\xi}_t\right). \tag{30}$$

Let ξ and $\tilde{\xi}$ be the generic random variables for ξ_t and $\tilde{\xi}_t$. The key point in our example is to choose properly these random variables, as well as the unbounded cost and discount factor functions. For instance, we set $\xi \equiv 1$, and for $\varepsilon > 0$,

$$\tilde{\xi} \equiv \tilde{\xi}_\varepsilon = 1 + \varepsilon \min \{\mathcal{H}(\varepsilon), \eta\},$$

where $\mathcal{H}(\varepsilon) := \exp\left(\frac{1}{\varepsilon^2}\right) - 1$, $\eta = |\zeta|$, and ζ is a random variable with density function $g_\zeta(y) = \dfrac{1}{\pi(1+y^2)}$, $y \in \Re$.

Observe that $E \min \{\mathcal{H}(\varepsilon), \eta\} \geq \dfrac{1}{2\pi\varepsilon^2}$. Therefore

$$E\tilde{\xi}_\varepsilon \to \infty, \quad \text{as } \varepsilon \searrow 0. \tag{31}$$

In addition, for each $\omega \in \Omega$, $\tilde{\xi}_\varepsilon(\omega) \to 1$ as $\varepsilon \to 0$, which implies that $\tilde{\xi}_\varepsilon$ converges weakly:

$$\tilde{\xi}_\varepsilon \Rightarrow \xi. \tag{32}$$

On the other hand, the one-stage cost and discount factor functions are defined as

$$c(x,a) = c(x) = \begin{cases} 1 & 0 \leq x \leq 1 \\ x & x > 1 \end{cases}, \tag{33}$$

and, for some $\bar{\alpha} < 1$,

$$\alpha(x) = \begin{cases} \bar{\alpha} & 0 \leq x < 0.5 \\ 2\bar{\alpha}(1-x) & 0.5 \leq x \leq 1 \\ 0 & x > 1 \end{cases}. \tag{34}$$

Under this setting, it is easy to verify that all assumptions in [22] are satisfied with $w(y) := y + \gamma$, for some $\gamma > 0$. Hence, we can ensure the existence of stationary optimal policies $f_*, \tilde{f}_* \in \mathbb{F}$ for the controlled processes (29) and (30), respectively.

Let us show that for all sufficiently small $\varepsilon > 0$, $\tilde{f}_*(0) = 0$. If we let $\tilde{f}_*(0) = 1$, then form (30) we have $\tilde{x}_1 = \tilde{\xi}_1$. Thus,

$$\tilde{V}(0, \tilde{f}_*) \geq 1 + \bar{\alpha} E\tilde{\xi}_\varepsilon \to \infty, \quad \text{as } \varepsilon \to 0.$$

Therefore, $\tilde{f}_*(0) = 0$.

Let $f' \in \mathbb{F}$ be a stationary policy with $f'(x) \equiv 0$, $x \in X$. Clearly, under f', we have $\tilde{x}_t = 0$ for all $t \in \mathbb{N}$; and moreover

$$\tilde{V}(0, f') = \sum_{t=0}^{\infty} \bar{\alpha}^t = \frac{1}{1-\bar{\alpha}}.$$

Now, applying $\tilde{f}_* \in \mathbb{F}$ to control the system (29), again we have $x_t = 0$ for all $t \in \mathbb{N}_0$, and

$$V(0, \tilde{f}_*) = \frac{1}{1-\bar{\alpha}}. \tag{35}$$

On the other hand, let $f \in \mathbb{F}$ be a stationary policy with $f(x) \equiv 1$, $x \in X$. Then, under this policy, we have that $x_t = t$ for all $t \in \mathbb{N}_0$, and from (33) and (34),

$$V(0, f) = 1 + \bar{\alpha}.$$

Finally, from (35) we have the following lower bound for the stability index

$$\Delta(0) \geq V(0, \tilde{f}_*) - V(0, f) = \frac{\bar{\alpha}^2}{1-\bar{\alpha}}. \tag{36}$$

Hence we conclude that, by choosing a suitable value of $\bar{\alpha} < 1$, the right-hand side of (36) can be made arbitrarily large, even when (32) holds.

4 Examples

4.1 Approximation by Means of Empirical Distributions

We assume that S is a Borel subset of $\Re$, and denote by G and $\tilde{G}$ the distribution functions of ξ and $\tilde{\xi}$, respectively. To illustrate the empirical approximation, we use the following well-known relation between the Dudley and Kantorovich metrics $\kappa(\cdot,\cdot)$ (see, for instance, [18]):

$$Dud(G, \tilde{G}) \leq \kappa(G, \tilde{G}) := \int_{-\infty}^{\infty} \left| G(x) - \tilde{G}(x) \right| dx. \tag{37}$$

Suppose now that G is unknown, but observations $\xi_1, \xi_2, ..., \xi_n$ are available. Let $\tilde{G} \equiv \hat{G}_n$ be the empirical distribution function. Then, using the results from [1], we find that if $E\left|\xi\right|^{2+\delta} \leq \gamma < \infty$ for some $\gamma, \delta > 0$, then

$$\kappa(G, \hat{G}_n) \leq M(\gamma, \delta) n^{-1/2}, \quad n = 1, 2, ...,$$

where $M(\gamma, \delta)$ is a explicitely computed constant. Therefore, under conditions of Theorem 2, if we assume that in the approximating controlled process (10) the random variable ξ_1 has the distribution $\hat{G}_n$, then (see (25)) for $\Delta \equiv \Delta_n$,

$$E \sup_{x \in X} \Delta_n(x) \leq \frac{\bar{K}}{2} M(\gamma, \delta) n^{-1/2}, \quad n \in \mathbb{N}.$$

Observe that inequality (28) is useless when approximating by empirical distributions.

4.2 A Water Reservoir Control Model

We consider a simple water reservoir control system with finite capacity M. The water stock process $\{x_t\}$ evolves according to the equation

$$x_t = \min\{x_{t-1} - a_t + \xi_t, M\}, \quad t \in \mathbb{N},$$

where the control a_t is the volume of water released during the stage t, whereas $\{\xi_t\}$ is a sequence of i.i.d. random variables representing water inflows with common distribution function G_ξ. In this case $X = [0, M]$, $A(x) = [0, x]$, $x \in X$, and $S = [0, \infty)$.

We assume that the one-step cost $c(x, a)$ satisfies Assumption 1(a). For instance, it could be $c(x, a) = c_0 a$, where c_0 is the cost per unit of water. Then, it is rather natural to admit that the discount may depend on stock levels in previous period (reflecting possible water shortage or overflows). We let the Assumption 1(c) is satisfied. It is easy to see the fulfilment of Assumptions 1(b), (e), and (f). On the other hand, simple calculations show that Assumption 1 (d) is satisfied if we suppose that the random variable ξ has a bounded density satisfying the Lipschitz condition. Therefore, considering the approximating process

$$\tilde{x}_t = \min\left\{\tilde{x}_{t-1} - \tilde{a}_t + \tilde{\xi}_t, M\right\}, \quad t \in \mathbb{N},$$

where $\left\{\tilde{\xi}_t\right\}$ is a sequence of i.i.d. random variables with distribution function $G_{\tilde{\xi}}$, inequalities (24) and (37) yield

$$\sup_{x \in X} \Delta(x) \leq \frac{K}{2} \int_0^\infty \left| G_\xi(x) - G_{\tilde{\xi}}(x) \right| dx.$$

4.3 A Radioactive Material Stock Control Model

Let $X = \Re$, $A(x) = A$ where A is a compact subset of $\Re$, and $S = \Re^2$. Consider the controlled process

$$x_t = \xi_t x_{t-1} + g(\eta_t, a_t), \quad t \in \mathbb{N}, \tag{38}$$

where (ξ_t, η_t), $t \in \mathbb{N}$, are i.i.d. random vectors with values in $\Re^2$.

The equation (38) represents the evolution of the stock of radioactive material, where ξ_t expresses the decay due to radioactivity, and $g(\eta_t, a_t)$ is a controlled supply, but randomly perturbed (see, e.g., [20, 21]). Note that if intervals between successive supplies are regular, then basically ξ_t is constant. The situation is different if such intervals are random.

We admit a one-step cost function c satisfying Assumption 2(a), a discount function $\alpha(x)$ as in Assumption 2(b), and suppose that the function g in (38) satisfies the Lipschitz condition. Also we assume that $\bar{\alpha} E|\xi| < 1$. Then, it is easy to see that Assumption 2 is satisfied, and therefore the stability inequality (22) can be applied when the unknown distribution of (ξ, η) is approximated.

5 Appendix

Proof of Theorem 1. Let V_* and $\tilde{V}_*$ be the value functions for the control models $\mathcal{M}$ and $\tilde{\mathcal{M}}$, respectively (see (6) and (10)). For each fixed $(x,a) \in \mathbb{K}$, we define:

$$H(x,a) := c(x,a) + \alpha(x) E\left[V_*\left(F\left(x,a,\xi\right)\right)\right] \tag{39}$$

and

$$\tilde{H}(x,a) := c(x,a) + \alpha(x) E\left[\tilde{V}_*\left(F\left(x,a,\tilde{\xi}\right)\right)\right]. \tag{40}$$

Now, let $\tilde{f}_* \in \mathbb{F}$ be the optimal policy for the model $\tilde{\mathcal{M}}$. Then, from (15), (17), (39), (40), and the Markov property, we get, for each $h_t \in \mathbb{H}_t, t \in \mathbb{N}$,

$$\begin{aligned} E_x^{\tilde{f}_*}\left[\alpha(x_{t-1}) V_*(x_t) | h_t\right] &= H(x_{t-1}, a_t) - c(x_{t-1}, a_t) \\ &\quad - \inf_{a \in A(x_{t-1})} H(x_{t-1}, a) + \inf_{a \in A(x_{t-1})} H(x_{t-1}, a) \\ &= \Lambda(x_{t-1}, a_t) - c(x_{t-1}, a_t) + V_*(x_{t-1}), \end{aligned} \tag{41}$$

where

$$\Lambda(x_{t-1}, a_t) := H(x_{t-1}, a_t) - \inf_{a \in A(x_{t-1})} H(x_{t-1}, a). \tag{42}$$

Then, for each $t \in \mathbb{N}$, from (41) (recall $\alpha(x_{-1}) := 1$ and $x_0 = x$):

$$\prod_{i=0}^{t-2} \alpha(x_i) c(x_{t-1}, a_t) = \prod_{i=0}^{t-2} \alpha(x_i) V_*(x_{t-1})$$

$$-E^{\tilde{f}_*}\left[\prod_{i=0}^{t-1} \alpha(x_i) V_*(x_t) | h_t\right] + \prod_{i=0}^{t-2} \alpha(x_i) \Lambda(x_{t-1}, a_t).$$

Hence, using properties of conditional expectation, for $n \in \mathbb{N}$, we have

$$\sum_{t=1}^{n} E_x^{\tilde{f}_*}\left[\prod_{i=0}^{t-2} \alpha(x_i) c(x_{t-1}, a_t)\right] =$$

$$\sum_{t=1}^{n}\left\{E_x^{\tilde{f}_*}\left[\prod_{i=0}^{t-2} \alpha(x_i) V_*(x_{t-1})\right] - E_x^{\tilde{f}_*}\left[\prod_{i=0}^{t-1} \alpha(x_i) V_*(x_t)\right]\right\}$$

$$+\sum_{t=1}^{n} E_x^{\tilde{f}_*}\left[\prod_{i=0}^{t-2} \alpha(x_i) \Lambda(x_{t-1}, a_t)\right]$$

$$= V_*(x) - E_x^{\tilde{f}_*}\left[\prod_{i=0}^{n-1} \alpha(x_i) V_*(x_n)\right] + \sum_{t=1}^{n} E_x^{\tilde{f}_*}\left[\prod_{i=0}^{t-2} \alpha(x_i) \Lambda(x_{t-1}, a_t)\right]. \tag{43}$$

Letting $n \to \infty$, the left-hand side of (43) converges to $\tilde{V}(x, \tilde{f}_*)$. Therefore, from (8),

$$\Delta(x) = V\left(x, \tilde{f}_*\right) - V_*(x) \le \sum_{t=1}^{\infty} \bar{\alpha}^{t-1}\left|E_x^{\tilde{f}_*} \Lambda(x_{t-1}, a_t)\right|. \tag{44}$$

In order to obtain the inequality (20), we proceed to analyze the term $\Lambda(x_{t-1}, a_t) \equiv \Lambda_t$. From (19), (40), (42), and the fact that the policy $\tilde{f}_*$ is optimal for the controlled process (10), we can write:

$$\begin{aligned} \Lambda_t = & H(x_{t-1}, a_t) - \tilde{H}(x_{t-1}, a_t) \\ & + \inf_{a \in A(x_{t-1})} \tilde{H}(x_{t-1}, a) - \inf_{a \in A(x_{t-1})} H(x_{t-1}, a). \end{aligned}$$

Hence (see (39), (40))

$$\begin{aligned} |\Lambda_t| & \leq 2 \sup_{a \in A(x_{t-1})} \left| H(x_{t-1}, a) - \tilde{H}(x_{t-1}, a) \right| \\ & \leq 2 \sup_{a \in A(x_{t-1})} \alpha(x_{t-1}) \left| E\left[V_*\left(F\left(x_{t-1}, a, \xi\right)\right)\right] - E\left[\tilde{V}_*\left(F\left(x_{t-1}, a, \tilde{\xi}\right)\right)\right] \right| \\ & \leq 2\bar{\alpha} \sup_{a \in A(x_{t-1})} \left| E\left[V_*\left(F\left(x_{t-1}, a, \xi\right)\right)\right] - E\left[V_*\left(F\left(x_{t-1}, a, \tilde{\xi}\right)\right)\right] \right| \\ & \quad + 2\bar{\alpha} \sup_{a \in A(x_{t-1})} \left| E\left[V_*\left(F\left(x_{t-1}, a, \xi\right)\right)\right] - E\left[\tilde{V}_*\left(F\left(x_{t-1}, a, \tilde{\xi}\right)\right)\right] \right|. \end{aligned} \tag{45}$$

On the other hand, using the contraction property of the operator $\tilde{T}$ (see Remark 1(i)), from (17) we have

$$\begin{aligned} \left\| V_* - \tilde{V}_* \right\| & \leq \left\| T V_* - \tilde{T} V_* \right\| + \left\| \tilde{T} V_* - \tilde{T} \tilde{V}_* \right\| \\ & \leq \bar{\alpha} \left\| V_* - \tilde{V}_* \right\| + \left\| T V_* - \tilde{T} V_* \right\|, \end{aligned}$$

which, in turns, implies (see (15), (16))

$$\begin{aligned} \left\| V_* - \tilde{V}_* \right\| & \leq \frac{1}{1 - \bar{\alpha}} \left\| T V_* - \tilde{T} V_* \right\| \\ & \leq \frac{\bar{\alpha}}{1 - \bar{\alpha}} \sup_{k \in \mathbb{K}} \left| E\left[V_*\left(F\left(k, \xi\right)\right)\right] - E\left[V_*\left(F\left(k, \tilde{\xi}\right)\right)\right] \right|. \end{aligned}$$

The last inequality provides the upper bound for the last term on the right-side of (45). Therefore,

$$|\Lambda_t| \leq \frac{2\bar{\alpha}}{1 - \bar{\alpha}} \sup_{k \in \mathbb{K}} \left| E\left[V_*\left(F\left(k, \xi\right)\right)\right] - E\left[V_*\left(F\left(k, \tilde{\xi}\right)\right)\right] \right|,$$

and in view of (44),

$$\Delta(x) \leq \frac{2\bar{\alpha}}{(1 - \bar{\alpha})^2} \sup_{k \in \mathbb{K}} \left| E\left[V_*\left(F\left(k, \xi\right)\right)\right] - E\left[V_*\left(F\left(k, \tilde{\xi}\right)\right)\right] \right|. \tag{46}$$

The next step is to show that the family of bounded functions

$$\{V_*\left(F\left(k, \cdot\right)\right) : k \in \mathbb{K}\} \tag{47}$$

is equi-Lipschitz. That is, there exists a constant $\tilde{L} > 0$ such that

$$|V_* [F(k,s)] - V_* [F(k,s')]| \le \tilde{L} d_S(s,s'), \qquad \forall s, s' \in S, k \in \mathbb{K}.$$

To this end, denote

$$g(x,a) := c(x,a) + \alpha(x) E\left(V_* [F(x,a,\xi)]\right),$$

where V_* is the value function (12) (see (17)). Then, for $k = (x,a)$, $k' = (x',a') \in \mathbb{K}$, from (11), Assumption 1 (d), and the definition of the total variation metric $\mathcal{V}_T$ we get

$$\begin{aligned} |g(x,a) - g(x',a')| &\le L_c d_{\mathbb{K}}(k,k') + L_\alpha d_X(x,x') \frac{b}{1-\bar{\alpha}} \\ &\quad + \frac{\bar{\alpha} b}{1-\bar{\alpha}} L_* d_{\mathbb{K}}(k,k') \\ &\le \left[L_c + L_\alpha \frac{b}{1-\bar{\alpha}} + \frac{\bar{\alpha} b}{1-\bar{\alpha}} L_* \right] d_{\mathbb{K}}(k,k'), \end{aligned}$$

where the last inequality is due to the fact $d_X(x,x') \le d_{\mathbb{K}}(k,k')$. Hence the function g is Lipschitz with constant

$$L = L_c + (L_\alpha + \bar{\alpha} L_*) \frac{b}{1-\bar{\alpha}}. \tag{48}$$

Then, under Assumption 1 (b), we can prove that the value function (see (17))

$$V_*(x) = \inf_{a \in A(x)} g(x,a), \quad x \in X \tag{49}$$

is Lipschitz with constant $(1 + L_A)L$ (see details in [8]). Thus, by Assumption 1(e),

$$\begin{aligned} |V_* [F(k,s)] - V_* [F(k,s')]| &\le (1 + L_A) L d_X \left(F(k,s), F(k,s')\right) \\ &\le (1 + L_A) L L_F d_S(s,s'), \quad \forall s, s' \in S, k \in \mathbb{K}. \end{aligned} \tag{50}$$

Therefore, combining (48) and (50), we prove that the family of functions (47) is equi-Lipschitz with constant

$$\tilde{L} := (1 + L_\alpha) L_F \left[L_c + (L_\alpha + \bar{\alpha} L_*) \frac{b}{1-\bar{\alpha}} \right].$$

Multiplying and dividing the right-hand side of (46) by $\frac{b}{1-\bar{\alpha}} + \tilde{L}$, the definition of the Dudley metric in (26) together with inequality (46), yield inequality (24).

Finally, using the fact that $Dud(\mu, \tilde{\mu}) \le 2\pi_r(\mu, \tilde{\mu})$, we get inequality (20) with the constant given in (21). ■

Proof of Theorem 2. Firstly, observe that to get inequality (46), we have explored only the parts of Assumption 1 which ensure the existence of stationary optimal policies related to the corresponding optimality equations. Then, in the context of Assumption 2, such an inequality holds . Therefore, our starting point for the proof of Theorem 2 is precisely inequality (46). In this sense, we should offer an alternative proof of Lipschitz property of the family of functions (47). See, e.g., [14] for similar arguments those given below.

Let $V_0 \equiv 0$, and with the operator T as in (15)

$$V_1(x) = TV_0(x) = \inf_{a \in A} c(x,a), \quad x \in X.$$

Then

$$\begin{aligned} |V_1(x) - V_1(x')| &\leq \sup_{a \in A} |c(x,a) - c(x',a)| \\ &\leq L_c d_X(x,x') \equiv L_1 d_X(x,x'), \quad x, x' \in X. \end{aligned}$$

Now let

$$V_2(x) := TV_1(x) = \inf_{a \in A} \{c(x,a) + \alpha(x) E\left(V_1\left[F\left(x,a,\xi\right)\right]\right)\}.$$

In view of Assumption 2 (a), (b) (c)

$$|V_2(x) - V_2(x')|$$

$$\begin{aligned} &\leq \sup_{a \in A} |c(x,a) + \alpha(x) E\left(V_1\left[F\left(x,a,\xi\right)\right]\right) - c(x',a) + \alpha(x') E\left(V_1\left[F\left(x',a,\xi\right)\right]\right)| \\ &\leq \bar{L}_c d_X(x,x') + L_\alpha d_X(x,x') \|V_1\| + \sup_{a \in A} \bar{\alpha} L_1 E\left[d_X\left(F\left(x,a,\xi\right), F\left(x',a,\xi\right)\right)\right] \\ &\leq \left[\bar{L}_c + L_\alpha \|V_1\| + \beta L_1\right] d_X(x,x') \\ &\leq \left[\bar{L}_c + L_\alpha \frac{b}{1-\bar{\alpha}} + \beta \bar{L}_c\right] d_X(x,x') =: L_2 d_X(x,x'), \qquad x, x' \in X. \end{aligned}$$

Defining $V_n = TV_{n-1}$, we find, by induction, that the function V_n satisfies the Lipschitz condition with the constant

$$L_n = \sum_{k=0}^{n-2} \beta^k \left[\bar{L}_c + L_\alpha \frac{b}{1-\bar{\alpha}}\right] + \beta^{n-1} \bar{L}_c, \quad n \geq 2.$$

Since $\|V_n - V_*\| \to 0$, as $n \to \infty$, the value function V_* satisfies the Lipschitz condition with constant

$$L^* = \frac{1}{1-\beta}\left[\bar{L}_c + L_\alpha \frac{b}{1-\bar{\alpha}}\right].$$

The remainder of the proof is similar to those given in proof of Theorem 1. ■

6 Acknowledgement

This research has been supported partially by Consejo Nacional de Ciencia y Tecnologia (CONACyT) under grants CB2015/254306 and CB2013/01222739.

References

1. Bobkov, S. and Ledoux, M.: One dimensional empirical measures, order statistics and Kantorovich transport distances, 2014, *Preprint.*
2. Dudley, R.M. *Real Analysis and Probability.* Cambridge University Press, Cambridge 2002.
3. Dynkin, E.B. and Yushkevich, A.A. *Controlled Markov Processes.* Springer-Verlag, NY, 1979.
4. González-Hernández, J., López-Martínez, R.R. and Minjárez-Sosa, J.A.: Approximation, estimation and control of stochastic systems under a randomized discounted cost criterion. *Kybernetika* **45** (2009), 737-754.
5. González-Hernández, J., López-Martínez, R.R., Minjárez-Sosa, J.A. and Gabriel-Arguelles, J.R.: Constrained Markov control processes with randomized discounted cost criteria: occupation measures and extremal points. *Risk and Decision Analysis,* **4** (2013) 163-176.
6. González-Hernández, J., López-Martínez, R.R., Minjárez-Sosa, J.A. and Gabriel-Arguelles, J.R.: Constrained Markov control processes with randomized discounted rate: infinite linear programming approach. *Optim. Control Appl. Meth.*, **35** (2014) 575-591.
7. Gordienko, E. and Minjárez-Sosa, J.A.: Adaptive control for discrete-time Markov processes with unbounded costs: discounted criterion. *Kybernetika* **34** (1998) 217–234.
8. Gordienko, E., Lemus-Rodríguez, E. and Montes-de-Oca, R.: Average cost Markov control processes: stability with respect to the Kantorovich metric. *Math. Methods Oper. Res.* **70** (2009) 13-33.
9. Gordienko, E. and Salem, F.: Estimates of stability of Markov control processes with unbounded costs. *Kybernetika* **36** (2000) 195-210.
10. Gordienko, E. and Yushkevich, A.: Stability estimates in the problem of average optimal switching of a markov chain. Math. Methods Oper. Res. **57** (2003), 354-365.
11. Hernández-Lerma, O. and Lasserre, J.B. *Discrete-Time Markov Control Processes: Basic Optimality Criteria.* Springer-Verlag, NY, 1996.
12. Hilgert, N. and Minjárez-Sosa, J.A.: Adaptive control of stochastic systems with unknown disturbance distribution: discounted criteria. *Math. Methods Oper. Res.* **63** (2006) 443-460.
13. Hinderer, K. *Foundations of Non-stationary Dynamic Programming with Discrete Time Parameter.* Lecture Notes Oper. Res. 33, Springer, NY, 1970.
14. Hinderer, K.: Lipschitz continuity of value functions in Markov decision processes. *Math. Methods Oper. Res.* **62** (2005) 3-22.
15. Horiguchi, M. and Piunovskiy, A.B.: Optimal stopping model with unknown transition probabilities. *Control and Cybernetics* **42** (2013) 593-612.
16. Minjárez-Sosa, J.A.: Approximation and estimation in Markov control processes under a discounted criterion. *Kybernetika* **40** (2004) 681-690.
17. Minjárez-Sosa, J.A.: Markov control models with unknown random state-action-dependent discount factors. *TOP* (2015). DOI 10.1007/s11750-015-0360-5.
18. Rachev, S.T. *Probability Metrics and the Stability of Stochastic Models.* Wiley, Chichester, 1991.
19. Schäl, M.: Conditions for optimality in dynamic programming and for the limit of n-stages optimal policies to be optimal, *Z. Wahrscheinlichkeitstheor Verwandte Geb.* **32** (1975) 179-196.

20. Uppuluri, V. R. R., Feder, P.I. and Shenton, L.R.: Random difference equations occurring in one-compartment models. *Math. Biosci.* **2** (1967) 143-171.
21. Vervaat, W.: On a stochastic difference equation and representation of non-negative infinitely divisible random variables. *Adv. Appl. Prob.* **11** (1979) 750-783.
22. Wei, Q. and Guo, X.: Markov decision processes with state-dependent discount factors and unbounded rewards/costs. *Oper. Res. Letters* **39** (2011) 369-374.

Elimination and Insertion Operations for Finite Markov Chains

Isaac M. Sonin[1] and Constantine Steinberg[2]

[1] University of North Carolina at Charlotte
Dept. of Mathematics Charlotte, NC 28223, USA
imsonin@uncc.edu

[2] Wells Fargo Securities
Charlotte, NC 28202, USA
stan.steinberg@wellsfargo.com

Abstract. A Markov chain (MC) observed only when it is outside of a subset D is again a MC with a well-known transition matrix P_D. This matrix can be obtained also in a few iterations, each requiring $O(n^2)$ operations, when the states from D are "eliminated" one at a time. We modify these iterations to allow for a state previously eliminated to be "reinserted" into the state space in one iteration. This modification sheds a new light on the relationship between an initial and censored MC, and introduces a new operation - "insertion" into the theory of MCs.
Keywords: Markov chain, censored Markov chain, State Elimination Algorithm.
AMS 2000 subject classification: Primary 60J10, Secondary 60J20, 60J22

1 Introduction

Let X be a finite state space, $|X| = n$, $P = \{p(x, y)\}$ a stochastic matrix indexed by elements of X, and (Z_n) a Markov chain (MC) defined by X, P with some initial distribution. Let $D \subset X$, $S = X \setminus D$. It is well known that a MC observed only at visits to the subset S is again a MC (sometimes called a censored or embedded MC) with a new transition matrix P_D, see formula (5) below. This formula can be found e.g. in the classical text [3]. We say that set D is "eliminated." Such a transition matrix has an especially simple form when D consists of only one point, $D = \{z\}$. In this case, if we denote $P_{\{z\}} = P'$, it is easy to see that

$$p'(x, y) = p(x, y) + p(x, z)n(z)p(z, y), x \in X, y \in S, \tag{1}$$

where $n(z) = \sum_{n=0}^{\infty} p^n(z, z) = 1/(1 - p(z, z))$. According to formula (1), each row-vector of the new stochastic matrix P' is a linear combination of two rows of P (with the z-column deleted). This transformation corresponds formally to one step of the Gaussian elimination and requires $O(n^2)$ operations. We call such a transformation of a matrix an *iteration.* If $|D| = k$ then the transition matrix P_D can be obtained in k iterations.

Censored MCs have had numerous applications in Probability Theory and Linear Algebra, see e.g. [11] and [4]. A relatively recent method for recursively calculating many important characteristics of MCs based on formula (1), is the so called *state reduction (SR) approach.* This was initiated by the papers [1] and [6], where the so-called GTH/S algorithm to calculate the invariant distribution for an ergodic Markov chain was introduced. See also [5], where similar ideas were analyzed mainly from the algebraic point of view.

Briefly this approach can be described as follows. If an initial Markov model $M = (X, P)$ is finite, $|X| = n$, and only one point is eliminated each time, then a sequence of stochastic matrices $(P_k), k = 1, ..., n-1$ can be calculated recursively on the basis of formula (1) or (5) below. Such a sequence of stochastic matrices provides an opportunity to calculate many characteristics of the initial Markov model M recursively starting from some reduced model $M_s, 1 < s \leq n$.

Another application of formula (1) in the area of the Optimal Stopping (OS) of MCs was started in [8], where the so called the State Elimination Algorithm was introduced. According to this algorithm all points that do not belong to an optimal stopping set are eliminated one by one, or more generally in the countable case at some steps a subset may be eliminated. The order in which states are eliminated is defined by some auxiliary procedure. Another algorithm based on the State Elimination algorithm was used to calculate the Generalized Gittins index in [9].

Recently in [10] we presented an algorithm for finding an optimal strategy and the value function for a Markov Decision Process (MDP) model where at each moment of discrete time a decision maker (DM) can apply one of three possible actions - *continue* when MC evolves according to the transition matrix P, *quit* when the evolution of a MC is stopped, and *restart* when the MC is moved to one of a finite number m of fixed "restarting" points. A decision at state x brings a corresponding reward, positive or negative, $c(x), q(x)$ or $r_i(x), i = 1, ..., m$. The goal of a DM is to maximize the total expected discounted reward. Such a model is a generalization of a model of Katehakis and Veinott in [2], where a restart to a unique point was allowed without any fee and the quit action was absent. Both models are related to the well-known Gittins index. For the case $m = 1$, a recursive algorithm to solve this model by performing $O(n^3)$ operations was proposed. An important part of this algorithm is a sequence of recursive steps when a transition matrix is transformed. It turns out that at some steps the points are eliminated but on other steps they need to be included (inserted) back. Assume, for example, that three points, z_1, z_2 and z_3, are subsequently eliminated. Reversing formula (1), it is easy to restore point z_3. This requires one iteration step. What if we wish to restore point z_1, i.e. to have only points z_2 and z_3 be eliminated? At first glance it seem that we either have to keep the matrix P_1 in memory and eliminate points z_2 and z_3 or restore (in three iteration steps) points z_3, z_2 and z_1 and then eliminate points z_2 and z_3. It turns out that we can insert point z_1 in one iteration if we will keep in memory a *nonstochastic* matrix W_D similar to matrix P_D. This matrix is also obtained by iterations. This result was given by Theorem 3 in [10]. The proof was somewhat tedious, part of it was given in Appendix and we wrote that "we fail to find a simpler proof though one likely exists". The main goals of this note are: first, to describe this new operation of *insertion* and to present corresponding formulas, and second, to give simpler, shorter and more transparent proof of this theorem, based on two new lemmas, keeping the formulation of the theorem the same. We also note that these "new elimination" steps allow us as a byproduct to obtain a recursive algorithm to calculate a fundamental matrix $N = N_D = (I - Q)^{-1}$ corresponding to any transient MC with substochastic matrix Q. In Section 3 we prove this theorem and we discuss the transformation of transition matrices of MCs under elimination and insertion. At the end we give two small numerical examples.

2 Censored MC and Elimination

An important and traditional tool for the study of Markov chains (MCs) is the notion of a Censored (Embedded) MC. Two operations on stochastic and related matrices are introduced in this section. They serve as building blocks for the algorithm in [10].

A pair $M = (X, P)$, where X is a state space and P is a stochastic matrix is called Markov model. Let us assume that a Markov model $M = (X, P)$ is given and $D \subset X$, $S = X \setminus D$. Then the matrix $P = \{p(x, y)\}$ can be decomposed as follows

$$P = \begin{bmatrix} Q & T \\ R & P_0 \end{bmatrix}, \tag{2}$$

where the substochastic matrix Q describes the transitions inside of D, P_0 describes the transitions inside of S and so on. Let us introduce the sequence of Markov times $\tau_0, \tau_1, ..., \tau_n, ...$, where $\tau_0 = 0$, and $\tau_n, n \geq 1$ are the times of first, and so on, *return* of the MC (Z_n) to the set S, i.e., $\tau_{n+1} = \min\{k > \tau_n, Z_k \in S\}$. Let us consider the random sequence $Y_n = Z_{\tau_n}, n = 0, 1, 2, ...$ and assume that $Z_0 = x \in S$. The strong Markov property and standard probabilistic reasoning imply the following basic lemma of the SR approach which should probably be credited to Kolmogorov and Doeblin.

Lemma 1. *[Elimination Lemma] (a) The random sequence (Y_n) is a Markov chain in a model $M'_D = (S, P'_D)$, where*

(b) the transition matrix $P'_D = \{p'(x, y), x, y \in S\}$ is given by the formula

$$P'_D = P_0 + RU = P_0 + RNT. \tag{3}$$

Here U is the matrix of the distribution of the MC at the time of the first visit to S starting from $x \in D$ and $N = N(D)$ is the fundamental matrix for the substochastic matrix Q, i.e. $N = \sum_{n=0}^{\infty} Q^n = (I - Q)^{-1}$, where I is the $|D| \times |D|$ identity matrix. This representation is given, for example, in the classical text [3].

The matrix $N = N(D)$ has the following well-known probabilistic interpretation, $N = \{n(x, y), x, y \in D\}$, where $n(x, y) = E_x \sum_{n=0}^{\tau} I_y(Z_n)$, and τ is the time of the *first visit* to S, i.e. $\tau = \min(n \geq 0 : Z_n \in S)$. Thus $n(x,y)$ is the expected number of visits to y starting from x until τ. The matrix N also satisfies the equalities

$$N = I + QN = I + NQ. \tag{4}$$

MC (Y_n) is called an *embedded* (*censored*) MC.

An important case is when the set D consists of one nonabsorbing point z. In this case formula (3) takes the form (1), where $n(z) = 1/(1 - p(z,z))$, is a "fundamental matrix".

In Lemma 1, the Markov model M'_D has a reduced state space S. Sometimes, it is more convenient to have all stochastic matrices of equal full size. Then we consider a MC (Y_n) with initial points x not only in S but also in D, though after the first step the MC (Y_n) is always in S. Lemma 1 remains true but now we obtain a Markov model $M_D = (X, P_D)$. In addition to (3) for $x, y \in S$, we have the equality $T + QNT = (I + QN)T = NT$ for $x \in D, y \in S$. The last equality is true by (4). Thus instead of (3) we have the following full size transition matrix

$$P_D = \begin{bmatrix} 0 & NT \\ 0 & P_0 + RNT \end{bmatrix}. \tag{5}$$

Note that the rows of matrix P_D give the distribution of MC (Z_n) at the time τ_1 of the *first return* to set S, i.e. $\tau_1 = \min\{n > 0 : Z_n \in S\}$ and $P(Z_{\tau_1} = y) = P_D(Y_1 = y)$ when $x \in X, y \in S$. For $x \in D$, the moment of the first return coincides with the moment of the first visit and this distribution is given by submatrix NT. For the points from $x \in S$ the corresponding distribution is given by submatrix $P_0 + RNT$. If $D = \{z\}$, i. e. when state z is eliminated, formula (5) is replaced by the one-state elimination formula, written here for columns, $(P_{\{z\}} = P')$,

$$p'(\cdot, z) = 0, \qquad p'(\cdot, y) = p(\cdot, y) + p(\cdot, z)\frac{p(z,y)}{1 - p(z,z)}, y \neq z. \tag{6}$$

We say that matrix P' is obtained from P in one *iteration.* Thus matrix P_D can be calculated directly by (5) or recursively using formula (6) in $|D|$ iterations.

3 Elimination vs Insertion

Suppose that the set $D = \{z_1, z_2, \ldots. z_k\}$ is eliminated in an initial model M, and P_D is the corresponding matrix obtained recursively by formula (6). Let $z \in D$, say $z = z_1$. How can one obtain the transition matrix $P_{D \setminus z}$, i.e. when only the points $z_2, \ldots. z_k$ are eliminated? Of course we can obtain this matrix starting from an initial matrix P and eliminating these points, performing $k-1$ iterations. Is there a way to obtain this matrix in just one iteration?

The answer for this rhetorical question is Yes. In this case we say that point z is inserted (restored). To do this, initially, instead of the stochastic matrices $P_1, P_2, \ldots, P_k = P_D$, we must calculate recursively similar but different non-stochastic (!) matrices $W_1, W_2, \ldots, W_k = W_D$. We do this by applying the second part of formula (6) to *all columns*, including previously eliminated states, i.e. when state z is eliminated using the formula

$$w'(\cdot, y) = w(\cdot, y) + w(\cdot, z)\frac{w(z,y)}{1 - w(z,z)}, y \in X. \tag{7}$$

Let us show immediately that this transformation of matrix W into W' can be reversed, i.e. the following statement holds

Lemma 2. *If matrix W' is obtained from matrix W by elimination formula* (7) *then matrix W can be obtained from matrix W' by insertion formula*

$$w(\cdot, y) = w'(\cdot, y) - w'(\cdot, z)\frac{w'(z,y)}{1 + w'(z,z)}, y \in X. \tag{8}$$

Proof. Using formula (7) for $x = z$ we obtain the equality $w'(z,y) = w(z,y)/(1 - w(z,z))$ and hence formula (7) can be written as

$$w(\cdot, y) = w'(\cdot, y) - w(\cdot, z)w'(z,y), y \in X. \tag{9}$$

Applying this formula for $y = z$ we obtain $w(\cdot, z) = w'(\cdot, z)/(1 + w'(z,z))$ and hence formula (9) can be written as (8). ■

Note that by the definition of matrices W_D and P_D their columns for $y \notin D$ coincide but the matrix P_D has zero columns for $y \in D$. The interpretation of the nonzero columns in W_D is given in Theorem 1.

Let us introduce the matrix $N^+ = N^+(D) = \{n^+(x, y|D)\}$, $x \in X$, $y \in D$, where $n^+(x, y|D)$ is the expected number of visits of a MC (Z_n) to state y *after the initial moment* until the moment of first *return* (visit) to set S. As with matrix $N = N(D)$ we usually will skip D. Note the differences between matrices N^+ and $N = \{n(x,y)\}, x, y \in D$: first, N^+ is the $|X| \times |D|$ matrix, N is the $|D| \times |D|$ matrix; second, $n^+(x,y)$ counts the number of visits to y after the initial moment, where as $n(x,y)$ including the initial moment. We also have obvious equalitiess: if $x, y \in D$ and $x \neq y$ then $n^+(x,y) = n(x,y)$, and $n^+(x,x) = n(x,x) - 1$. If $y \in D.x \notin D$ then $n^+(x,y) = n(x,y) = \sum_{z \in D} p(x,z)n(z,y)$.

To describe the structure of matrix N^+ it is convenient to introduce also an auxiliary matrix, $P(D)$. It consists of the first D columns of matrix P, see (2), i.e. has dimension $|X| \times |D|$ and contains blocks Q and R

Lemma 3. *(One point Lemma). Let $D \subset X$, and N, N^+, $P(D)$ are matrices defined above. Then*

a) matrix $N^+ = P(D)N$, i.e.

$$N^+ = \begin{bmatrix} QN \\ RN \end{bmatrix} = \begin{bmatrix} N - I \\ RN \end{bmatrix}, \tag{10}$$

b) the columns of N^+ can also be obtained by formula

$$n^+(\cdot, y|D) = \frac{p_{D\setminus y}(\cdot, y)}{1 - p_{D\setminus y}(y,y)} = p_{D\setminus y}(\cdot, y)n(y, y|D), y \in D. \tag{11}$$

First, we show that it is sufficient to consider only the case when D contains only one point, $D = \{y\}$. This explains the name for this lemma. The reason is that all points in D except y one can be eliminated without changing $n(y, y|D)$ or $n^+(\cdot, y|D)$. More precisely

Proposition 1. *Let $D \subseteq X, y \in D$. Then*

$$n(\cdot, y|D) = n_{D\setminus y}(\cdot, y), n^+(\cdot, y|D) = n^+_{D\setminus y}(\cdot, y).$$

Proof. Intuitively this statement is almost obvious: the expected number of visits to state y starting in y before exit to S remains the same if model M is transformed to model $M_{D\setminus y}$. The strict proof of this and more general statement about trajectories in an initial and reduced model was given in [8]. ∎

Now we can prove Lemma 3.

Proof of Lemma 3. Let $D = \{y\}$. Then $p_{D\setminus y}(\cdot, y) = p(\cdot, y)$ and $n(y, y)$ is the expectation of a geometric random variable with the P(success)= P(exit from D) $= 1 - p(y, y)$ and hence $n(y, y) = 1/(1 - p(y, y)$. Correspondingly $n^+(y, y) = n(y, y) - 1 = p(y, y)/(1 - p(y, y)$ and $n^+(x, y) = p(x, y)n(y, y)$. ∎

Before formulating our main theorem, recall that both matrices W_D and P_D are defined recursively, starting from the same stochastic matrix P and their columns for $y \notin D$ are the same, i.e. P_D is a part of W_D. The rows of P_D have a simple probabilistic interpretation as the distribution of MC (Z_n) at the moment of the first return to set $S = X \setminus D$. Thus there is no question whether the matrix P_D for $D = \{z_1, ...z_k\}$ depends on the order of states in D, it does not. For the y-columns of matrix W_D for $y \in D$ this is initially an open question, but formula (12) (or (14)) shows that as with P_D the order is irrelevant. Let us denote by P^0_D a matrix which consists only of non zero columns of matrix P_D, see (5), and thus has dimension $|X| \times |S|$ and contains blocks NT and $P_0 + RNT$, and denote by $p_{D\setminus y}(\cdot, y)$ the columns of matrix $P_{D\setminus y}$.

Theorem 1. *Let $D \subset X, S = X \setminus D$, the transition matrix P is decomposed as in (2) and N, N^+, P^0_D are matrices defined above. Then*

a) the matrix $W_D = [N^+|P^0_D]$, i.e. the first $|D|$ columns of W_D for $y \in D$ coincide with columns of matrix $N^+ = N^+(D)$ and the remaining columns, for $y \in S$ coincide with columns of matrix P^0_D, i.e.

$$W_D = \begin{bmatrix} QN & NT \\ RN & P_0 + RNT \end{bmatrix}, \tag{12}$$

b) given matrix W_D, any point $z \in D$ can be inserted in one iteration and matrix $W_{D\setminus z}$ can be obtained by the formula (8) with $W = W_{D\setminus z}, W' = W_D$, i.e.

$$w_{D\setminus z}(\cdot, y) = w_D(\cdot, y) - w_D(\cdot, z)\frac{w_D(z, y)}{1 + w_D(z, z)}, y \in X. \tag{13}$$

Corollary 1. *The equality* (13) *(the y-th column of matrix W_D for $y \in D$), can be described also by the formula*

$$w_D(\cdot, y) = n^+(\cdot, y|D) = \frac{p_{D\setminus y}(\cdot, y)}{1 - p_{D\setminus y}(y, y)}, y \in D, \tag{14}$$

This corollary follows immediately from point a) of Theorem 1 and formula (11).

Before to prove the theorem note that by formula (4) the submatrix QN in formula (12) coincides with $N - I$. Since matrix W_D is calculated recursively, it means that our theorem implies immediately

Corollary 2. *For any set $D \subset X$ and corresponding substochastic matrix Q the fundamental matrix $N = N(D)$ can be also calculated recursively as a part of matrix W_D. It means also that the insertion of one point into fundamental matrix can be also obtained in one iteration.*

To prove the theorem we need the following key lemma.

Lemma 4. *(Two point Lemma). Let $D \subset X, y \in D, z \notin D, G = D \cup \{z\}$, then*

$$n^+(\cdot, y|G) = n^+(\cdot, y|D) + n^+(\cdot, z|G)n^+(z, y|D). \tag{15}$$

Proof. Similarly to the proof of (3), using Proposition 1, it is sufficient to consider only the case when $D = \{y\}, G = \{y, z\}$, thus the name of this lemma. In this case $p_{D\setminus y}(\cdot, \cdot) = p(\cdot, \cdot)$. The proofs of (15) for all three possible cases $x = z, x = y, x \notin \{y, z\}$ are very similar so we consider only the more difficult case $x = z$. In this case $n^+(z, y) = n(z, y)$ and formula (11) applied separately for the sets D and G, and correspondingly for point $x = z$ and columns y and z, implies the equalities

$$n^+(z, y|D) = p(z, y)n(y, y)|D) = \frac{p(z, y)}{1 - p(y, y)},$$

$$n^+(z, z|G) = \frac{p_D(z, z)}{1 - p_D(z, z)}.$$

Then the right side of formula (15) with $x = z$ can be written as equality

$$p(z, y)n(y, y)|D)(1 + n^+(z, z|G)) = \frac{p(z, y)}{(1 - p(y, y))(1 - p_D(z, z))}. \tag{16}$$

Thus to prove lemma we need to prove that the left side of (15), i.e. $n^+(z, y|G) \equiv n(z, y)$ coinsides with (16).

Using the second equality in (4) the left side of (15) can be represented (skipping G) as $n(z, y) = n(z, y)p(y, y) + n(z, z)p(z, y)$ and hence $n(z, y) = n(z, zz)p(z, y)/(1 - p(y, y))$. By point b) of (3) we have $n(z, z) \equiv n(z, z|G) = 1/(1 - p_D(z, z))$. Thus $n(z, y)$ coincides with the right side of (16). Lemma is proved. ■

Now we can prove Theorem 1.

Proof of Theorem 1. To prove point a) is equivalent to prove (14). For the case when $D = \{y\}$ by formula (7) with $W' = P_D, W = P$ and $z = y$ we have

$$w_D(\cdot, y) = p(\cdot, y) + p(\cdot, y)\frac{p(y, y)}{1 - p(y, y)} = \frac{p(\cdot, y)}{1 - p(y, y)}, y \in D,$$

i.e. formula (14) is valid. Suppose that (14) holds for a set $D \subset X$. Let us prove that then this equality holds for any set $G = D \cup z, z \notin D$. To simplify notation let us denote matrices and columns

$$W_D = W_1, W_G = W_2, p_{D\backslash y}(\cdot, y) = s_1(\cdot, y), p_{G\setminus y}(\cdot, y) = s_2(\cdot, y).$$

Thus our goal is to prove the first equality in the following formula

$$w_2(\cdot, y) = n^+(\cdot, y|G) = \frac{s_2(\cdot, y)}{1 - s_2(y, y)}, y \in G, \tag{17}$$

where the second equality holds by (11).

If formula(14) holds for a set $D \subseteq X$, then two equivalent statements are true

$$w_1(\cdot, y) = n^+(, y|D) = \frac{s_1(\cdot, y)}{1 - s_1(y, y)}, s_1(\cdot, y) = \frac{w_1(\cdot, y).}{1 + w_1(y, y)}, y \in D. \tag{18}$$

The second formula follows from the first one if we apply the first formula for $x = y$ to obtain the equality $1 - s_1(y, y) = 1/(1 + w_1(y, y))$.

For $y = z$ formula (17) can be checked directly: formula (7) with $W' = W_2, W = W_1$, state z eliminated, and with $y = z$ implies $w_2(\cdot, z) = w_1(\cdot, z)/(1 - w_1(z, z))$; since $z \notin D$ by definition of s_2 we have $s_2(\cdot, z) = p_{G\setminus z}(\cdot, z) = p_D(\cdot, z) = s_1(\cdot, z)$, and by definition of s_1 we have $s_1(\cdot, z) = w_D(\cdot, z)$.

For $y \in D$ by definition of W_2, i.e. by formula (7) with $W' = W_2, W = W_1$, state z eliminated, using the first formula in ((18)), the equalities $w_D(\cdot, z) = p_D(\cdot, z), w_D(z, z) = p_D(z, z)$ (since $z \notin D$), and the first formula in (18) for $x = z$, we have

$$w_2(\cdot, y) = n^+(\cdot, y|D) + \frac{p_D(\cdot, z)}{1 - p_D(z, z)} n^+(z, y|D), y \in D, z \notin D. \tag{19}$$

By formula (11) in Lemma 3 applied to the case when set D is replaced set G and y is replaced by z, we have

$$\frac{p_D(\cdot, z)}{1 - p_D(z, z)} = n^+(\cdot, z|G).$$

Therefore applying lemma 4, i.e. formula (15) the right side of (19) coincides with $n^+(\cdot, z|G)$, i.e. formula (17) is proved. ∎

Remark 1. In the MDP problems with a current reward function $c(x)$ it is possible to introduce a transformation of the cost function $c(x)$ (or any function $f(x)$) defined on X into the cost function $c'_D(x)$ under transition from model M to model M'_D or correspondingly into function $c_D(x)$ under transition to model M_D. This formula was used first in [7] in the context of MDP. A corresponding transformation can be obtained for the insertion situation as well.

Table 1. The elimination of states #1, #2, #3.

0.4000	**0.3000**	**0.2000**	**0.1000**	**0.0000**
0.3000	**0.5000**	**0.1000**	**0.1000**	**0.0000**
0.2000	**0.3000**	**0.4000**	**0.1000**	**0.0000**
0.1000	**0.1000**	**0.1000**	**0.5000**	**0.2000**
0.1000	**0.1000**	**0.2000**	**0.2000**	**0.4000**

0.6667	**0.5000**	**0.3333**	**0.1667**	**0.0000**
0.5000	**0.6500**	**0.2000**	**0.1500**	**0.0000**
0.3333	**0.4000**	**0.4667**	**0.1333**	**0.0000**
0.1667	**0.1500**	**0.1333**	**0.5167**	**0.2000**
0.1667	**0.1500**	**0.2333**	**0.2167**	**0.4000**

1.3810	1.4286	**0.6190**	**0.3810**	**0.0000**
1.4286	1.8571	**0.5714**	**0.4286**	**0.0000**
0.9048	1.1429	**0.6952**	**0.3048**	**0.0000**
0.3810	0.4286	**0.2190**	**0.5810**	**0.2000**
0.3810	0.4286	**0.3190**	**0.2810**	**0.4000**

3.2188	3.7500	2.0313	**1.0000**	**0.0000**
3.1250	4.0000	1.8750	**1.0000**	**0.0000**
2.9688	3.7500	2.2813	**1.0000**	**0.0000**
1.0313	1.2500	0.7188	**0.8000**	**0.2000**
1.3281	1.6250	1.0469	**0.6000**	**0.4000**

Table 2. The insertion of states #2 and #3.

0.8750	**0.7500**	0.6250	**0.2500**	**0.0000**
0.6250	**0.8000**	0.3750	**0.2000**	**0.0000**
0.6250	**0.7500**	0.8750	**0.2500**	**0.0000**
0.2500	**0.2500**	0.2500	**0.5500**	**0.2000**
0.3125	**0.3250**	0.4375	**0.2750**	**0.4000**

0.6667	**0.5000**	**0.3333**	**0.1667**	**0.0000**
0.5000	**0.6500**	**0.2000**	**0.1500**	**0.0000**
0.3333	**0.4000**	**0.4667**	**0.1333**	**0.0000**
0.1667	**0.1500**	**0.1333**	**0.5167**	**0.2000**
0.1667	**0.1500**	**0.2333**	**0.2167**	**0.4000**

4 Example

Example shows elimination and insertion for the Markov chain with 5 states. Initially, the first 3 states are eliminated one by one, then states #2 and #3 are inserted.

All calculations are performed with double floating-point precision (16 decimal significant digits), the values in tables are rounded to 4 digits after decimal point.

The original transition matrix and the result of consecutive elimination of states #1, #2, #3 is shown in Table 1. The Table 2 shows result of consecutive insertion of states #2 and #3. Bold columns correspond to non-zero columns in P_D, these columns form a stochastic matrix. Note, that the last matrix in Table

2 coincides with second matrix in Table 1 (with eliminated state #1), which is expected.

References

1. Grassmann, W.K., Taksar, M.I and Heyman, D.P.: Regenerative analysis and steady state distributions for Markov chains. *Operations Research*, **33** (1985) 1107-1116.
2. Katehakis, M.N and Veinott, A.F.: The multi-armed bandit problem: decomposition and computation. *Mathematics of Operations Research*, **12** (1987) 262-268.
3. Kemeny, J.G. and Snell, J.L. *Finite Markov Chains.* Springer, Princeton, NJ, 1976.
4. Langville, A,N. and Meyer, C.D.: Updating Markov chains with an eye on Google's PageRank. *SIAM J. Matrix Anal. Appl.*, **27** (2005) 968-987.
5. Meyer, C.D.: Stochastic complementation, uncoupling Markov chains, and the theory of nearly reducible systems. *SIAM Rev.*, **31** (1989) 240-272.
6. Sheskin, T.J.: Technical note—a Markov chain partitioning algorithm for computing steady state probabilities. *Operations Research*, **33** (1985) 228-235.
7. Sheskin, T.J.: State reduction in a Markov decision process. *Internat. J. Math. Ed. Sci. Tech.*, **30** (1999) 167-185.
8. Sonin, I.M.: The elimination algorithm for the problem of optimal stopping. *Mathematical Methods of Operations Research*, **49** (1999) 111-123 .
9. Sonin, I.M.: A generalized Gittins index for a Markov chain and its recursive calculation. *Statistics & Probability Letters*, **78** (2008) 1526 - 1533.
10. Sonin, I.M. and Steinberg, C.: Continue, quit, restart probability model. *Annals of Operations Research* (2012) 1-24 .
11. Stewart, W. *Probability, Markov Chains, Queues, and Simulation: The Mathematical Basis of Performance Modeling.* Princeton University Press, 2009.

Whittle's Indexation Approach to and Applications of Bi-Objective Two-State Binary-Action Markov Decision Processes

Peter Jacko[1] and Vladimír Novák[2]

[1] Department of Management Science, Lancaster University,
Lancaster, LA1 4YX, UK
p.jacko@lancaster.ac.uk

[2] CERGE-EI, a joint workplace of Charles University in Prague and the Economics Institute of the Czech Academy of Sciences, Politickych veznu 7, 111 21 Prague, Czech Republic
vladimir.novak@cerge-ei.cz

Abstract. In this chapter we present how the Whittle's indexation approach, originally developed for the restless bandits problem, can be used to address multi-objective stochastic optimization problems. For clarity, we focus on Markov decision processes with two objectives, two states and two actions, and provide an optimal solution in terms of Whittle indices. We then broadly discuss the applications of our solution in optimal scheduling of queueing systems with abandonments and optimal choices of venture capitalists investments. We believe there is a need to develop the Whittle index theory further in order to tackle more general multi-objective problems.
Keywords: Bi-objective optimization, Markov decision processes, Whittle index
AMS 2000 subject classification: Primary 90C29, 90C40, Secondary 90B36, 90C31

1 Introduction

In this chapter we consider a fundamental stochastic bi-objective optimization problem. Multi-objective optimization is an important and challenging area with many applications in practice where one needs to find the best solution in the presence of trade-offs between several conflicting objectives. On the other hand, stochastic optimization addresses the ubiquitous decision making under uncertainty. Even though many problems in practice are both stochastic and multi-objective, research on problems lying at this intersection is relatively scarce. For an excellent motivating introduction and survey of existing research see [6].

For the sake of clarity, in this chapter we consider the simplest such problem: optimizing two objectives (bi-objective) in the model of a two-state and binary-action Markov decision process (MDP). Apart from naturally arising in practice (see e.g., [13]), such problems appear as subproblems in more complex problems, or in the solution procedure of constrained stochastic optimization problems when these are relaxed using the Lagrangian approach (see e.g., [11]).

Traditionally, multi-objective optimization formulations and solution approaches are divided into (i) scalarized, where the objectives are combined into a single weighted objective, leading to a unique solution, and (ii) non-scalarized, where objectives are kept separately, leading to a set (or frontier) of Pareto efficient solutions. For our stochastic multi-objective optimization problem we use the approach of [17], which, is based on solving a parametrized optimization problem and, surprisingly, leads to both scalarized and non-scalarized solutions, by solving the problem for all values of the weight parameters at the same time.

To motivate, we briefly describe applications where special cases of our formulation appeared in the literature. Our model appears as a subproblem in the multi-armed restless bandit problem and in the framework of dynamic and stochastic resources allocation problem, see [7]. In particular, we will discuss in Section 5 the following applications:

- scheduling of queueing systems with customer abandonment, see, e.g., [1–3, 9], [15, Section 7];
- optimal decisions for venture capitalist investments [12].

A related MDP model was also addressed in [10, Section 6.1]. We further note that two-state MDPs provide a fundamental model in case of partially observable systems, which are typically modelled using Bayesian approaches, see e.g., [12, 16]. Optimal priority index policies were extensively studied, see e.g., [5].

2 Problem Description and MDP Formulation

We consider a generic problem as described below. There is a system characterized at every time by its current state. The decision maker must decide at every time, which action to apply in order to affect the system's evolution. The action decided when a particular state is observed leads to a certain reward, but requires certain amount of work. It is necessary to pay a cost per unit of work. The rewards and costs are discounted by a geometric discount factor $0 < \beta < 1$ per unit of time. There are two (typically conflicting) objectives: to maximize the total expected discounted reward and to minimize the total expected discounted cost of work carried out.

We formulate the above problem as a discrete time Markov Decision Process (MDP). Consider the time slotted into epochs $t \in \mathcal{T} := \{0, 1, 2, \ldots\}$ at which decisions can be made. The time epoch t corresponds to the beginning of the time period t. We will assume that one of the states is absorbing. We will further consider only two states and two actions.

2.1 System

The system is defined by the tuple $\left(\mathcal{N}, \mathcal{A}, (W_n^a)_{n \in \mathcal{N}}^{a \in \mathcal{A}}, (R_n^a)_{n \in \mathcal{N}}^{a \in \mathcal{A}}, (\boldsymbol{P}^a)^{a \in \mathcal{A}}\right)$, where

- $\mathcal{N} := \{0, 1\}$ is the *state space*, where state 0 represents a service already completed or abandoned, and state 1 means that the service is uncompleted and not abandoned;
- $\mathcal{A} := \{0, 1\}$ is the *action space*, where action $a = 0$ means to be passive and action $a = 1$ means to be active;
- R_n^a is the expected one-period *revenue* earned at state n if action a is decided at the beginning of a period;
- W_n^a is the expected one-period *work* required at state n if action a is decided at the beginning of a period; for simplicity, we assume that for any $n \in \mathcal{N}$, the amount of work carried out does not depend on state, i.e.,

$$W_n^1 := W^1, \qquad\qquad W_n^0 := W^0;$$

- $\boldsymbol{P}^a := \left(p_{n,m}^a\right)_{n,m \in \mathcal{N}}$ is the (stationary) one-period *state transition probability matrix* if action a is decided at the beginning of a period, i.e., $p_{n,m}^a$ is the probability of moving to state m from state n under action a; in particular, we can simplify the notation to

$$\boldsymbol{P}^1 := \begin{matrix} & 0 & 1 \\ 0 & \\ 1 & \end{matrix}\!\!\!\!\!\! \begin{pmatrix} 1 & 0 \\ p^1 & 1 - p^1 \end{pmatrix}, \quad \boldsymbol{P}^0 := \begin{matrix} & 0 & 1 \\ 0 & \\ 1 & \end{matrix}\!\!\!\!\!\! \begin{pmatrix} 1 & 0 \\ p^0 & 1 - p^0 \end{pmatrix}.$$

The dynamics of the system is captured by the *state process* $X(\cdot)$ and the *action process* $a(\cdot)$, which correspond to state $X(t) \in \mathcal{N}$ and action $a(t) \in \mathcal{A}$ at all time epochs $t \in \mathcal{T}$. As a result of deciding action $a(t)$ in state $X(t)$ at time epoch t, the system consumes the work, earns the revenue, and evolves its state for the time epoch $t + 1$.

2.2 Objective

We consider the set of policies $\Pi_{X,a}$, where every $\pi \in \Pi_{X,a}$ is a randomized and non-anticipative policy depending on the state-process $X(\cdot)$ and deciding the action-process $a(\cdot)$. Let $\mathbb{E}_\tau^\pi$ denote the expectation over the state process $X(\cdot)$ and over the action process $a(\cdot)$, conditioned on the state-process history $X(0), X(1), \ldots, X(\tau)$, on the action-process history $a(0), a(1), \ldots, a(\tau - 1)$, and on policy π.

It is necessary to pay a cost ν per unit of work. We therefore study the problem

$$\max_{\pi \in \Pi_{X,a}} \mathbb{E}_0^\pi \left[\sum_{t \in \mathcal{T}} \beta^t R_{X(t)}^{a(t)} \right] - \nu \, \mathbb{E}_0^\pi \left[\sum_{t \in \mathcal{T}} \beta^t W_{X(t)}^{a(t)} \right]. \tag{1}$$

3 Solution

In this section we identify an optimal policy π^* for problem (1). Problem (1) fits the framework of *restless bandits* and can be optimally solved by assigning a set of Whittle index values ν_n to each state $n \in \mathcal{N}$ under certain conditions [11], which, as we prove below, are satisfied for our problem.

To rule out impractical cases, we assume $W^1 > W^0 \geq 0$.

3.1 Optimal Solution via Whittle Index

Let us denote by

$$\nu_0 := \frac{R_0^1 - R_0^0}{W^1 - W^0}, \tag{2}$$

$$\nu_1 := \begin{cases} \dfrac{R_1^1(1-\beta+\beta p^0) - R_1^0(1-\beta+\beta p^1) + R_0^0\beta(p^1-p^0)}{(W^1-W^0)(1-\beta+\beta p^0)}, & \text{if } \nu > \nu_0, \\ \dfrac{R_1^1(1-\beta+\beta p^0) - R_1^0(1-\beta+\beta p^1) + R_0^1\beta(p^1-p^0)}{(W^1-W^0)(1-\beta+\beta p^1)}, & \text{if } \nu \leq \nu_0. \end{cases} \tag{3}$$

We will call ν_n the Whittle index value of state $n \in \mathcal{N}$.

Theorem 1. *For problem* (1), *the following holds (where, in the case of equality, both actions are optimal):*

1. *it is optimal to be active in state* 1 *if and only if* $\nu \leq \nu_1$;
2. *it is optimal to be active in state* 0 *if and only if* $\nu \leq \nu_0$;

Proof. The proof follows the survey [11], and so is based on establishing indexability of the problem and characterizing the Whittle index values. Indexability is satisfied because any two-state MDP is indexable.

Let us denote the optimal value function by V_n for state n. The Bellman equation for state 0 is

$$V_0 = \max\{R_0^1 - \nu W^1 + \beta\,[1 \cdot V_0 + 0 \cdot V_1]\,; R_0^0 - \nu W^0 + \beta\,[1 \cdot V_0 + 0 \cdot V_1]\}. \tag{4}$$

The previous equation shows that it is optimal to be active in state 0 if

$$R_0^1 - \nu W^1 \geq R_0^0 - \nu W^0$$

what gives us a condition for optimality of being active in state 0

$$\frac{R_0^1 - R_0^0}{W^1 - W^0} \geq \nu \tag{5}$$

Statement 2. then follows immediately. Next we prove 1., dividing the proof into two cases.

Case $\nu \leq \frac{R_0^1 - R_0^0}{W^1 - W^0}$. First, it is straightforward to obtain that $V_0 = \frac{R_0^1 - \nu W^1}{1-\beta}$. We want to derive when it is optimal to be active and when it is not optimal to be passive in state 1. The Bellman equation in this case is

$$\begin{aligned} V_1 = \max\{&R_1^1 - \nu W^1 + \beta\left[p^1 V_0 + (1-p^1)V_1\right]; \\ &R_1^0 - \nu W^0 + \beta\left[p^0 V_0 + (1-p^0)V_1\right]\} \end{aligned} \tag{6}$$

If we are active in state 1, we get

$$V_1 = R_1^1 - \nu W^1 + \beta \left[p^1 V_0 + (1 - p^1) V_1 \right]$$

and after substituting V_0, it is straightforward to obtain

$$V_1 = \frac{R_1^1(1-\beta) + \beta p^1 R_0^1 - \nu \left(W^1(1-\beta) + \beta p^1 W^1 \right)}{(1-\beta)(1-\beta+\beta p^1)}$$

When we substitute V_0 and V_1 in (6), where it holds that if the first terms is greater or equal to the second one, then it is optimal to be active. After rearranging, we obtain the following condition for optimality of being active in state 1

$$\frac{R_1^1(1-\beta+\beta p^0) - R_1^0(1-\beta+\beta p^1) + R_0^1 \beta (p^1 - p^0)}{(W^1 - W^0)(1-\beta+\beta p^1)} \geq \nu$$

Case $\nu > \frac{R_0^1 - R_0^0}{W^1 - W^0}$. We proceed similarly as in the previous case. We first obtain $V_0 = \frac{R_0^0 - \nu W^0}{1-\beta}$. If we are active in state 1 and we substitute V_0, we get

$$V_1 = \frac{R_1^1(1-\beta) + \beta p^1 R_0^0 - \nu \left(W^1(1-\beta) + \beta p^1 W^0 \right)}{(1-\beta)\,(1-\beta+\beta p^1)}$$

Now we can substitute V_0 and V_1 in (6). After rearranging, we obtain the following condition for optimality of being active in state 1:

$$\frac{R_1^1(1-\beta+\beta p^0) - R_1^0(1-\beta+\beta p^1) + R_0^0 \beta (p^1 - p^0)}{(W^1 - W^0)(1-\beta+\beta p^0)} \geq \nu.$$

■

3.2 Whittle Index under the Long-Run Average Criterion

Letting $\beta \to 1$ in expressions (2) and (3) gives the optimal policy for the problem under the long-run average criterion. We present below the Whittle index under the assumption $p^a > 0$ for any $a \in \mathcal{A}$.

While ν_0 remains the same, ν_1 simplifies to

$$\nu_1 = \begin{cases} \dfrac{R_1^1 p^0 - R_1^0 p^1 + R_0^0 (p^1 - p^0)}{(W^1 - W^0) p^0}, & \text{if } \nu > \nu_0, \\ \dfrac{R_1^1 p^0 - R_1^0 p^1 + R_0^1 (p^1 - p^0)}{(W^1 - W^0) p^1}, & \text{if } \nu \leq \nu_0. \end{cases} \tag{7}$$

Under the long-run average criterion, the most common situation is when $R_0^1 = R_0^0 = R_0$, in which case we can rewrite the above as

$$\nu_0 = 0, \tag{8}$$

$$\nu_1 = \begin{cases} \dfrac{R_1^1 p^0 - R_1^0 p^1 + R_0 (p^1 - p^0)}{W^1 - W^0} \cdot \dfrac{1}{p^0}, & \text{if } \nu > 0, \\ \dfrac{R_1^1 p^0 - R_1^0 p^1 + R_0 (p^1 - p^0)}{W^1 - W^0} \cdot \dfrac{1}{p^1}, & \text{if } \nu \leq 0. \end{cases} \tag{9}$$

Note that $1/p^1$ and $1/p^0$ appearing in (9) are the expected times spent in state 1 until the transition to state 0 under the active and passive action, respectively.

4 Interpretation of the Whittle Index Values

Let us denote by

$$C^- := \frac{R_1^1(1-\beta+\beta p^0) - R_1^0(1-\beta+\beta p^1) + R_0^1\beta(p^1-p^0)}{(1-\beta+\beta p^1)(1-\beta+\beta p^0)}$$

and

$$C^+ := \frac{R_1^1(1-\beta+\beta p^0) - R_1^0(1-\beta+\beta p^1) + R_0^0\beta(p^1-p^0)}{(1-\beta+\beta p^1)(1-\beta+\beta p^0)}$$

so the Whittle index value of state 1 can be rewritten as

$$\nu_1 := \begin{cases} \frac{C^+}{W^1-W^0}(1-\beta+\beta p^1), & \text{if } \nu > \frac{R_0^1-R_0^0}{W^1-W^0}, \\ \frac{C^-}{W^1-W^0}(1-\beta+\beta p^0), & \text{if } \nu \le \frac{R_0^1-R_0^0}{W^1-W^0}, \end{cases} \tag{10}$$

Note that $\frac{1}{1-\beta+\beta p^a}$ appearing in (10) is the expected total discounted time spent in state 1 if action a is always used.

Proposition 1.

1. *C^- is the difference between the expected total discounted revenue if being always active and between the expected total discounted revenue if being passive in state* 1 *and being active in state* 0*;*
2. *C^+ is the difference between the expected total discounted revenue if being active in state* 1 *and being passive in state* 0 *and between the expected total discounted revenue if being always passive.*

Proof. It is straightforward to obtain that the expected total revenue if being always active is

$$\frac{R_1^1 + \beta p^1 \frac{R_0^1}{1-\beta}}{1-\beta+\beta p^1}, \tag{11}$$

the expected total revenue if being passive in state 1 and being active in state 0 is

$$\frac{R_1^0 + \beta p^0 \frac{R_0^1}{1-\beta}}{1-\beta+\beta p^0}, \tag{12}$$

the expected total revenue if being active in state 1 and being passive in state 0 is

$$\frac{R_1^1 + \beta p^1 \frac{R_0^0}{1-\beta}}{1-\beta+\beta p^1}, \tag{13}$$

and the expected total revenue if being always passive is

$$\frac{R_1^0 + \beta p^0 \frac{R_0^0}{1-\beta}}{1 - \beta + \beta p^0}. \tag{14}$$

In order to prove 1., we subtract (12) from (11). After rearranging we obtain

$$\frac{R_1^1(1-\beta+\beta p^0) - R_1^0(1-\beta+\beta p^1) + R_0^1\beta(p^1-p^0)}{(1-\beta+\beta p^0)(1-\beta+\beta p^1)} = C^-.$$

Similarly we establish 2. After subtraction of (14) from (13) and rearranging we have

$$\frac{R_1^1(1-\beta+\beta p^0) - R_1^0(1-\beta+\beta p^1) + R_0^0\beta(p^1-p^0)}{(1-\beta+\beta p^0)(1-\beta+\beta p^1)} = C^+.$$

■

In the practically relevant case when $R_0^1 = 0$ and $R_0^0 = 0$, this interpretation simplifies as follows.

Proposition 2. *Suppose that* $R_0^1 = 0$ *and* $R_0^0 = 0$*. Then,* $C^- = C^+$*, and it is the difference between the expected total discounted revenue if being always active and between the expected total discounted revenue if being always passive.*

5 Applications

The Whittle index values (2) and (3) presented above provide a generalization of the indices obtained in several recent papers. In this section we will briefly describe how MDP formulations differ in these model and how they fit our formulation given in Section 2.

All of the these models have the following common features: $W^1 := 1, W^0 := 0, R_0^1 := 0$ and $R_0^0 := 0$. The remaining parameters are specific to each model and are described below. Note that the problem as considered in previous sections materializes in all these models after the Lagrangian relaxation and decomposition, and cost ν appears as the Lagrange multiplier.

5.1 Scheduling of Customers

The discrete-time model in [7, 8] addresses the problem of optimal scheduling in a multi-class queueing system and defines every customer of class k as

$$R_1^1 := -c_k(1-\mu_k), \qquad R_1^0 := -c_k, \qquad p^1 := \mu_k, \qquad p^0 := 0,$$

where c_k is the waiting cost accrued at the end of each period if the customer is in the system and μ_k is the probability that the customer's service is completed in one period if in service.

After substitution of these parameters in (2) and (3), we obtain Whittle index values

$$\nu_{k,0} = 0, \qquad \nu_{k,1} = \begin{cases} \dfrac{c_k \mu_k}{1-\beta}, & \text{if } \nu > 0, \\ \dfrac{c_k \mu_k}{1-\beta+\beta\mu_k}, & \text{if } \nu \leq 0. \end{cases}$$

The static priority policy which uses the $c_k \mu_k$ as priorities of each class k is a well-known policy, so-called the $c\mu$-rule, and was shown to be optimal in [14, 4]. These priorities are recovered in the natural case $\nu > 0$ as the scaled Whittle index values $\nu_{k,1}(1-\beta)$.

5.2 Scheduling of Customers with Abandonment I

The model in [2] addresses the problem of optimal scheduling in a multi-class queueing system in which customers can abandon, a.k.a. discrete-time multi-class $M/M/S+M$ system, where *at most* S customers are served at any time, and the arrivals and departures (from service and abandonment) are Markovian (geometrically distributed). The paper defines every customer of class k as

$$R_1^1 := -c_k(1-\mu_k), \quad R_1^0 := -c_k(1-\theta_k) - d_k\theta_k, \quad p^1 := \mu_k, \quad p^0 := \theta_k,$$

where c_k is the waiting cost accrued at the end of each period if the customer is in the system, d_k is the abandonment penalty incurred if the customer abandons, μ_k is the probability that the customer's service is completed in one period if in service, and θ_k is the probability that the customer abandons in one period if not in service. It is further assumed that $c_k(\mu_k - \theta_k) + d_k\theta_k(1-\beta+\beta\mu_k) \geq 0$.

After substitution of these parameters in (2) and (3), we obtain Whittle index values

$$\nu_{k,0} = 0, \qquad \nu_{k,1} = \frac{c_k(\mu_k - \theta_k) + d_k\theta_k(1-\beta+\beta\mu_k)}{1-\beta+\beta\theta_k}.$$

The authors in [2] showed in computational experiments with a single server ($S = 1$) and Bernoulli arrivals under the long-run average criterion that the static priority policy which uses the undiscounted Whittle index values $\nu_{k,1} = \frac{c_k(\mu_k-\theta_k)+d_k\theta_k\mu_k}{\theta_k}$ as priorities of each class k performs nearly optimally and often outperforms other proposed schedulers.

5.3 Scheduling of Customers with Abandonment II

The authors in [3] also addressed the problem of optimal scheduling in the multi-class $M/M/S+M$ system, however in the continuous-time version with Markovian (Poisson distributed) arrival and departure rates. After uniformization and discretization, the paper defines every customer of class k as

$$R_1^1 := -c_k', \qquad R_1^0 := -c_k' - \beta d_k \theta_k', \qquad p^1 := \mu_k', \qquad p^0 := \theta_k',$$

where c'_k, μ'_k, and θ'_k are the uniformized versions of the waiting cost rate c_k, service completion rate μ_k, and abandonment rate θ_k, respectively. Both rates follow the Poisson distribution.

After substitution of these parameters in (2) and (3), we obtain Whittle index values

$$\nu_{k,0} = 0,$$
$$\nu_{k,1} = \begin{cases} C'_k(1-\beta+\beta\mu'_k), & \text{if } \nu > 0, \\ C'_k(1-\beta+\beta\theta'_k), & \text{if } \nu \leq 0, \end{cases}$$

where C'_k is given as follows

$$C'_k = \beta \frac{c'_k(\mu'_k - \theta'_k) + d_k\theta'_k(1-\beta+\beta\mu'_k)}{(1-\beta+\beta\theta'_k)(1-\beta+\beta\mu'_k)}.$$

See [3] for the transformation of these expressions back to the continuous-time model in the general (discounted) case. The (long-run average) Whittle index rates are

$$\xi_{k,1} := \begin{cases} C_k\mu_k, & \text{if } \nu > 0 \\ C_k\theta_k, & \text{if } \nu \leq 0 \end{cases}$$

where $C_k = d_k - c_k\,(1/\mu_k - 1/\theta_k)$. Computational experiments with 1 and 2 servers and Poisson arrivals presented in [3] indicated that the static priority policy which uses the (long-run average) Whittle index rates as priorities of each class k in the following sense

$$\xi_{k,1} := \begin{cases} C_k\mu_k, & \text{if } C_k \geq 0 \\ C_k\theta_k, & \text{if } C_k < 0 \end{cases}$$

performs nearly optimally and systematically outperforms other proposed schedulers in system with *exactly* S customers in service. The policy was also shown to perform well when virtual idling options are introduced with $\xi_{k,1} = 0$, causing customers with negative priority never being served, representing thus a system with *at most* S customers in service at any time. Note that now the priorities do not depend on ν, as this parameter is not available in the scheduling problem.

5.4 Scheduling of Customers with Abandonment III

The continuous-time model in [1] also addresses the problem of optimal scheduling in the multi-class $M/M/S+M$ system. However, it uses a different approach and does not include an MDP formulation of the problem. As shown in [3], there is an equivalent MDP formulation, which after uniformization and discretization, leads to defining every customer of class k as

$$R_1^1 := 0, \qquad R_1^0 := -c'_k - \beta d_k\theta'_k, \qquad p^1 := \mu'_k, \qquad p^0 := \theta'_k,$$

where the parameters have the same interpretation as in the previous subsection. The main difference in this model is that the waiting cost is not accrued while the customer is being served (i.e., under action 1).

After substitution of these parameters in (2) and (3), we obtain Whittle index values

$$\nu_{k,0} = 0,$$
$$\nu_{k,1} = \begin{cases} \dfrac{(c'_k + \beta d_k \theta'_k)(1 - \beta + \beta\mu'_k)}{1 - \beta + \beta\theta'_k}, & \text{if } \nu > 0, \\ c'_k + \beta d_k \theta'_k, & \text{if } \nu \leq 0. \end{cases}$$

Using the same transformation of these expressions back to the continuous-time model gives the (long-run average) Whittle index rates

$$\xi_{k,1} := \begin{cases} \dfrac{(c_k + d_k\theta_k)\mu_k}{\theta_k}, & \text{if } \nu > 0 \\ c_k + d_k\theta_k, & \text{if } \nu \leq 0 \end{cases}$$

In [1], the authors considered the long-run average criterion and proved that the static priority policy which uses the following priorities of each class k

$$\frac{(c_k + d_k\theta_k)\mu_k}{\theta_k}$$

is asymptotically optimal in the many-server fluid-scaling limit regime under overload conditions (i.e., $\sum_k \lambda_k/\mu_k > S$, where λ_k is the arrival rate of class k).

5.5 Scheduling of Customers with Abandonment IV

The continuous-time model in [15, Section 7] also addresses the problem of optimal scheduling in a multi-class queueing system, in a slightly more general model. After uniformization and discretization, we can define every customer of class k as

$$R_1^1 := -\widetilde{c}'_k - \beta\widetilde{d}_k\widetilde{\theta}'_k, \qquad R_1^0 := -c'_k - \beta d_k\theta'_k, \qquad p^1 := \mu'_k + \widetilde{\theta}'_k, \qquad p^0 := \theta'_k,$$

where the parameters have the same interpretation as in the previous subsection. The main difference in this model is that it is possible to abandon while the customer is being served (i.e., under action 1) and that the waiting cost rates, abandonment penalties and abandonment rates may differ under the two actions.

After substitution of these parameters in (2) and (3), and the transformation of these expressions back to the continuous-time model gives the (long-run average) Whittle index rates

$$\xi_{k,1} := \begin{cases} \dfrac{(c_k + d_k\theta_k)(\mu_k + \widetilde{\theta}_k)}{\theta_k} - (\widetilde{c}_k + \widetilde{d}_k\widetilde{\theta}_k), & \text{if } \nu > 0 \\ (c_k + d_k\theta_k) - \dfrac{(\widetilde{c}_k + \widetilde{d}_k\widetilde{\theta}_k)\theta_k}{\mu_k + \widetilde{\theta}_k}, & \text{if } \nu \leq 0 \end{cases}$$

It was proved in [15] that the policy defined below is asymptotically optimal in the many-server fluid-scaling limit regime under the long-run average criterion with the constraint that *at most* S customers are served at any time. The static priority policy uses the following priorities of each class k

$$\frac{(c_k + d_k\theta_k)(\mu_k + \widetilde{\theta}_k)}{\theta_k} - (\widetilde{c}_k + \widetilde{d}_k\widetilde{\theta}_k)$$

as long as these priorities are positive; customers of classes whose priorities are negative are never served.

5.6 Venture Capitalists Investments

In [12, Section 2.2], the author considers a venture capitalist investor who periodically reassesses investment opportunities into a number of entrepreneurial companies and decides which few of them to actively collaborate with, invest in and push to IPO or acquisition. The rest of the companies are only passively monitored.

An investment opportunity k is characterized by cost of investment c_k, success probability μ_k, and bankruptcy probability θ_k. There are also bankruptcy penalty d_k^1, which describes damages connected to the bankruptcy of the company such as investor's reputation loss, waste of prepared processes and strategies for that particular company, etc.; and lost-opportunity penalty d_k^0, which describes loss in reputation because of an unnoticed opportunity to invest into a successful company. If the investment is successful, the investor gains reward r_k, which represents the net present value of an annuity of dividends. We assume that all the rewards and costs are accrued at the end of each period.

Thus, investment opportunity k is defined as

$$R_1^1 := -c_k(1 - \mu_k - \theta_k) + r_k\mu_k - d_k^1\theta_k, \quad R_1^0 := -d_k^0\mu_k, \quad p^1 := p^0 := \mu_k + \theta_k.$$

After substitution of these parameters in (2) and (3), we obtain Whittle index values

$$\nu_{k,0} = 0, \qquad \nu_{k,1} = -c_k(1 - \mu_k - \theta_k) + r_k\mu_k - d_k^1\theta_k + d_k^0\mu_k.$$

Note that this solution is independent of the discount factor β, and is inherently myopic, as $\nu_{k,1} = R_1^1 - R_1^0$.

6 Conclusion

In this chapter we have presented a detailed solution using the Whittle's indexation approach to the simplest bi-objective stochastic optimisation problem. We believe this approach can be developed for more general multi-objective stochastic problems. However, this brings up several theoretical (and modelling) challenges that need to be addressed in future work.

We have further shown that our two-state model covers and generalizes several existing models. It provides useful results for a couple of application areas, some of which were obtained using different approaches in previous literature. Another interesting model to investigate in detail would be a three-state MDP, which has attracted limited attention so far, but is useful in practice.

References

1. Atar, R., Giat, C. and Shimkin, N.: The $c\mu/\theta$ rule for many-server queues with abandonment. *Operations Research*, **58(5)** (2010) 1427–1439.
2. Ayesta, U., Jacko, P. and Novak, V.: A nearly-optimal index rule for scheduling of users with abandonment. *Proceedings of IEEE INFOCOM*, (2011) 2835-2843.
3. Ayesta, U., Jacko, P. and Novak, V.: Scheduling of multi-class multi-server queueing systems with abandonments. *Journal of Scheduling*, DOI: 10.1007/s10951-015-0456-7, (2015), in press.
4. Buyukkoc, C., Varaiya, P. and Walrand, J.: The $c\mu$ rule revisited. *Advances in Applied Probability*, **17(1)** (1985) 237-238.
5. Gittins, J. C. *Multi-Armed Bandit Allocation Indices.* J. Wiley & Sons, New York, 1989.
6. Gutjahr, W. J. and Pichler, A.: Stochastic multi-objective optimization: a survey on non-scalarizing methods. *Annals of Operations Research*, DOI: 10.1007/s10479-013-1369-5, (2013), in press.
7. Jacko, P.: Adaptive greedy rules for dynamic and stochastic resource capacity allocation problems. *Medium for Econometric Applications*, **17(4)** (2009) 10-16.
8. Jacko, P.: Restless bandits approach to the job scheduling problem and its extensions. In *Modern Trends in Controlled Stochastic Processes: Theory and Applications*, (A. B. Piunovskiy Ed.), Luniver Press, United Kingdom, 2010, 248-267.
9. Larrañaga, M., Ayesta, U., and Verloop, I. M.: Dynamic fluid-based scheduling in a multi-class abandonment queue. *Performance Evaluation*, **70(10)** (2013) 841-858.
10. Niño-Mora, J.: Restless bandits, partial conservation laws and indexability. *Advances in Applied Probability*, **33(1)** (2001) 76-98.
11. Niño-Mora, J.: Dynamic priority allocation via restless bandit marginal productivity indices. *TOP*, **15(2)** (2007) 161-198.
12. Novak, V.: *Learning in Finance (Master Thesis).* Comenius University Bratislava, Slovakia, 2013.
13. Piunovskiy, A. B.: Bicriteria optimization of a queue with a controlled input stream. *Queueing Systems*, **48** (2004) 159-184.
14. Smith, W. E.: Various optimizers for single-stage production. *Naval Research Logistics Quarterly*, **3(1-2)** (1956) 59-66.
15. Verloop, I. M.: Asymptotically optimal priority policies for indexable and non-indexable restless bandits. *Annals of Applied Probability*, (2015) to appear.
16. Villar, S. S.: Indexability and optimal index policies for a class of reinitialising restless bandits. *Probability in the Engineering and Informational Sciences*, DOI: 10.1017/S026996481500025X, (2015), in press.
17. Whittle, P.: Restless Bandits: Activity Allocation in a Changing World. *A Celebration of Applied Probability, J. Gani (Ed.), Journal of Applied Probability*, **25A** (1988) 287-298.

Completely Mixed Stochastic Games with Small Unfixed Discount Factor

Konstantin Avrachenkov and Anastasiia Varava

Inria Sophia Antipolis, France and KTH, Sweden
k.avrachenkov@inria.fr; varava@kth.se

Abstract. Motivated by uncertainty in the value of the interest rate, we study discounted zero-sum stochastic games with unfixed discount factor. Our general goal is to obtain a power series expansion of the value of the game with respect to the discount factor around its nominal value. We consider a specific but important class of stochastic games – completely mixed stochastic games. As an illustrative example we take tax evasion model.
Keywords: discounted stochastic games, completely mixed games, unfixed discount factor, power series, Shapley-Snow kernel, tax evasion model.
AMS 2000 subject classification: Primary 91A15, Secondary 49L20

1 Introduction

The present work is motivated by uncertainty in the value of the interest rate. We consider the discounted stochastic game and pose a question what happens with the value of the game if the interest rate, or equivalently, the discount factor, deviates from its nominal value. In the spirit of the perturbation analysis [2], we try to find efficient algorithms for computation of some initial coefficients of the power series of the value of the game with respect to the discount factor.

The perturbation analysis of stochastic games with respect to the discount factor appears to be very challenging in its full generality (see e.g., [7, 11]). Therefore, in this work we limit ourselves to a specific but important class of stochastic games – completely mixed stochastic games. In particular, in the case of completely mixed stochastic games the value of the game has a Taylor series expansion at the vicinity of zero discount factor.

Our approach is based on generalization of the Shapley value iterations [9] from the field of real numbers to the field of power series. Such an approach was successfully used before for Blackwell optimality in Markov decision processes [6], for singularly perturbed Markov decision processes [1] and for Blackwell equilibrium [3] in stochastic games with perfect information or switching controller games [4]. It is interesting to observe that in the present setting each Shapley value iteration produces an exact new coefficient in the Taylor series of the game value.

The structure of the paper is as follows: In the next section we define the discounted stochastic game and provide necessary background material on stochastic and matrix games. In Section 3 we study the case of completely mixed stochastic games. In Section 4 we provide an illustrative example of the tax evasion model. Finally, in Section 5 we give conclusions and discuss open problems.

2 Background on Discounted Stochastic Games and Matrix Games

The notion of stochastic game was first introduced by Shapley in 1953 [9]. Following [9], we consider two-person zero-sum stochastic games on infinite time horizon and discounted payoff. The game has a finite set of positions, called states. For each state there are two finite sets of actions for the first and the second player, respectively. Each pair of actions corresponding to the same state defines the immediate reward for both players as well as the probabilities of transitions to the other states. At each step the players simultaneously choose actions and receive corresponding rewards. After that the system immediately moves to the next state with respect to the probability distribution defined by chosen pair of actions. Let us next formally define the two-person zero-sum stochastic game (an interested reader can find much more information on stochastic games in the book by Filar and Vrieze [5]).

Definition 1. *A system with the following structure is called two-person zero-sum stochastic game Γ:*

1. *there are two players, P_1 and P_2 (also called "the first player" and "the second player", respectively);*
2. *$S = \{1, 2, .., n\}$ is a finite set of states of the game;*
3. *$A^i(s) = \{1, 2, .., m_i(s)\}$ represent sets of actions of ith player with respect to the current state $s \in S$;*
4. *the function $r(s, i, j)$ represents immediate rewards for player 1, and $-r(s, i, j)$ is the immediate reward for player 2. Here $s \in S$, $i \in A^1(s), j \in A^1(s)$, which means that the game is currently in state s and the players choose actions i and j, respectively. One usually denotes the matrix of immediate rewards in state s by $R(s)$.*
5. *Transition probabilities $p(s'|s, i, j) : s, s' \in S, i \in A^1(s), j \in A^2(s)$ where $p(s'|s, i, j)$ is the probability of transition from state s to state s' given that players 1 and 2 choose actions $i \in A^1(s), j \in A^2(s)$, respectively. It is assumed that the transition probabilities and the immediate rewards are known to both players.*

A strategy for a player is a rule of selecting an action at each step of the game. In general, strategies can depend on complete history of the game until the current stage. Such strategies are called behavioural strategies. We are looking at the simpler class of strategies called stationary strategies which depend only on the current state s and not on how s has been reached.

Definition 2. *A stationary strategy of a player is a function from the state space to the set of probability distributions on player's action set. A strategy is called pure or deterministic when for each state of the game the player deterministically choose exactly one action with probability 1.*

In his paper [9], Shapley has shown that it is enough to consider only stationary strategies for the discounted stochastic games, so we restrict ourselves to them.

Discounted payments are accumulating throughout the game. The first player aims to maximize the β-discounted payoff, whereas the second player aims to minimize it.

Definition 3. *Given an initial state s_0, a pair of stationary strategies (f, g) of players 1 and 2, resp., and a discount factor $\beta \in [0, 1)$, we define β-discounted payoffs as follows:*

$$[J_\beta(f, g)](s_0) = \sum_{t=0}^{\infty} \beta^t E_{s_0}^{f,g}[r_t],$$

where t corresponds to discrete moments of time and r_t is an immediate payoff on a corresponding (t^{th}) step of the game with respect to the initial state s_0 and strategies of the players.

Under this payoff one can define an equilibrium pair of strategies and the value vector of the game.

Definition 4. *A pair (f^*, g^*) such that*

$$[J_\beta(f, g^*)](s) \le [J_\beta(f^*, g^*)](s) \le [J_\beta(f^*, g)](s),$$

for all f and g and for all $s \in S$, is called a pair of equilibrium strategies. The vector $J_\beta(f^*, g^*)$ is called equilibrium value vector, or game value vector and is denoted by $\mathbf{v}$.

One can view a zero-sum discounted stochastic game $\Gamma(\beta)$ as a generalization of static matrix game to a multistate and multistage situation. Indeed, assume that $|S| = N$ and $m_1(s) = |A^1(s)|, m_2(s) = |A^2(s)|$ for each $s \in S$. Then we can naturally define N matrix games that are in one-to-one correspondence with the states of $\Gamma(\beta)$:

$$R(s) = [r(s, a_1, a_2)]_{a_1=1, a_2=1}^{m_1(s), m_2(s)}$$

Now we can think of each state of a game as of a simple matrix game $R(s)$ with corresponding action sets $A^1(s), A^2(s)$. Notice that actions of players determine not only their rewards (as in the classical matrix game), but also probability transitions $p(s'|s, a_1, a_2)$ to the matrix game $R(s')$ that can be played at the next step.

In his original paper [9], Shapley has shown that under the discounted payoff criterion, there always exist an equilibrium pair of strategies, and the equilibrium value vector $\mathbf{v}(f^*, g^*)$ is unique. More precisely, the following theorem takes place:

Theorem 1. *(Shapley, 1953) The discounted, zero-sum, stochastic game $\Gamma(\beta)$ possesses the value vector* $\mathbf{v}$ *that is unique solution of the equations*

$$v(s) = \mathrm{val}[R(s, \mathbf{v})]$$

for all $s \in \mathbf{S}$*, where* $\mathbf{v}^T = (v(1), v(2), ..., v(N))^T$ *and*

$$R(s, \mathbf{v}) = \left[r(s, a_1, a_2) + \beta \sum_{s' \in \mathbf{S}} p(s'|s, a_1, a_2) v(s') \right].$$

We call the following set of equations N (one equation per state of the game) "Shapley equations":

$$v(s) = \mathrm{val} \left[r(s, a_1, a_2) + \beta \sum_{s' \in \mathbf{S}} p(s'|s, a_1, a_2) v(s') \right], \quad s \in S$$

In addition, we call the following map "Shapley operator":

$$\mathrm{T} : \mathbb{R}^N \to \mathbb{R}^N, \quad \mathrm{T}(\mathbf{x}_s) = \mathrm{val}\, R(s, \mathbf{x}_s)$$

Shapley [9] has shown that this operator is a contraction with coefficient β. In the above mentioned paper he has also shown how one can reduce the process of solving stochastic game to solving several static matrix games. The idea of this algorithm is based on the fact that the operator T is a contraction. The value of the game is the unique fixed point of this operator, or, in other words, the unique solution of the equation

$$\mathbf{x} = \mathrm{T}\,\mathbf{x}.$$

By simple iterations one can find the approximated value of the game. Let $\mathbf{v}_0$ be an arbitrary initial approximation. Then, by Banach fixed point theorem, the sequence of approximations $\mathbf{v}_k$ converges to the exact solution $\mathbf{v}$ of the game:

$$\mathbf{v}_{k+1} = \mathrm{T}\,\mathbf{v}_k, \quad k = 0, 1, ...$$

Notice that at each iteration we have to solve N matrix games. As we mentioned in the previous section, the problem of solving a matrix game is of polynomial complexity (e.g., is solvable by linear programming). We will call this algorithm Shapley value iteration.

Let us also recall some useful facts about static matrix games. Each static matrix game can be presented as a matrix M by identifying rows with pure strategies of player 1 and columns with pure strategies of player 2. The element $M[a_1, a_2]$ of a matrix represents the reward $r(a_1, a_2)$. Clearly, the first player aims to maximize his payoff, whereas the second player aims to minimize his cost. We assume that both players are rational. It is known that there always exists a pair of equilibrium strategies (f^*, g^*) such that:

$$\forall f, g : \; r(f, g^*) \leq r(f^*, g^*) \leq r(f^*, g).$$

The value $v = r(f^*, g^*)$ is called the value of the game and is known to be unique. By solving a game one usually means finding its value (and, possibly, optimal strategies). Solving a matrix game is not a trivial problem. There are several approaches to it, and one of them is due to Shapley and Snow [10].

Theorem 2. *(Shapley and Snow, 1950) If A is a matrix game and* $\operatorname{val} A \neq 0$, *then A has a square invertible submatrix $\hat{A}$, called a Shapley-Snow kernel, such that:*

- $\operatorname{val} A = \operatorname{val} \hat{A} = \dfrac{\det \hat{A}}{\sum_{ij} \operatorname{adj} \hat{A}[i,j]}$;
- *There is a pair of equilibrium strategies $(\hat{x}, \hat{y})$ for $\hat{A}$ which are also equilibrium strategies for A (after inserting zeroes at corresponding entries) that satisfy*

$$(\hat{x})^T = (\operatorname{val} A)\mathbf{1}^T \hat{A}^{-1}, \quad \hat{y} = (\operatorname{val} A)\hat{A}^{-1}\mathbf{1},$$

where $\mathbf{1} = (1, .., 1)^T$.

Without loss of generality, we assume that all rewards of the game are strictly positive and hence the value is positive as well, and so Shapley-Snow kernels are always defined.

Following [11], let us call Shapley-Snow kernels, which are completely mixed, *cmv-kernels.* It has been shown in [11] that they always exist in arbitrary matrix game. From now we will consider only these kernels.

Despite the fact that this theorem has a theoretical value, in practice matrix games are usually solved in other ways, e.g., by linear programming. In particular, this implies that a matrix game can be solved in polynomial time.

3 Completely Mixed Stochastic Games

Consider a special class of stochastic games – completely mixed stochastic games.

Definition 5. *Stochastic game is called completely mixed, if for each state $s \in S$, the Shapley matrix*

$$R(s, \mathbf{v}(\beta), \beta) = \left[r_{s,i,j} + \beta \textstyle\sum_{l=1..N} p^l_{s,i,j} v_l(\beta)\right]_{i,j=1}^{|A^1(s)| \times |A^2(s)|}, \quad \forall s \in S$$

is completely mixed.

In other words, for each state s the cmv-kernel of the Shapley matrix $R(s, \mathbf{v}(\beta), \beta)$ is the whole matrix per se. Clearly, in this case for each $s \in S$ it is required that $|A^1(s)| = |A^2(s)|$.

We can propose an easily verifiable condition to check if a stochastic game is completely mixed for small values of the discount factor.

Assumption 1. *Assume that we are given a game $\Gamma(\beta)$ in which the static matrix game at each state $s \in S$ is completely mixed.*

The above assumption implies useful structural properties of the stochastic game. Namely, we have

Lemma 1. *Let Assumption 1 hold. Then, the stochastic game $\Gamma(\beta)$ is completely mixed for the values of the discount factor β in some interval $[0,\delta)$ and the value vector of the game possesses a Taylor series expansion in the vicinity of $\beta_0 = 0$ with the convergence radius $R \leq \delta$.*

Proof. Since the set of completely mixed matrix games is open in the space of all matrix games of the corresponding dimension [11], there exist a neighborhood $\beta \in \omega = [0,\delta)$, such that for all $\beta \in \omega$ and all states $s \in S$, the games defined by $R(s, \mathbf{v}(\beta), \beta)$ are completely mixed. It then implies that for all $\beta \in \omega$ the stochastic game $\Gamma(\beta)$ is completely mixed as well.

It follows from the arguments in the proof of Lemma 4.1 in [11] that in this case the value function $\mathbf{v}(\beta)$ of the stochastic game $\Gamma(\beta)$ is analytic on $\beta \in \omega$. More precisely, the fact that the cmv-kernels of Shapley matrices do not change in some neigbourhood of $\beta = 0$ is crucial here. In its turn, it is a consequence of the completely-mixed assumption on the payoff matrices.

Thus, the value of the game can be represented as a Taylor series expansion around zero for each state s, with the convergence radius $0 < R \leq \delta$. ∎

Our goal is to find an approximation of the value function $\mathbf{v}(\beta)$ given by the first m terms of its power series expansion. We take a constant vector as an initial approximation, e.g., $\mathbf{v_0}(\beta) := 0$. By $\mathbf{v_k}(\beta)$ we denote the kth approximation of $\mathbf{v}(\beta)$. Then the next approximation can be obtained as follows:

$$v_{k+1,s}(\beta) = \frac{\det\left(R(s, \mathbf{v_k}(\beta), \beta)\right)}{\sum_{i,j=1}^{r} \operatorname{adj}\left(R(s, \mathbf{v_k}(\beta), \beta)\right)[i,j]}, \qquad \forall s \in S. \tag{1}$$

Here $\operatorname{adj}\left(R(s, \mathbf{v_k}(\beta), \beta)\right)$ is the adjugate matrix of $R(s, \mathbf{v_k}(\beta), \beta)$. Recall that each entry of $R(s, \mathbf{v_k}(\beta), \beta)$ looks as follows:

$$R(s, \mathbf{v_k}(\beta), \beta)[i,j] = r_{s,i,j} + \beta \sum_{l=1..N} p^l_{s,i,j} v_{k,l}(\beta).$$

Each $v_{k,l}(\beta)$ is a rational function, continuous (and analytical) for $\beta \in \omega$. Thus, we can present it as a power series expansion

$$v_{k,l}(\beta) = \sum_{i=0}^{\infty} a^i_{k,l} \beta^i.$$

Substituting this representation into the Shapley matrix, we obtain

$$\begin{aligned} R(s, \mathbf{v_k}(\beta), \beta)[i,j] &= r_{s,i,j} + \beta \sum_{l=1..N} p^l_{s,i,j} \left(\sum_{i=0}^{\infty} a^i_{k,l} \beta^i \right) = \\ & r_{s,i,j} + \beta \sum_{l=1..N} p^l_{s,i,j} \left(a^0_{k,l} + a^1_{k,l}\beta + a^2_{k,l}\beta^2 + ... \right) = \\ & r_{s,i,j} + \beta \sum_{l=1..N} p^l_{s,i,j} a^0_{k,l} + \beta^2 \sum_{l=1..N} p^l_{s,i,j} a^1_{k,l} + \beta^3 \sum_{l=1..N} p^l_{s,i,j} a^2_{k,l} + ... \end{aligned} \tag{2}$$

Let us define a reduction $\mod \beta^m$ on the power series:

$$a(\beta) = \sum_{i=0}^{\infty} a_i \beta^i;$$

$$a(\beta) \mod \beta^m = \sum_{i=0}^{\infty} a_i \beta^i \mod \beta^m = \sum_{i=0}^{m} a_i \beta^i.$$

It can be easily seen that if

$$a(\beta) \mod \beta^m = b(\beta) \mod \beta^m, \quad c(\beta) \mod \beta^m = d(\beta) \mod \beta^m$$

then

$$(a(\beta) \pm c(\beta)) \mod \beta^m = (b(\beta) \pm d(\beta)) \mod \beta^m, \tag{3}$$

$$(a(\beta) \cdot c(\beta)) \mod \beta^m = (b(\beta) \cdot d(\beta)) \mod \beta^m. \tag{4}$$

Now let us look at the formula of the value of a completely mixed matrix game $R(s, \mathbf{v_k}(\beta), \beta)$:

$$\operatorname{val}(R(s, \mathbf{v_k}(\beta), \beta)) = \frac{\det(R(s, \mathbf{v_k}(\beta), \beta))}{\sum_{i,j=1}^{r} \operatorname{adj}(R(s, \mathbf{v_k}(\beta), \beta))[i,j]}. \tag{5}$$

Firstly consider the numerator. Let us use the series representation of the entries of the Shapley matrix. Then from the way of computing the determinant it is easy to see that the numerator can be written as follows:

$$\det(R(s, \mathbf{v_k}(\beta), \beta)) = c_{k,s}^0 + c_{k,s}^1 \beta + c_{k,s}^2 \beta^2 + c_{k,s}^3 \beta^3 + \dots$$

So, it can be represented as a power series expansion.

Now consider the denominator. Similarly, the sum of the elements of the adjugate matrix is the sum of cofactors of the initial matrix, which are (up to sign) determinants of its $r-1$ submatrices. So the denominator can be written as

$$\sum_{i,j=1}^{r} \operatorname{adj}(R(s, \mathbf{v_k}(\beta), \beta))[i,j] = d_{k,s}^0 + d_{k,s}^1 \beta + d_{k,s}^2 \beta^2 + d_{k,s}^3 \beta^3 + \dots$$

Note that from the definition of a completely-mixed game we have $c_{k,s}^0 \neq 0$, $d_{k,s}^0 \neq 0$. Now $v_{k+1}^s(\beta)$ can be expressed as a ratio of two power series:

$$v_{k+1}^s(\beta) = \operatorname{val}(R(s, \mathbf{v_k}(\beta), \beta)) = \frac{c_{k,s}^0 + c_{k,s}^1 \beta + c_{k,s}^2 \beta^2 + c_{k,s}^3 \beta^3 + \dots}{d_{k,s}^0 + d_{k,s}^1 \beta + d_{k,s}^2 \beta^2 + d_{k,s}^3 \beta^3 + \dots} \tag{6}$$

This is one of the key observations in our method. From now we will use the fact that $v_{k+1}^s(\beta)$ can be represented as

$$v_{k+1,s}(\beta) = \frac{P_{k,s}(\beta)}{Q_{k,s}(\beta)},$$

where $P_{k,s}(\beta)$ and $Q_{k,s}(\beta)$ are power series of β:

$$P_{k,s}(\beta) = c^0_{k,s} + c^1_{k,s}\beta + c^2_{k,s}\beta^2 + c^3_{k,s}\beta^3 + ...,$$

and

$$Q_{k,s}(\beta) = d^0_{k,s} + d^1_{k,s}\beta + d^2_{k,s}\beta^2 + d^3_{k,s}\beta^3 + ...$$

We shall show that if at some moment K we have calculated the value function expansion at $\beta = 0$ up to the m-th term, then at the next iteration we get the next $(m+1)$-st term of the expansion. Formally, we can formulate the following statement.

Theorem 3. *Let $\Gamma(\beta)$ be a discounted stochastic game with unfixed discount parameter $\beta \in [0, 1)$ and let $\mathbf{v}(\beta)$ be its value function. Let Assumption 1 hold. Then starting from $\mathbf{v_0} = \underline{0}$, we can obtain K first terms of the Taylor series expansion of $\mathbf{v}(\beta)$ at $\beta_0 = 0$ after K iterations given by (1).*

Proof. Notice that when $\beta = 0$ we have

$$v_{k+1,s}(0) = \frac{c^0_{k,s}}{d^0_{k,s}}.$$

In particular, if $k = 0$ and $\mathbf{v_0} = \underline{0}$, we have that

$$v_{1,s} = \text{val}(R(s, \underline{0}, \beta)) = \text{val}(R(s)) = v_s(0). \tag{7}$$

Let us now consider the derivative of order m of the value function approximation

$$v_{k+1,s}(\beta) = \frac{P_{k,s}(\beta)}{Q_{k,s}(\beta)}.$$

We can think of $v_{k+1,s}(\beta)$ as a rational function of $P_{k,s}(\beta)$ and $Q_{k,s}(\beta)$. Similarly, the derivative $v^{(m)}_{k+1,s}(0)$ is a rational function of $P_{k,s}(0)$, $P^{(1)}_{k,s}(0)$, ..., $P^{(m)}_{k,s}(0)$ and $Q_{k,s}(0)$, $Q^{(1)}_{k,s}(0)$, ..., $Q^{(m)}_{k,s}(0)$.

On the other hand, notice that $P^{(q)}_{k,s}(0) = q! \cdot c^q_{k,s}$ and $Q^{(q)}_{k,s}(0) = q! \cdot d^q_{k,s}$. So, we can say that $v_{k+1,s}{}^{(m)}(0)$ is a rational function of $c^q_{k,s}$ and $d^q_{k,s}$, for $q \in \{0, .., m\}$. Therefore, the first $m+1$ terms of the Taylor series expansion of $v_{k+1,s}(\beta)$ are completely defined by $c^q_{k,s}$ and $d^q_{k,s}$, for $q \in \{0, .., m\}$.

The latter means that if for some $K \in \mathbb{N}$:

$$\forall s \in S,\ \forall o, p > K:$$
$$P_{o,s}(\beta) \mod \beta^m = P_{p,s}(\beta) \mod \beta^m,$$
$$Q_{o,s}(\beta) \mod \beta^m = Q_{p,s}(\beta) \mod \beta^m, \tag{8}$$

then the first $m+1$ terms of the value function series expansion will not change at all on the subsequent iterations:

$$\forall s \in S,\ \forall o, p > K+1: \ v_{o,s}(\beta) \mod \beta^m = v_{p,s}(\beta) \mod \beta^m \tag{9}$$

On the other hand, recall that $c_{k,s}^q$ and $d_{k,s}^q$ for $q \in \{0,..,m\}$ are defined by the corresponding Shapley matrix approximation, $R(s, \mathbf{v}_k(\beta), \beta)[i,j]$. Thus, if K terms are comuted exactly, the $(K+1)$-st term will be exactly derived at the next iteration. Invoking the principle of mathematical induction together with the base (7), we conclude the proof of the theorem. ∎

The complexity of the proposed approach depends on the number of states, on the number of possible actions at each state and, obviously, on the number of terms of the expansion that we are looking for. Assume that we have N states, at each state $s \in \{1,..,N\}$ both of the players have r_s possible actions, and we want to compute K terms of the series expansion of each of the value component. Observe that for each state s at each iteration we have to compute the determinant and the adjugate of an $r_s \times r_s$ matrix. We can do this performing $O(r_s^3)$ operations per iteration using Gaussian elimination. This means that in general our algorithm requires $O(K \cdot \sum_{s=1}^{N} r_s^3)$ operations.

4 Example: Tax Evasion Model

As it has been already mentioned, stochastic games is a powerful tool for modeling different real-life situations. They have applications in economics, evolutionary biology, computer networks etc. In this section we consider a simple model of tax evasion as a simple example of a stochastic game application. This model is inspired by [8], but we slightly simplify it in our work, since this particular application is not the main objective of the present work. We propose a numerical example, which is a two-states completely mixed game, and we find an approximation in terms of power series using our technique described above.

We consider a situation where there are two agents with the opposite interests: the taxpayer and the auditor. The objective of the former is to pay as less as possible, while the latter aims to collect as much money as possible. Each time slot, say, each month, the taxpayer has to decide whether declare his income honestly or not. In his turn, each month the auditor can decide whether to trust the taxpayer or to audit. Hence both of the agents have two possible behaviors. To motivate taxpayers to be honest, the auditor introduces a system of penalties and rewards. Furthermore, there are two different possible states. Normally, the taxpayer is assumed to be honest. In such a situation penalties are lower, and rewards are higher. However, if the auditor suspects (based on the experience of previous months) that the taxpayer is a cheater, then rewards are lower and penalties are higher. Hence we have two different states: when the taxpayer is presumed to be honest and when he is assumed to be suspicious. Since the taxpayer can be "short-sighted" and the auditor can be oblivious or subject to staff mobility, we feel that our setting of small unfixed discount factor is particularly relevant in this model. Formally, we can describe the model as the following stochastic game:

Player 1: auditor;

Player 2: taxpayer;

Pure strategies of Player 1: to audit or not to audit (to trust);

Pure strategies of Player 2: to declare the income honestly or to cheat;

States of the game: State 1 (Good) – the taxpayer is presumed to be honest; State 2 (Bad) – the taxpayer is presumed to cheat;

Payoffs: the amount of money that the taxpayer pays; it is defined by 3 constants: n – normal tax; p – penalty for cheating; r – reward for being honest.

The payoff is accumulated throughout the game with a discount parameter $\beta \in [0, 1)$.

We assume that in a "good" state the taxpayer receives a reward for being honest in case if the auditor decides to check his income. He pays the penalty, if he is found to be guilty. Moreover, in this case the auditor becomes more severe and the game moves to the "bad" state, where the penalty is higher and the reward for being honest is lower. However, if after moving to the "bad" state the taxpayer becomes honest and the auditor notices this, the game moves back to the "good" state. The payoffs and the transition probabilities can be expressed with the help of the following tables:

Table 1. State 1 (Good)

	Be honest	**Cheat**
Audit	$n - r$ (1, 0)	$n + p$ (0, 1)
Trust	n (1, 0)	0 (1, 0)

Let us now consider concrete numerical example. Let $n = 5$, $p = 2$, $r = 2$. Then we have the following Shapley equations:

$$v_1(\beta) = \text{val} \begin{bmatrix} 3 + \beta \cdot v_1(\beta) & 7 + \beta \cdot v_2(\beta) \\ 5 + \beta \cdot v_1(\beta) & 0 + \beta \cdot v_1(\beta) \end{bmatrix},$$

$$v_2(\beta) = \text{val} \begin{bmatrix} 4 + \beta \cdot v_1(\beta) & 9 + \beta \cdot v_2(\beta) \\ 5 + \beta \cdot v_2(\beta) & 0 + \beta \cdot v_2(\beta) \end{bmatrix}$$

It is easy to check that when β is close to zero, both of the games at the right-hand side are completely-mixed. On the one hand, the exact solution of this game is given by the following system of equations:

$$v_1(\beta) = \frac{(3 + \beta v_1(\beta))\beta v_1(\beta) - (7 + \beta v_2(\beta))(5 + \beta v_1(\beta))}{\beta v_1(\beta) - \beta v_2(\beta) - 9},$$

Table 2. State 2 (Bad)

	Be honest	**Cheat**
Audit	$n - r/2$ (1, 0)	$n + 2 \cdot p$ (0, 1)
Trust	n (0, 1)	0 (0, 1)

$$v_2(\beta) = \frac{(4 + \beta v_2(\beta))\beta v_2(\beta) - (9 + \beta v_2(\beta))(5 + \beta v_2(\beta))}{\beta v_1(\beta) - \beta v_2(\beta) - 10}. \quad (10)$$

We see that even in this simple example the exact solution is not easy to compute and to represent analytically.

On the other hand, we can apply our technique to easily find the solution as a series expansion in the neighbourhood of zero. Using the system for symbolic computations *Maple*, we obtain the following segments of the series:

$$v_1(\beta) = 35/9 + 2890/729\beta + 940969/236196\beta^2 + 95458552/23914845\beta^3 + O(\beta^4),$$

$$v_2(\beta) = 9/2 + 1499/360\beta + 593543/145800\beta^2 + 762448589/188956800\beta^3 + O(\beta^4).$$

5 Conclusion and Discussion

Motivated by uncertainty in the value of the interest rate, we study discounted zero-sum stochastic games with unfixed discount factor. Our general goal is to obtain a power series expansion of the value of the game with respect to the discount factor around its nominal value. Even though we could not solve the problem in its full generality, we considered a specific but important class of stochastic games – completely mixed stochastic games. In this class of games we show that the value vector can be expanded as a Taylor series near zero discount factor and provide a generalization of the Shapley iterations to compute an initial segment of the Taylor series. It is very interesting that iterations subsequently produce exact values of the Taylor series coefficients. We illustrate our technique on tax evasion model.

There is a number of very interesting open research questions. In the case of completely mixed games, can the generalized Shapley iterations be adapted to some nominal values of the discount factor different from zero? Our numerical experiments indicate that such generalization is likely to be possible but there is no any more nice term-by-term convergence. Then, how to estimate the radius of convergence of the obtained power series? If the game is not completely mixed, how one can compute the Puiseux series expansion of the value of the game? In the general case, one needs to deal with the fractional power Puiseux series expansions instead of the Taylor series.

Acknowledgement

This work is partly supported by EU Project Congas FP7-ICT-2011-8-317672.

References

1. Altman, E. Avrachenkov K.E. and Filar, J.A.: Asymptotic linear programming and policy improvement for singularly perturbed Markov decision processes. *Math. Meth. Oper. Res.*, **49(1)** (1999) 97–109.
2. Avrachenkov, K.E. Filar J.A. and Howlett, P.G. *Analytic Perturbation Theory and its Applications.* SIAM, 2013.
3. Avrachenkov, K. Cottatellucci, L. and Maggi, L.: Algorithms for uniform optimal strategies in two-player zero-sum stochastic games with perfect information. *Operations Research Letters*, **40(1)** (2012) 56–60.
4. Filar, J.A.: Order field property of stochastic games when the player who controls the transition changes from state to state. *J. Optim. Theory Appl.*, **34** (1981) 505–517.
5. Filar, J.A. and Vrieze, K. *Competitive Markov Decision Processes*, Springer, 1997.
6. Hordijk, A. Dekker, R. and Kallenberg, L.C.M.: Sensitivity-analysis in discounted Markovian decision problems. *Operations-Research-Spektrum*, **7(3)** (1985) 143–151.
7. Neyman, A.: Real algebraic tools in stochastic games. In *Stochastic games and applications*, NATO Science Series C, Mathematical and Physical Sciences, **570**, 2003, 57–75.
8. Raghavan, T.E.S.: A stochastic model of tax evasion. *Annals of the International Society of Dynamic Games*, **8** (2006) 397–420.
9. Shapley, L.S.: Stochastic games. *Proceedings of the National Academy of Sciences*, **39** (1953) 1095–1100.
10. Shapley, L.S. and Snow, R.N.: Basic solutions of discrete games. *Annals of Mathematics Studies*, **24** (1950) 27–35.
11. Szczechla, W. Connell, S. Filar, J. and Vrieze, K.: On the Puisuex series expansion of the limit discount equation of stochastic games. *SIAM J. Control Optim.*, **35(3)** (1994) 860–875.

Real Life Optimization Problems: A CMDP-Based Approach

Alexander Zadorojniy

IBM Research - Haifa
Mount Carmel, Haifa 31905, Israel
zalex@il.ibm.com

Abstract. This work provides a hands-on approach for the application of Markov Decision Processes (MDP) with constraints to real-life problems. We give an overview of the state-of-the-art applications, emphasizing modeling difficulties and algorithmic approaches. We also discuss the differences between theoretical and practical MDP research, and summarize with recommendations on what methods should be used in different circumstances, to provide value to industries and simplify the MDP application process.
Keywords: MDP, Constraints, Real-Life Applications, Modeling.
AMS 2000 subject classification: Primary 90C40, Secondary 80M50

1 Introduction

In this article we consider real-life applications of constrained Markov Decision Process (CMDP) framework. The CMDP framework, especially its non-constrained counterpart (MDP), is a well-known and well-researched framework [2, 21]. MDP first appeared in Shapley's paper in 1953 [22] but received attention following Bellman's publication in 1957 [3, 4]. There has been tremendous theoretical progress over the last six decades [2, 12, 21]. However, in real-life application of MDPs/CMDPs the situation is completely different.

In the 1980s, White [24, 25] published a two-paper survey that noted several real-life MDP applications such as

1. Maintenance problems (road condition vs. requirements).
2. Power generation (production cost vs. sales income).
3. Repair or replace problem (repair vs. replacement costs).

Over time, these applications became iconic in Operations Research [14]. Since then, technology, especially computing power, has progressed dramatically, raising the interesting question of whether an MDP/CMDP framework is useful today for real-life application. Surprisingly, it appears that there are not that many MDP/CMDP real-life applications (see Definition 1). We discuss three of them [26, 23, 18], identify the differences between theoretical approaches and practical ones, and conclude with recommendations that we hope will enable wider CMDP usage in real-life applications.

Note, we do not cover real-life applications in planning and asset management that were successfully treated by the Approximate Dynamic Programming (ADP) technique, which is related to MDP. Powell provides the details about ADP in his book [20].

We use the following definition for real-life application:

Definition 1. *A real-life MDP/CMDP application is an application such that an entity such as a factory, bank, company, etc., either used or is using MDP/CMDP-based solutions on a regular basis.*

This definition short-lists only the MDP/CMDP applications that were proven to bring value to industries, for example in cost reduction, quality improvement, operational management simplification, etc.

In this paper we do not aim to survey all such applications (some of them may be confidential for proprietary reasons). We provide an overview of successful applications that shed light on how to apply a MDP/CMDP framework in real-life.

1.1 Organization

In Section 2 we briefly introduce the definitions of MDP and CMDP. In Section 3 we discuss modeling difficulties that practitioners face and theoreticians typically ignore. In Sections 4, 5, and 6 we summarize CMDP applications for wastewater treatment plants, banking, and tax collection, respectively. We provide recommendations for CMDP application to real-life problems in Section 7. In Section 8, we discuss open problems.

2 CMDP Framework

2.1 Definition of MDP

An MDP [21] is a 4-tuple $\langle X, U, P, c\rangle$, where $X = \{0, \ldots, n-1\}$ is a finite set of *states*, $U = \{0, \ldots, k-1\}$ is a finite set of *actions*, $P : X^2 \times U \to [0,1]$ is a *transition probability function*, and $c : X \times U \to \mathbb{R}$ is a *cost function*. The probability of transition from state x to state y when action u is chosen is specified by the function P and denoted by $P(y|x,u)$. The cost associated with selecting the action u when in state x equals $c(x,u)$. We denote initial states by x_0.

Time is discrete, and in each time unit t, let x_t denote the random variable that equals the state at time t. Similarly, let u_t denote the random variable that equals the action selected at time t. A non-stationary policy is a function $\pi : X \times U \times t \to [0,1]$, such that $\sum_u \pi(x,t,u) = 1$ for every $x \in X$ for each time unit t. A stationary policy is a function $\pi : X \times U \to [0,1]$, such that $\sum_u \pi(x,u) = 1$ for every $x \in X$. A policy controls the action selected in each state as follows: the probability of selecting action u in state x equals $\pi(x,u)$. A policy can be either randomized or deterministic. A randomized policy is a policy with a state x_i for which $\pi(x_i,u) > 0$ for more than one action u. A deterministic policy is a policy where for all states $x \in X$, there is exactly one action $u \in U$ such that $\pi(x,u) = 1$. The initial state together with a policy determine a probability measure on states and actions. The goal is to find a policy that minimizes the cost $C(\pi)$ defined below. We consider a discounted cost model with infinite horizon throughout the paper.

Discounted cost model. In the discounted cost model, the parameter $\beta \in (0,1)$ specifies the rate by which future costs are reduced. Let $P^\pi(x_t = x, u_t = u)$ denote the probability of the event $x_t = x$ and $u_t = u$ when the initial state equals x_0 (once set, remains unchanged and omitted from the notation) and the policy is π. The infinite horizon discounted expected cost $C(\pi)$ is defined by

$$C(\pi) \triangleq (1-\beta) \cdot \sum_{t=0}^{\infty} \beta^t \cdot E^\pi[c(x_t,u_t)]. \tag{1}$$

Occupation measures. Every policy π induces a probability measure over the state-action pairs. We call this probability measure the *occupation measure* corresponding to π and denote it by ρ^π such that $\rho^\pi(x,u) \triangleq (1-\beta)\cdot\sum_{t=0}^{\infty} \beta^t \cdot P^\pi(x_t = x, u_t = u)$ (for simplicity we will omit π from the denotation of ρ).

Given an occupation measure $\rho(x,u)$ over $X \times U$, the policy π^ρ induced by ρ is defined by $\pi^\rho(x,u) \triangleq \rho(x,u)/\sum_{u'} \rho(x,u')$. (Note that if $\sum_{u'} \rho(x,u') = 0$, then one may define $\pi^\rho(x,u)$ arbitrarily as long as $\sum_u \pi^\rho(x,u) = 1$.) A cost can be rewritten using occupation measure notations such as

$$C(\pi) \triangleq \sum_{x \in X, u \in U} c(x,u) \cdot \rho^\pi(x,u). \tag{2}$$

2.2 Constrained Markov Decision Processes (CMDPs)

A constrained MDP [2] is an MDP with additional input consisting of a set of constraints L ($|L| = l$), associated with a cost function vector $\bar{d} : X \times U \to \mathbb{R}^l$ and a vector of parameters $\bar{\alpha}$. The cost $D^i(\pi)$ of π for any constraint $i \in L$ is defined similar to $C(\pi)$, $E^\pi[d^i(s_t,u_t)] = (1-\beta)\cdot\beta^t \sum_{x \in X, u \in U} d^i(s,u)\cdot P^\pi(s_t = s, u_t = u)$. The additional input defines the constraints $\bar{D}(\pi) \le \bar{\alpha}$ that a feasible policy must satisfy. The optimization problem in $CMDP(\bar{\alpha})$ is to find a policy π that minimizes $C(\pi)$ subject to the constraints $\bar{D}(\pi) \le \bar{\alpha}$, where the inequality is interpreted componentwise.

2.3 Linear Programming(LP) Formulation

Publications from the 1960s [9–11, 17] proved that MDP and CMDP can be formulated as an LP problem. They also proved that there is a stationary optimal deterministic policy for MDP and a stationary optimal randomized policy for CMDP. Below 3-7 we show an LP dual formulation for CMDP that appears to be useful in real-life applications. To formulate the LP, we switch to vectorized representation such that c and ρ are vectors of length of $|X| \cdot |U|$, P is a transition probabilities matrix with $|X|$ rows and $|X| \cdot |U|$ columns, I is an identity matrix, $(1-\beta, 0, 0, 0 \ldots, 0)$ is a vector that represents the initial states distribution where $1 - \beta$ corresponds to state x_0, $\bar{d}$ is a matrix with l columns and $|X| \cdot |U|$ rows.

$$\min_{\rho} c^T \cdot \rho \tag{3}$$

$$s.t. \tag{4}$$

$$(I - \beta \cdot P) \cdot \rho = (1 - \beta, 0, 0, 0 \ldots, 0)^T \tag{5}$$

$$\rho \geq 0 \tag{6}$$

$$\bar{d}^T \cdot \rho \leq \bar{\alpha} \tag{7}$$

3 Modeling Difficulties

As opposed to theoretical research, where knowledge of an underlying process is assumed to be known and the problem size is normally not an obstacle [20], in practice state space choice and transitions probability estimation is difficult to resolve. Moreover, problem size is extremely important (e.g., hundreds of variables might be required for state representation) and dealing with this may require a model reduction approximation instead of an optimal solution. In addition, there is a limitation of data from the perspective of both quality and quantity (e.g., the number of sensors is limited and their quality is not ideal). Since most of the practical problems require constraints, this further complicates the problem. Constrained MDPs, even in theory, are less studied and understood.

4 Waste Water Treatment Plant (WWTP) Application

In this section, we discuss a CMDP application from Zadorojniy et al. [26]. They describe a WWTP application [26] of CMDP that includes modeling and solution aspects. The goal of a WWTP is to accept sewage, treat it, and deliver the cleaned water to the closest river or lake. Solid remains are loaded on trucks and taken away from the plant. There are three main parts: the liquid line (LL), the sludge line (SL), and the gas line (GL), as depicted in Figure 1. The LL accepts the sewage, provides initial treatment, and delivers cleaned water that satisfies regulatory requirements to the river or lake. The most important part of the LL is the aeration tank. The aeration tank is primarily regulated by the following controllers: dissolved oxygen (DO) concentration level, chemical dosing (e.g., ferric chloride) for phosphorus removal, internal recycle rate (how much water is fed from the output to the input of the aeration tank), and waste-activated sludge pump rate, which specifies the rate of the water from the LL to the SL. Moreover, these controllers impact the overall process in the plant. The sludge line carries out secondary treatment for the more solid parts of the water. Part of the treated water from the SL output feeds back to the plant's input for additional treatment; another part is loaded on trucks and taken away from the plant. A byproduct of the process is methane gas, which feeds into the GL. This gas is used as fuel for the gas engine in the GL to produce self-generated electricity, which is used as an alternative source of energy.

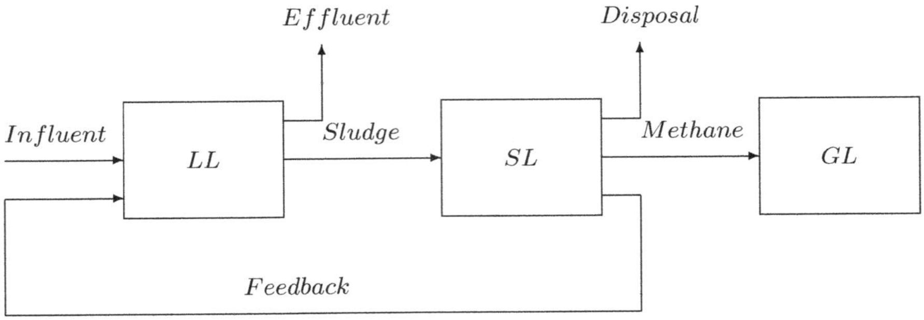

Fig. 1. WWTP Overview. Cost: Electricity, sludge, gas, chemicals; Actions: Blowers speed, various feedbacks, chemical dosing, etc.; Constraints: Effluent quality.

To apply a CMDP framework, a 5-tuple of $\langle X, U, P, c, \bar{d} \rangle$ must be defined. The simplest and most straightforward input to define is the actions space U. In the current application, this is a list of existing controllers that the client is ready to change.

Action Variables:

1. Dissolved Oxygen Set Point: a variable controlling a dissolved oxygen level in the WWTP reactor.
2. Waste Activated Sludge Flow: a variable controlling a pump that regulates a WAS flow from the LL to the SL.
3. Internal Recycle Flow: a variable controlling a pump that regulates a feedback flow inside the LL block.

The action is a vector that consists of the above three variables. Each of the above three variables can get $3-4$ discrete values.

The set of state variables is much trickier. First, it can be either too big (e.g., hundreds of variables) or too small (e.g., only a few variables can be estimated in real-time). Thus, a compromise between the two must be achieved. Zadorojniy et al. [26] showed empirically that the list of variables below is suitable. The state is a vector that consists of all the variables below. The set of all possible realizations of the vector is a state space X. The values for each one of the state variables follow the ':'.

State Variables:

1. Feedback: On and Off.
2. Influent Flow: Discretized influent flow, $3-4$ values.
3. Cost: Discretized total cost, $3-4$ values. Note, this is not standard. The total cost variable was added to reduce the state space (calculation of total cost requires several variables).
4. Electricity Time Period: There are 3 different price intervals over the course of the day. Intervals and prices are fixed per month.
5. Effluent Total Nitrogen (E_TN): Discretized effluent total nitrogen, $3-4$ values.
6. Effluent Total Phosphorus (E_TP): Discretized effluent total phosphorus, $3-4$ values.

Immediate Cost - c. Immediate cost is defined as a sum of electricity consumed, sludge disposal, and chemical dosing minus electricity produced, for each pair of state and action (estimated using the simulator [6]). The immediate cost is defined as an action independent. This is because the actions have no immediate impact, and thus, the last time interval cost is a reasonable approximation for the current immediate cost.

Constraints - $\bar{d}$. There are two conflicting constraints in the model: effluent total phosphorus and total nitrogen daily average concentration levels. Note, that there are more constraints in the permit. However, if total effluent phosphorus and nitrogen constraints are satisfied, the rest of the constraints are satisfied with large margins (based on the simulation results and historical data analysis). Thus, there is no reason to model them explicitly. Let E_TN^i and E_TP^i be the $i's$ values of E_TN and E_TP variables, respectively. Let $X_{E\_TN^i}$ to be $x \in X : x\_E\_TN = E\_TN^i$ and $X_{E\_TP^i}$ to be $x \in X : x\_E\_TP = E\_TP^i$. Thus,

1. $\sum_i d(E\_TN^i) \cdot \rho(E\_TN^i) \leq tn\_avg\_level[mg/l]$,
2. $\sum_i d(E\_TP^i) \cdot \rho(E\_TP^i) \leq tp\_avg\_level[mg/l]$,

where $\rho(E\_TN^i) = \sum_{x' \in X_{E\_TN^i}} \sum_{u \in U} \rho(x', u)$, $d(E\_TN^i) = E\_TN^i$,
where $\rho(E\_TP^i) = \sum_{x' \in X_{E\_TP^i}} \sum_{u \in U} \rho(x', u)$, $d(E\_TP^i) = E\_TP^i$.

Transitions Probability Matrix Generation. The last input that has to be estimated is the transitions probability matrix for the reduced states model defined above. Zadorojniy et al. used a WWTP simulator and the plant simulation model, which was generated for this simulator. They then used a heuristic that drives the simulator to go over as many as possible different states using all available actions; at every state the least used action so far was applied. Since this process does not guarantee the estimation of all required transitions, they also applied an interpolation (see Appendix in Zadorojniy et al. [26]).

CMDP Dual LP Formulation. After defining the required input, Zadorojniy et al. used Dual LP formulation 3-7 for CMDP representation. This formulation is very convenient, since it allows insertion of constraints and a polynomial run-time that is guaranteed to solve it [15, 5].

Actual Usage. The system using the above CMDP was deployed at a medium size European city in the middle of November, 2014. As of August 2015 it remains in use. The state value is estimated using on site sensor readings and regression formulas derived from the simulation model. The recommendations for new controller values are provided every 2 hours, on a 24/7 basis. Initial results showed significant cost reduction, effluent water quality improvement, and better controllability of the process [26].

5 Banking Application

In this section, we discuss a CMDP application from Trench et al. [23], for more efficient credit card management by Bank One Card Services, Inc. In most developed countries, credit cards are widely used and people frequently have more than one such card. These credit cards are usually issued by different banks and each bank is interested in increasing customer usage of its respective card. We use a CMDP framework to do this in the most efficient manner. The idea behind the problem modeling was to capture customer behavior while the bank applies various control strategies, as shown in Fig. 2.

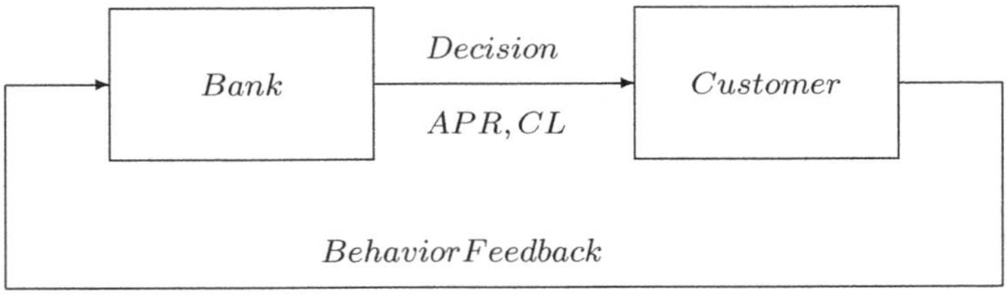

Fig. 2. Banks Credit Cards Management.

As in the previous application, to apply a CMDP a 5-tuple of $\langle X, U, P, c, \bar{d} \rangle$ must be defined.

There are two action variables that the bank can change.

Action Variables:

1. Annual percentage rate (APR).
2. Credit Line Level (CL).

The action is a vector that consists of the above two variables.

State Variables: To obtain state variables, the authors used a regression tree analysis. For proprietary reasons they do not provide the exact variables list they used, only the number of variables, which was six. However, they do explain how this list was generated. The variables list represents customer behavior with respect to risk, card usage, purchasing, payments, etc. The state is a vector, which consists of these variables. The set of all possible realizations of the vector is a state space X. Note, each customer (customer account) is treated independently. The same action is applied for each customer at the same state and the same time.

Both action and state variables were discretized with 5 to 10 values for actions and 2 to 4 values for states variables.

Immediate Cost - c. Immediate cost is defined as a net cash flow (NCF) for a state and action pair.

Constraints - $\bar{d}$. The problem has risk-related constraints (e.g., the bank will refuse to increase the credit line of high-risk customers who may not pay the money back to the bank in the future).

Transitions Probability Matrix Generation. The transitions probability matrix was built based on the real data of 3 million accounts over a period of 18 months. Similar to the previous section application, it was not enough to estimate all the required entries of the matrix. Thus, a heuristic was used to interpolate the missing data. The details of the heuristic are proprietary, therefore they are not revealed in the paper.

Note, that one of the possible interpolation heuristics was presented in the Appendix of Zadorojniy et al. [26].

CMDP Formulation. The problem was formulated as a finite horizon (36 months), discounted cost MDP problem, where constraints were not modeled as a part of the problem. It was solved by a value iteration algorithm (dynamic programming). After the solution was found, the policy was adjusted heuristically to satisfy the constraints (e.g., decreasing credit line will decrease the risk). Obviously, there is no optimality guarantee for the solution. Moreover, the policy can be non-stationary.

Actual Usage. The bank began to use the method in November 2001.

Possible Simplification and Improvement. We think that an infinite horizon discounted CMDP formulation 3-7 should be considered for such a long horizon (36 months). Constraints can be modeled using occupation measures terminology as follows:

1. $\sum_i d(risk^i) \cdot \rho(risk^i) \leq risk\_level.$

By increasing or decreasing the $risk\_level$ parameter a bank can obtain a policy with more or less risk. An optimal policy will be found for MDP with risk constraints satisfied after solving LP program 3-7. Moreover, the optimal policy is stationary.

6 Tax Collection Application

In this section, we discuss a CMDP application from Miller at el. [18]. The objective of this CMDP application is to assist the New York State Department of Taxation and Finance (NYS DTF) to increase tax collection from delinquent tax payers. To collect these taxes, the DTF has to follow legal rules (e.g., what can be done and at what process collection stage), and business rules (e.g., resource limitations to perform certain actions). The DTF acts (e.g., warrant) based on the process state (e.g., available to levy) of the tax collection from a certain customer. The state of the process is pending customer response to DTF actions (e.g., ready to pay after a phone call from DTF) as shown in Figure 3.

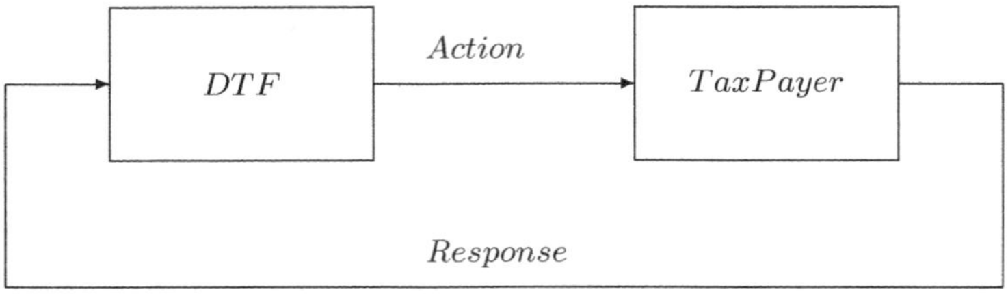

Fig. 3. Tax Collection Process.

As in the previous applications to apply CMDP a 5-tuple of $\langle X, U, P, c, \bar{d} \rangle$ must be defined.

Below are four action variables with possible values after the ':' that are in use by the DTF [18] :

Action Variables:

1. Collections: Contact taxpayer by mail, Contact taxpayer by telephone, Create warrant, Create income execution, Create levy.
2. Movement: Move to district office, Move to high-value team, Move to collection vendors, Move to individual case enforcement.
3. Organization-specific: Perform field visit.
4. Do nothing: Take no action.

The action is a vector that consists of the above variables.

States in this application are stages in the collection process. Although the authors did not use the variables notation, this is convenient for representation. We assign a variable for every stage in the tax collection process. Information after the ':' shows possible realizations of the corresponding variable.

State Variables:

1. Stage 1: Assessment initiated.
2. Stage 2: Assigned to call center.
3. Stage 3: Available to warrant, Install payment.
4. Stage 4: Available to levy, Find financial source, Complete.

The state is a vector consisting of the four above variables. The set of all possible realizations of the vector is a state space X.

The authors note that the actual states size is much bigger and that the above list of states is just exemplary.

Below are two variables with their possible realizations for constraints formulation.

Constraints - $\bar{d}$:

1. Contact rule: A collection letter should not be sent to a taxpayer whose mailing address is invalid. A contact action should only occur for a taxpayer with at least one open mature assessment. A contact by mail must not be made for a taxpayer with an active promise to pay within 30 days.
2. Levy rule: A levy is not allowed for a taxpayer unless the taxpayer has at least one perfected warrant.

CMDP Formulation. The authors formulated the problem as an infinite horizon, discounted cost problem with constraints. They used constrained reinforcement learning approximation to solve the problem [1, 18]. Note, the policy obtained might not be stationary.

Actual Usage. The system with the above CMDP engine became operational in December 2009.

7 Discussion and Recommendations

We gave an overview of real-life applications such as wastewater treatment, banking, and tax collection domains. They were all formulated as CMDPs and solved by different algorithms. The common difficulty was that all the authors had to invest significant effort to resolve modeling and constraint-related issues. The following two paragraphs summarize our recommendations. We hope they will simplify the task of CMDP application to real-life problems.

'Classical' Approach. First, we need to understand the problem (e.g., plant, organization, working concept, and what the customer wants to improve). Second, define the objective function, including what actions impact it, and which of them are available for optimization. Third, note any restrictions (business, legal, etc., constraints). Fourth, we need to understand what state variables are needed and which can be estimated based on available readings. Aggregation of several

variables to one should be considered (cost variable example in Zadorojniy et. al. [26]). Fifth, identify what information is available and what can be obtained from historical data, simulators, theoretical models, and others to be used for transitions probability matrix estimation. To summarize:

1. Define action space.
2. Define state space (the most important variables for actions impact evaluation).
3. Define immediate cost function.
4. Estimate transitions probability matrix.
5. Define constraints.

Given a discounted cost model with infinite horizon (probably the most practical combination), there is a stationary optimal policy. Namely, the optimal policy can be calculated off-line in the back-end and stored at hash well before it is actually applied. (The hash key is a state and hash value is an optimal action for this state.) Then, at the time of application on site it will require $O(1)$ time (average/practical performance) to find the optimal action for any plant state. Since we are considering practical application, the optimal action for a new state value will be provided almost immediately. Moreover, using LP formulation, very efficient existing SW packages can be used [7, 8].

Decomposition and Reinforcement Learning Approaches. When state space is too big even for a reduced model, then decomposition [13, 16, 19, 26] techniques can be considered. The idea of decomposition is to break the problem into smaller ones that can be solved within a reasonable time, speed, and quality to obtain an approximate solution for the original problem. There is no general decomposition technique for MDPs, even without constraints, and no guarantee that an appropriate technique exists or can be easily developed for the desired problem. Therefore, another alternative such as reinforcement learning can be considered [18]. The disadvantages of reinforcement learning are sub-optimal performance, and specific application-driven assumptions, which may be reasonable for one problem but unacceptable in another [1].

8 Open Problems

One of the most crucial and difficult tasks is generation of the transition probability matrix. This process frequently consists of model reduction, obtaining statistics for this model using available data and simulators, and finally the interpolation of missing information to complete the matrix estimation. The important question at the end of the process is: how good is this matrix and how can its quality be measured? Let's begin with the definition of a metric for quality. Let π^*, π^r_{Pideal}, π^r_P be optimal policies with respect to the original model, reduced model (*Pideal* represents an ideally calculated transitions probability matrix for this model) and a reduced model described by the matrix P, and which quality we would like to estimate. Let $\epsilon_q(\pi)$ denote the quality of the matrix P with respect to policy π. Then, we define two quality measures below

Definition 2. $\epsilon_q(\pi^*) = C(\pi^r_P) - C(\pi^*)$,
$\epsilon_q(\pi^r_{Pideal}) = C(\pi^r_P) - C(\pi^r_{Pideal})$

Two questions are of interest: (i) What are the algorithms for generating transitions probability matrix P such that a reasonable upper bound for $\epsilon_q(\pi)$ can be provided? and (ii) Given matrix P, what can we say about its quality?

$\epsilon_q(\pi^*)$ provides the discrepancy of the model actually used with respect to the ideal one, while $\epsilon_q(\pi^r_{Pideal})$ provides the discrepancy of the matrix P compared to the ideal one in the reduced model.

9 Acknowledgment

The author would like to thank Sergey Zeltyn, Segev Wasserkrug, Zohar Feldman, Marco Laumanns, and Adam Shwartz for fruitful discussions.

References

1. Abe, N., Melville, P., Pendus, C., Reddy, C K., Jensen, D L., Thomas, V P., Bennett, J J., Anderson, G F., Cooley, B R., Kowalczyk, M. et al.: Optimizing debt collections using constrained reinforcement learning. In *Proceedings of the 16th ACM SIGKDD International Conference on Knowledge Discovery and Data Mining*, ACM, 2010, 75-84.
2. Altman, E. *Constrained Markov Decision Processes.* Chapman & Hall/CRC Press, Boca Raton, FL, 1999.
3. Bellman, R. *Dynamic Programming.* Princeton University Press, 1957.
4. Bellman, R.: A Markovian decision process. *Technical Report*, DTIC Document, 1957.
5. Bertsimas, D. and Tsitsiklis, J.N. *Introduction to Linear Optimization.* Athena Scientific Belmont, MA, 1997.
6. BIOWIN. *http://envirosim.com/.*
7. CLP. *http://www.coin-or.org/Clp/.*
8. CPLEX. *https://en.wikipedia.org/wiki/CPLEX.*
9. de Ghellinck, G.: Les problemes de decisions sequentielles. *Cahiers Centre Etudes Rech. Operationnelle*, **2** (1960) 161-179.
10. d'Epenoux, F.: A probabilistic production and inventory problem. *Management Science*, 1963, 98-108.
11. Derman, C.: On sequential decisions and Markov chains. *Management Science*, 1962, 16-24.
12. Feinberg, E A and Shwartz, A. *Handbook of Markov Decision Processes: Methods and Applications.* Springer Science & Business Media, 2012.
13. Hauskrecht, M., Meuleau, N., Kaelbling, L P., Dean, T., and Boutilier, C.: Hierarchical solution of Markov decision processes using macro-actions. In *Proceedings of the Fourteenth Conference on Uncertainty in Artificial Intelligence*, Morgan Kaufmann Publishers Inc., 1998, 220-229.

14. Hillier, F S. and Lieberman, G J. *Introduction to Operations Research, 4th ed.* Holden-Day, Inc. San Francisco, CA, USA, 1986.
15. Khachiyan, L G.: A polynomial time algorithm in linear programming. *Soviet Math. Dokl.*, (1979) 191-194.
16. Laroche, P., Boniface, Y., and Schott, R.: A new decomposition technique for solving Markov decision processes. In *Proceedings of the 2001 ACM Symposium on Applied Computing*, ACM, 2001, 12-16.
17. Manne, A S.: Linear programming and sequential decisions. *Management Science*, **6(3)** (1960) 259-267.
18. Miller, G., Weatherwax, M., Gardinier, T., Abe, N., Melville, P., Pendus, C., Jensen, D., Reddy, C K., Thomas, V., Bennett, J. et al.: Tax collections optimization for New York state. *Interfaces*, **42(1)** (2012) 74–84.
19. Parr, R.: Flexible decomposition algorithms for weakly coupled Markov decision problems. In *Proceedings of the Fourteenth Conference on Uncertainty in Artificial Intelligence*, Morgan Kaufmann Publishers Inc., 1998, 422–430.
20. Powell, W B. *Approximate Dynamic Programming: Solving the Curses of Dimensionality.* John Wiley & Sons, 2007.
21. Puterman, M L. *Markov Decision Processes: Discrete Stochastic Dynamic Programming.* John Wiley & Sons, Inc. New York, NY, USA, 1994.
22. Shapley, L S.: Stochastic games. *Proceedings of the National Academy of Sciences of the United States of America*, **39(10)** (1953) 1095-1100.
23. Trench, M S., Pederson, S P., Lau, E T., Ma, L., Wang, H., and Nair, S K.: Managing credit lines and prices for bank one credit cards. *Interfaces*, **33(5)** (2003) 4-21.
24. White, D J.: Real applications of Markov decision processes. *Interfaces*, **15(6)** (1985) 73-83.
25. White, D J.: Further real applications of Markov decision processes. *Interfaces*, **18(5)** (1988) 55-61.
26. Zadorojniy, A., Shwartz, A., Wasserkrug, S., and Zeltyn, S.: Operational optimization of wastewater treatment plants: A CMDP based decomposition approach. *Submitted*, 2015.

Bayesian Inference in Markov Decision Processes

Masayuki Horiguchi

Kanagawa University
Department of Mathematics and Physics, Faculty of Science, Hiratsuka, Kanagawa, 259-1293, Japan
horiguchi@kanagawa-u.ac.jp

Abstract. Multivariate Bayesian control chart is introduced by Makis [12] in which the model is formulated as Markov decision processes associated with partially observable two states. The change of state of system from in-control state to out-of-control state occurs at the exponentially distributed time. At each step of equally intervals the probability of the system being out-of-control is estimated by a sample of multivariate data set. The process of this probability is a continuous time piecewise-deterministic Markov process. In this talk, we consider the unknown case of transition parameter and show that, under the principle of estimation and control, we have the existence of average optimal adaptive policy of control-limit type.
Keywords: Markov decision model, multivariate Bayesian control, Bayesian inference, unknown parameter of transition.
AMS 2000 subject classification: Primary 90C40, Secondary 62F15

1 Introduction

In this paper, we consider the method of quality control based on Bayesian priori-posteriori analysis and the contents are sequel to our preceding study [17].

There are many preceding studies on adaptive quality control by using Bayesian inference (cf.[2, 15, 21]), the effectiveness of those quality control methods is reported in actual practice.

In quality control model based on Bayesian inference, the information accumulated by sampling is used by the decision maker to change the control limit, sample size and sampling intervals so as to adapt the decision to the change of state of system(cf. [21]).

When we design quality control chart, it is important to place an emphasis on either statistical standpoint or economic one, but it is known that they have both merits and demerits each other(cf. [25]).

Therefore, it is considered appropriate for making the control chart based on both statistic and economic standpoint. To realize this management, it needs to consider the model as sequential decision process and reflect their analysis data to control chart. There are many studies on this type (cf. [1, 8, 13, 20, 22, 23]). Recently, Makis[12] considered this approach under the assumption that the state moving from in-control(state 0) to out-of-control (state 1) is occurred with the exponentially distribution and known parameter θ. He consider the multivariate control model as Markov decision process, MDP(cf. [3, 14]) and under the long term average expected cost criterion he derived the existence of optimal control policy, and by this result, he proposed a method of multivariate control charts. The process of the probability of failure(out-of-control) is descrived as continuous time piecewise-deterministic Markov processes([4, 5]).

In [17], we consider "Makis model[12]" with unkown parameter θ, and discussed adaptive optimal policies under the two criteria of the average cost and discounted total expected cost respectively. At the time, in section 2 of that paper, by using MLE we take point estimation to unknown parameter θ. Moreover applying so-called principle of estimation and control(cf. [9, 18]), we proposed a method of adaptive control chart.

There is another approach of our recent work[10] to Markov decision model with unknown transition probability, in which optimal stopping problem is considered by using Bayesian approach.

In this paper, by using the method of Bayesian priori-posteriori analysis we compute a posteriori distribution on the basis of both the observed information of each steps and updating of priori distribution of unknown parameter θ. Applying the limit theorem for posteriori distribution we construct useful adaptive policy and utilize it in order to make control charts to control the system. In section 2, we explain Makis' control model and summarize our results in [17]. In section 3, we consider the optimization problem for adaptive model according to Bayesian priori-posteriori analysis.

2 Bayasian Control Model

In this section, we describe a quality control model treated in this paper and formulate equivalent Bayesian model to the original problem.

Let in-control state of the system be "0" and out-of-control "1". We assume that a transition time from state "0" to "1" is exporonentially distributed with parameter θ. The parameter θ is unknown and $\theta \in \Theta = \{\theta_1, \theta_2, \ldots, \theta_r\}$, where each θ_i is different and positive. Let the random variable of prior distribution of θ be $\widetilde{\theta}$. We denote by X_t the state of system at time $t (t \geqq 0)$.

For each constant $h > 0$, we obtain information on the system by sampling data (q-dimensional data of sample size n) at h time intervals. By this information, the decision maker select a action whether to continue the operation of system (action "0") or to stop the operation and search the system in failure or not(action "1"). When the decision "to stop and search" is taken, we can know whether the state of system is in-cntrol or out-of-control exactly and if the state is known as out-of-control the system is renewaled instantaneously and the process restart from in-control state.

It is noted that by the information of sampling data the decision maker choices either an action "0" or "1" at each decision epoch $ih (i = 1, 2, \ldots)$.

Let the samples of size n and q-dimensional data be

$$Y_i = \begin{bmatrix} \mathbf{y}_1^i \\ \mathbf{y}_2^i \\ \vdots \\ \mathbf{y}_n^i \end{bmatrix}, \quad \mathbf{y}_j^i = \left(y_{j1}^i, y_{j2}^i, \ldots, y_{jq}^i\right) \quad j = 1, 2, \ldots, n. \tag{1}$$

By this data, we have the information about estimation of state of system.

We assume the following:
If $X_{ih} = 0$ (or 1), $\mathbf{y}_1^i, \mathbf{y}_2^i, \ldots, \mathbf{y}_n^i$ are mutually independent and each $\mathbf{y}_j^i$ is identically distributed random variables of $N_q(\mu_0, \Sigma)$ (or $N_q(\mu_1, \Sigma)$), where $N_q(\mu_0, \Sigma)$ and $N_q(\mu_1, \Sigma)$ are q-dimensional normal distribution and Σ is variance covariance matrix(positive) and $\mu_0 = (\mu_{01}, \mu_{02}, \ldots, \mu_{0q}), \mu_1 = (\mu_{11}, \mu_{12}, \ldots, \mu_{1q})$ are mean vector respectively. For M-distance d_1 between μ_1 and μ_0 we assume the following:

$$d_1 := \left[(\mu_1 - \mu_0)\, \Sigma^{-1} (\mu_1 - \mu_0)\right]^{\frac{1}{2}} > 0. \tag{2}$$

Each cost is described as follows:

- investigation cost $A > 0$ to stop and search the system whether there is failure or not.
- renewal cost $R \geqq 0$ from state "1"(out-of-control) to "0"(in-control).
- operating cost $M > 0$ per unit time remaining the system in out-of-control.
- observation cost $b + nc$ $(b, c \geqq 0)$ of sample size n.

This decision process is regarded as partially observable Markov decision process. Therefore, we extend the state space to $S = [0, 1]$ whose state element $p \in S$ is the probability of $X_t = 1$(state of out-of-control at time t) and by Bayes theorem we transfer a priori probability distribution to a posteriori one which describes transition from in-control state to out-of-control state. Then, we have the equivalent completely observable Bayesian model(cf. [9, 24]).

For each decision epoch ih $(i = 1, 2, \ldots)$ of alternatives, we consider the Markov decision process as follows:

$$\begin{cases} S = [0, 1]\text{: state space,} \\ A = \{0, 1\}\text{: action space,} \\ \Theta \subset (-\infty, \infty)\text{: parameter space,} \\ c(p, a)\text{: immediate cost when } p \in S, a \in A. \end{cases}$$

Let sample space be $\overline{\Omega} = \Theta \times \Omega, \Omega = S \times (A \times S)^\infty$ and random variables of process $\tilde{\theta}, \tilde{p}_0, \tilde{a}_0, \tilde{p}_1, \tilde{a}_1, \ldots$. That is, for $\omega \in \overline{\Omega} = (\theta, p_0, a_0, p_1, a_1, p_2, \ldots)$, we have

$\tilde{\theta}(\omega) = \theta, \tilde{p}_0(\omega) = p_0, \tilde{a}_0(\omega) = a_0, \tilde{p}_1(\omega) = p_1, \ldots$, where we set $p_0 = 0$ without loss of generality.

When the state is $\tilde{p}_m = p$ at epoch mh, an action $\tilde{a}_m = 0$(or 1, respectively) is selected and at the next epoch $(m+1)h$ the information $Y_{m+1} = y^{m+1}$ is obtained and at epoch $(m+1)h$ the state moves to

$$\tilde{p}_{m+1} = T(p, y^{m+1}, 0) \text{ (or } T(p, y^{m+1}, 1) \text{ respectively) }, \tag{3}$$

where by Bayes theorem priori-posteriori Bayesian operator T is defined as (Lemma 1 in [12]):

$$\begin{cases} T(p,z,0) = (1-(1-p)e^{-\theta h})h_1(z)/h(z|p), \\ T(p,z,1) = T(0,z,0), \\ \text{where,} \\ z = 2\sum_{j=1}^{n}(\mathbf{y}_j - \mu_0)\Sigma^{-1}(\mu_0 - \mu_1)^T, \\ y = \begin{bmatrix} \mathbf{y}_1 \\ \mathbf{y}_2 \\ \vdots \\ \mathbf{y}_n \end{bmatrix}, \mathbf{y}_j = (y_{j1}, y_{j2}, \ldots, y_{jq}), \\ h(z|p) = (1-(1-p)e^{-\theta h})h_1(z) + (1-p)e^{-\theta h}h_0(z), \\ h_0(z) = N_1(0, 4nd_1^2), h_1(z) = N_1(-2nd_1^2, 4nd_1^2). \end{cases} \tag{4}$$

It is noted that z is sufficient statistics and the transition of states depends only on the value z. When $X_{ih} = 0$(or 1 respectively), z is a sample from one-dimensional normal distribution with mean 0(or $-2nd_1^2$ respectively) and variance $4nd_1^2$.

When θ is true parameter the costs are given below:

$$\begin{cases} c(p,0) = ME(\int_0^h I_{\{X_s=1\}}\,ds) + b + nc = M\left[h - \frac{1-p}{\theta}(1-e^{-\theta h})\right] + b + nc, \\ c(p,1) = c_1(p) + c(0,0), \end{cases} \tag{5}$$

where, $c_1(p) = A + Rp$.

The case of continuously discounted cost with rate $\alpha > 0$ is given below in the similar fashion.

$$\begin{cases} c_\alpha(p,0) = ME(\int_0^h I_{\{X_s=1\}}e^{-\alpha s}\,ds) + e^{-\alpha h}(b+nc) = MB_\theta + \beta(b+nc), \\ c_\alpha(p,1) = c_1(p) + c_\alpha(0,0), \end{cases} \tag{6}$$

where,

$$\begin{cases} B_\theta = \frac{1}{\alpha}\left\{\frac{\theta}{\theta+\alpha}\left(1-e^{-(\theta+\alpha)h}\right) - \beta(1-e^{-\theta h})\right\}, \\ \beta = e^{-\alpha h}. \end{cases} \tag{7}$$

$\pi = (\pi_0, \pi_1, \ldots)$ denotes the policy and $H_m = (\tilde{p}_0, \tilde{a}_0, \tilde{p}_1, \ldots, \tilde{p}_m)$ the history up to epoch mh. The set of policies on H_m at epoch mh is denoted by $\pi_m(H_m) \in A = \{0,1\}(m \geqq 0)$. We denote by Π the set of all policies.

Let $\mathcal{P}(\Theta)$ be the set of all probabilities defined on Θ. For any policy $\pi = (\pi_0, \pi_1, \ldots) \in \Pi$, we define the sequence of $\tau = (\tau_0, \tau_1, \tau_2, \ldots)$ as below:

$$\tau_0 = 0, \tau_k = \min\{k | \pi_k(H_k) = 1, k > \tau_{k-1}\}.$$

It is easily seen that policy π corresponds to the sequence of stopping times τ. Therefore, we identify the policy with corresponding sequence of stopping times in the subsequent description as necessary.

Z_m denotes random variable of observation data at epoch mh. If the state of system $X_{mh} = 0$(in-control), Z_m is a sample from a population with pdf $h_0(z) = N_1(0, 4nd_1^2)(z)$ and if $X_{mh} = 1$(out-of-control), Z_m a sample from a population with pdf $h_1(z) = N_1(-2nd_1^2, 4nd_1^2)(z)$.

Under the condition $\tilde{p}_0 = 0, \tilde{\theta} = \theta$ and

$$a_0 = a_1 = \cdots = a_m = 0, Z_1 = z_1, Z_2 = z_2, \cdots, Z_m = z_m,$$

we denote posteriori probability of failure by δ_m. We show the following lemma.

Lemma 1. *We have*

$$\delta_m = \frac{h_1(z_1, z_2, \ldots, z_m)}{h(z_1, z_2, \ldots, z_m)} \quad \text{and } \tilde{p}_m = \delta_m \ (m \geqq 1), \tag{8}$$

where,

$$\begin{cases} h_1(z_1, \ldots, z_m) = \sum_{l=1}^{m}(e^{-(l-1)\theta h} - e^{-l\theta h})h_0(z_1)\cdots h_0(z_{l-1})h_1(z_l)\cdots h_1(z_m), \\ h(z_1, \ldots, z_m) = h_1(z_1, \ldots, z_m) + e^{-m\theta h}h_0(z_1)\cdots h_0(z_m). \end{cases} \tag{9}$$

Proof. Let the joint pdf of $Z_1, Z_2, \ldots, Z_m$ be $h(z_1, z_2, \ldots, z_m)$. Since the probability of occurring the event of failure between $((l-1)h, lh]$ is $\int_{(l-1)h}^{lh} \theta e^{-\theta t}\, dt = e^{-(l-1)\theta h} - e^{-l\theta h}$, equation (9) holds. By Bayes theorem we obtain δ_m in (8). The latter part of Lemma 1, we show by induction as follows. It holds $\tilde{p}_1 = \delta_1$ obviously. Assume equation (8) holds when $m-1$. From equation (3), we have

$$\tilde{p}_m = \frac{\{\tilde{p}_{m-1}h_1(z_m) + (1-\tilde{p}_{m-1})(1-e^{-\theta h})h_1(z_m)\}}{h(z_m|\tilde{p}_{m-1})}, \tag{10}$$

where

$$h(z_m|\tilde{p}_{m-1}) = \tilde{p}_{m-1}h_1(z_m) + (1-\tilde{p}_{m-1})(1-e^{-\theta h})h_1(z_m) + (1-\tilde{p}_{m-1})e^{-\theta h}h_0(z_m). \tag{11}$$

Substituting $\tilde{p}_{m-1} = \delta_{m-1}$ for equation (11), it holds that

$$
\begin{aligned}
&h(z_1,\ldots,z_{m-1})h(z_m|\tilde{p}_{m-1})\\
&=\sum_{l=1}^{m-1}(e^{l-1}\theta h-e^{-l\theta h})h_0(z_1)\cdots h_0(z_{l-1})h_1(z_l)\cdots h_1(z_{m-1})h_1(z_m)\\
&\quad+(e^{-(m-1)\theta h}-e^{-m\theta h})h_0(z_1)\cdots h_0(z_{m-1})h_1(z_m)\\
&\quad+e^{-m\theta h}h_0(z_1)\cdots h_0(z_m)\\
&=h(z_1,z_2,\ldots,z_m).
\end{aligned}
\tag{12}
$$

Similarly, from equation (10) we have

$$h(z_1,\ldots,z_{m-1})h(z_m|\tilde{p}_{m-1})\tilde{p}_m=h_1(z_1,z_2,\ldots,z_m).$$

Therefore, it is proved that

$$\tilde{p}_m=\frac{h_1(z_1,\ldots,z_m)}{h(z_1,\ldots,z_m)}$$

holds. ■

Average expected cost $\varphi(\pi|\theta,p_0)$ given by $\tilde{\theta}=\theta\in\mathcal{P}(\Theta)$ and initial state distribution $p_0=p\in S$ is defined as follows:

$$\varphi(\pi|\theta,p_0)=\overline{\lim_{k\to\infty}}\frac{1}{E_\pi(\tau_k)}E_\pi\left[\sum_{m=0}^{\tau_k}c(\tilde{p}_m,\tilde{a}_m)\mid\theta,p_0\right], \tag{13}$$

where, $\pi=(\tau_0,\tau_1,\tau_2,\ldots)$ and $E_\pi\left[\cdot|\theta,p\right]$ is expectation with probability measure $P_\pi(\cdot|\theta,p)$ on $\overline{\Omega}$ given by parameters θ,p and policy π.

In addition, define discounted total expected cost $v(\pi|\theta,p_0)$ as follows:

$$v(\pi|\theta,p_0)=\sum_{m=0}^{\infty}\beta^m E_\pi\left[c_\alpha(\tilde{p}_m,\tilde{a}_m)|\theta,p_0\right], \tag{14}$$

where, $\beta=e^{-\alpha h}$ denotes discount factor. Each policy $\pi\in\Pi$ which minimize $\varphi(\pi|\theta,p),v(\pi|\theta,p)$ respectively call θ-average optimal and θ-discounted optimal respectively. We have the following theorems.

Theorem 1 (V. Makis[12]). *If $A+R<\dfrac{M}{\theta}$, there exists θ-average optimal policy $\pi^*$of the control-limit type. That is, there exists $p^*_\theta\in(0,1)$ such that control policy following the decision function $f_\theta:S\to A$ as below is θ-average optimal.*

$$f_\theta(p)=\begin{cases}0 & \text{if } p<p^*_\theta,\\ 1 & \text{if } p\geqq p^*_\theta.\end{cases} \tag{15}$$

Theorem 2 ([17]). *There exists θ-discounted optimal policy of control-limit type, that is, there exists $\overline{p}_\theta \in (0,1)$ such that optimal decision function $g_\theta : S \to A$ is given as below:*

$$g_\theta(p) = \begin{cases} 0 & \text{if } p < \overline{p}_\theta, \\ 1 & \text{if } p \geqq \overline{p}_\theta. \end{cases} \tag{16}$$

3 Adaptive Control Model

In this section, we consider an approach to control the system under Bayesian priori-posteriori analysis. We describe our control model in detail. If an action $a_m = 1$(to stop and search) is taken at m-th epoch, the system is investigated in more detail whether the state is failure or not. If necessarily the system replace renewal and the process resume with the probability of failure state of system is 0. We regard a period between the time of taking "stop" and the next "stop" actions as one cycle. In the process the sates of system repeat these cycles. Hence it is convenient that we consider the sequential approach of calculating a posteriori probability from the information about unknown parameter θ as sequences of units of batch which is done during one cycle.

For the sequences $Z_0, Z_1, \ldots$ of observation data, we assume that the stopping time σ is bounded above, i.e., there exists $M > 0$ such that $P(1 \leqq \sigma \leqq M) = 1$.

When the decision maker selects an action "stop", the value of $X_{\sigma h}$ is known so that the full information between 1st epoch and σth epoch is given by $(Z_1, Z_2, \ldots, Z_\sigma, X_{\sigma h})$. If $\tilde{\theta} = \theta$, the joint pdf for $Z_1 = z_1, Z_2 = z_2, \ldots, Z_\sigma = z_\sigma, X_{\sigma h} = x$ is

$$f(z_1, z_2, \ldots, z_\sigma, x|\theta) = f_0(z_1, z_2, \ldots, z_\sigma|\theta) I_{\{0\}}(x) + f_1(z_1, z_2, \ldots, z_\sigma|\theta) I_{\{1\}}(x), \tag{17}$$

where,

$$f_0(z_1, z_2, \ldots, z_\sigma|\theta) = \int_{\sigma h}^{\infty} \theta e^{-\theta t}\, dt h_0(z_1) \cdots h_0(z_\sigma) = e^{-\sigma\theta h} h_0(z_1) \cdots h_0(z_\sigma), \tag{18}$$

$$\begin{aligned} f_1(z_1, z_2, \ldots, z_\sigma|\theta) &= \sum_{l=1}^{\sigma} \int_{(l-1)h}^{lh} \theta e^{-\theta t}\, dt h_0(z_1) \cdots h_0(z_l) h_1(z_{l+1}) \cdots h_1(z_\sigma) \\ &= \sum_{l=1}^{\sigma} \left(e^{-(l-1)\theta h} - e^{-l\theta h}\right) h_0(z_1) \cdots h_0(z_l) h_1(z_{l+1}) \cdots h_1(z_\sigma). \end{aligned} \tag{19}$$

For the simplicity of notation, we define followings:
For policy of sequence of stopping times $\tau = (\tau_1, \tau_2, \ldots)$,

$$\sigma_l := \tau_l - \tau_{l-1} (l \geqq 1), \tag{20}$$

$$D_{\sigma_l} := (Z_{\tau_{l-1}+1}, Z_{\tau_{l-1}+2}, \ldots, Z_{\tau_l}, X_{\tau_l h}), \tag{21}$$

$$= (Z_{\tau_{l-1}+1}, Z_{\tau_{l-1}+2}, \ldots, Z_{\tau_{l-1}+\sigma_l}, X_{(\tau_{l-1}+\sigma_l)h}), \tag{22}$$

$$D'_{\sigma_l} := (Z_{\tau_{l-1}+1}, \ldots, Z_{\tau_l}). \tag{23}$$

Then, we express $D_{\sigma_l} = (D'_{\sigma_l}, X_{\tau_l h})$.

From equation (17), we have the following.

$$f(D_{\sigma_l}|\theta) = f_0(D_{\sigma_l}|\theta) I_{\{0\}}(X_{\tau_l h}) + f_1(D_{\sigma_l}|\theta) I_{\{1\}}(X_{\tau_l h}).$$

The possible value of D_{σ_l} is denoted by $d_{\sigma_l} = (z_{\tau_{l-1}+1}, z_{\tau_{l-1}+2}, \ldots, z_{\tau_l}, x_{\tau_l h})$. The likelihood function of θ by the possible value $(d_{\sigma_1}, d_{\sigma_2}, \ldots, d_{\sigma_k})$ up to the end of k-th cycle is as follows:

$$p(\theta|d_{\sigma_1}, d_{\sigma_2}, \ldots, d_{\sigma_k}) = \prod_{l=1}^{k} f(d_{\sigma_l}|\theta). \tag{24}$$

Let ρ be the priori distribution of θ. Then, posteriori distribution ρ_k for distribution ρ of $\tilde{\theta}$ is given by following:

$$\rho_k(\theta|d_{\sigma_1}, d_{\sigma_2}, \ldots, d_{\sigma_k}) \propto p(\theta|d_{\sigma_1}, d_{\sigma_2}, \ldots, d_{\sigma_k}) p(\theta), \quad (k \geqq 1) \tag{25}$$

It is obviously that the following recurrence formula holds

$$\begin{cases} \rho_k(\theta|d_{\sigma_1}, d_{\sigma_2}, \ldots, d_{\sigma_k}) \propto f(d_{\sigma_k}|\theta) \rho_{k-1}(\theta|d_{\sigma_1}, d_{\sigma_2}, \ldots, d_{\sigma_{k-1}}), & (k \geqq 2) \\ \rho_1(\theta|d_{\sigma_1}) \propto f(d_{\sigma_1}|\theta) \rho(\theta). \end{cases} \tag{26}$$

For any $k \geqq 1$, we denote by $\overline{\theta}_k$ the mean with ρ_k which is posteriori distribution of θ up to k cycle. For $0 < a < b$, we set $\Theta = [a, b]$ and denote by $\mathcal{P}[a, b]$ the set of all probability distribution on $[a, b]$.

For bounded stopping time σ, we assume separation property as below:

$$P_\tau(f(Z_1, Z_2, \ldots, Z_\sigma, X_{\sigma h}|\theta_1) \neq f(Z_1, Z_2, \ldots, Z_\sigma, X_{\sigma h}|\theta_2)) > 0, \quad (\theta_1 \neq \theta_2), \tag{27}$$

where, P_τ is the probability distribution over the process given by a policy τ. Then, we have the following theorem.

Lemma 2 (cf. Th. 2.4 in [24]). *Let $\rho \in \mathcal{P}[a, b]$. For any bounded measurable function s on $[a, b]$, we have* *(i)* $\lim_{k\to\infty}\int_a^b s(\theta)\rho_k(d\theta) = s(\tilde{\theta}),\quad P_\tau\text{-}a.s.,$

(ii) $\lim_{k\to\infty}\overline{\theta}_k = \lim_{k\to\infty}\int_a^b \theta\rho_k(d\theta) = \tilde{\theta}.\ \ P_\tau\text{-}a.s.$

Let the priori distribution ρ of $\tilde{\theta}$ be $\rho \in \mathcal{P}[a,b]$. We call a policy $\overline{\pi}$ is average optimal adaptive if the following equation holds ρ-a.s.:

$$\varphi(\overline{\pi}|\tilde{\theta},p_0) \leqq \varphi(\pi|\tilde{\theta},p_0),\quad \pi \in \Pi. \tag{28}$$

We denote by $\overline{\pi}^*$ the policy taken $\overline{\theta}_k$-average optimal stopping time $\tau^*(\overline{\theta}_k)$ in $(k+1)$ cycle, where, $\tau^*(\theta)$ is θ-average optimal stopping time derived by the result from Theorem 1. Applying Lemma 2, we have the following:

Theorem 3. *$\overline{\pi}^*$ is average optimal adaptive policy.*

The proof is presented in Appendix (Section 5).

For $k \geqq 1$, we define the following notation:

$$v_k(\pi|\theta,p_0) = E_\pi\left[\sum_{t=k}^{\infty}\beta^{t-k}c(\tilde{p}_t,\tilde{a}_m)\mid p_0,\theta\right].$$

We call $\tilde{\pi} = (\tilde{\pi}_0,\tilde{\pi}_1,\tilde{\pi}_2,\ldots)$ asymptotically discount optimal if the following holds ρ-a.s.:

$$v_{\tilde{\tau}_k}(\tilde{\pi}|\tilde{\theta},p_0) \to v(\tilde{\theta}|p_0)\ \ (k\to\infty), \tag{29}$$

where, π^* is θ-discounted optimal policy.

For any $k \geqq 1$, as similar definition with average optimal stopping time, we denote by $\tilde{\pi}^*$ the policy in $(k+1)$ cycle is taken by $\overline{\theta}_k$-discounted optimal stopping time, where stopping time $\tilde{\tau}(\theta)$ is depending on θ and derived from the results of Theorem 2. Then, we have the following.

Theorem 4. *The policy $\tilde{\pi}^*$ is asymptotically discount optimal.*

The proof is presented in Appendix (Section 5).

Concluding Remarks. An adaptive model treated in this paper is difficult to construct the equivalent general Markov decision process with complete known transition matrices(cf. [16]) since the model haven't Markov property in the relationship to priori-posteriori analysis between each decision epoch. Therefore, in this paper there is noting a method of effective control chart except the method applying the principle of estimation and control.

4 Acknowledgement

The author is thankful to Prof. Alexey B. Piunovskiy for giving me this opportunity, and also to EPSRC whose organization supported our stay and permitting us to valuable time of discussion and improving our study.

5 Appendix

Proof of Theorem 3. We select any positive integer $K > 0$ and fix it. Let C_K be the set of all stopping times for sequence of random variables $Z_1, Z_2, \dots$ and bounded from above by K. For the sequence of failure probability $\tilde{p}_0 = 0, \tilde{p}_1, \tilde{p}_2, \dots$, we define as follows:

$$X_k = c(\tilde{p}_0) + c(\tilde{p}_1) + \cdots + c(\tilde{p}_{k-1}) + c(\tilde{p}_k), \tag{30}$$

where,

$$\begin{cases} c_\alpha(p) = M\left[h - \dfrac{1-p}{\theta}(1 - e^{-\theta h})\right] + b + nc, \\ c_\alpha(p) = A + Rp, \\ \tilde{p}_0 = 0, \tilde{p}_l = T(\tilde{p}_{l-1}, z_{l-1}, 0), l = 1, 2, \dots, \\ \text{the pdf of } z_l \text{ is } h(z|\tilde{p}_{l-1}). \end{cases} \tag{31}$$

Let $\eta(\sigma|\theta) := \dfrac{E(X_\sigma|\theta, p_0)}{E(\sigma|\theta, p_0)}$, where $E(\cdot|\theta, p_0)$ is the expectation given by $p_0 = 0$ and θ is true. Let $\eta(\theta) = \inf_{\sigma \in C_K} \eta(\sigma|\theta)$. Then, for σ^* which satisfies $\eta(\theta) = \eta(\sigma^*|\theta)$, the policy $(\sigma^*, \sigma^*, \dots)$ is θ-average optimal over stopping times on restricted set C_K.

By following recurrence formula, we define the sequence of functions $\{U_m, m \geqq 0\}$.

$$\begin{cases} V_0(p, \lambda|\theta) = -A - RP, \\ V_{m+1}(p, \lambda|\theta) = \max\{-A - RP, \lambda h - (b + mc) \\ \qquad\qquad -M\left[h - \dfrac{1-p}{\theta}(1 - e^{-\theta h})\right] \\ \qquad\qquad + \displaystyle\int V_m(T(p, z, 0), \lambda|\theta) h(z|p) dz\}. \end{cases} \tag{32}$$

We consider all stopping times are restricted on C_K so we define the function $V(\cdot)$ as follows:

$$V(p, \lambda|\theta) := V_K(p, \lambda|\theta). \tag{33}$$

From equation (32), it can be proved by induction that $V(p, \lambda|\theta)$ is continuous in p, λ, θ and monotone increasing in λ. Moreover, we have $V(p, \lambda|\theta) < 0 (\lambda \to -\infty)$, $V(p, \lambda|\theta) \to \infty (\lambda \to \infty)$. Then, there exists $\overline{\lambda}(\theta)$ such that $V(p_0, \overline{\lambda}(\theta)|\theta) =$

0. By λ-maximization technique(cf.[7, 12]) we have $\overline{\lambda}(\theta) = \eta(\theta)$ so that $\overline{\lambda}(\theta)$ is defined uniquely.

Under the condition $a \leqq \theta \leqq b, 0 \leqq p \leqq 1$, by uniform continuity of $V(p, \lambda|\theta)$ and unique solution of $\overline{\lambda}(\theta)$, if $\overline{\theta}_k \to \theta$ then we have $\overline{\lambda}(\overline{\theta}_k \to \theta) = \eta(\overline{\theta}_k) \to \overline{\lambda}(\theta) = \eta(\theta)$. Therefore, for any $\varepsilon > 0$ there exists $k_\varepsilon > 0$ such that the followings holds:

$$\eta(\overline{\theta}_k) \leqq \eta(\theta) + \varepsilon, k \geqq k_\varepsilon. \tag{34}$$

By the definition of average cost criteria(see (13) in Section 2), we have

$$\varphi(\overline{\pi}^*|\theta, p_0) \leqq \eta(\theta) + \varepsilon. \tag{35}$$

Then as $\varepsilon \to 0$, we have

$$\varphi(\overline{\pi}^*|\theta, p_0) \leqq \eta(\theta). \tag{36}$$

Therefore, we get $\varphi(\overline{\pi}^*|\theta, p_0) = \eta(\theta)$ and show that $\overline{\pi}^*$ is average optimal adaptive policy. ■

Proof of Theorem 4. (The proof is done similarly as the proof of Theorem 3): Define

$$D_k = d(\tilde{p}_0) + \beta d(\tilde{p}_1) + \cdots + \beta^{k-1} d(\tilde{p}_{k-1}) + \beta^k d_1(\tilde{p}_k) \quad (\tilde{p}_0 = p), \tag{37}$$

where, $d(p), \tilde{p}_l (l \geqq 1)$ is given by the followings.

$$\begin{cases} d_p = c_\alpha(p, 0) \text{ given by } 5\ , d_1(p) = A + Rp, \\ \tilde{p}_0 = 0, \tilde{p}_l = T(\tilde{p}_{l-1}, z_{l-1}, 0), l = 1, 2, \ldots, \\ \text{the pdf of } z_l \text{ is } h(z|\tilde{p}_{l-1}). \end{cases} \tag{38}$$

For the stationary policy $(\sigma, \sigma, \ldots) = (\sigma)^\infty, \sigma \in C_K$, it holds that

$$v((\sigma)^\infty|\theta, p_0) = E\left[D_\sigma + \beta^\sigma v((\sigma)^\infty|\theta, p_0)\right]. \tag{39}$$

From this equation, we have

$$v((\sigma)^\infty|\theta, p_0) = \frac{E\left[D_\sigma|\theta, p_0\right]}{E\left[1 - \beta^\sigma|\theta, p_0\right]}. \tag{40}$$

Let

$$v((\sigma)^\infty|\theta, p_0) = \inf_{\sigma \in C_K} \frac{E\left[D_\sigma|\theta, p_0\right]}{E\left[1 - \beta^\sigma|\theta, p_0\right]}, \tag{41}$$

then, $v(\theta|p_0)$ is θ-optimal discounted total expected cost.

Now we apply λ-maximization technique to equation (41). We define the sequence $\{v_m, m \geqq 0\}$ by following recurrence:

$$\begin{cases} v_0(p, \lambda|\theta) := -A - Rp, \\ v_{m+1}(p, \lambda|\theta) := \max\{-A - Rp, \\ \qquad \lambda(1 - \beta) - C_\alpha(p, 0) + \beta \displaystyle\int v_{m-1}(T(p, z, 0), \lambda|\theta) h(z|p) dz\}. \end{cases} \tag{42}$$

Let $v(p,\lambda|\theta) := v_K(p,\lambda|\theta)$, then, $v(p,\lambda|\theta)$ is continuous in p,λ,θ and monotone increasing in λ. Moreover, it holds that $v(p,\lambda|\theta) < 0 \quad (\lambda \to -\infty), v(p,\lambda|\theta) \to \infty \quad (\lambda \to \infty)$. Consequently, there exists $\tilde{\lambda}(\theta)$ such that $v(p_0, \tilde{\lambda}(\theta)|\theta) = 0$. Also it holds that $\tilde{\lambda}(\theta)$ is defined uniquely by $\tilde{\lambda}(\theta) = v(\theta|p_0) = 0$. By the uniform continuity of $v(p_0,\lambda|\theta)$, $\overline{\theta}_k \to \theta$ implies that

$$\tilde{\lambda}(\tilde{\theta}_k) = v(\tilde{\theta}_k|p_0) \to \tilde{\lambda}(\theta) = v(\theta|p_0) \quad (k \to \infty). \tag{43}$$

This shows Theorem 4 holds. ■

References

1. Bather, J. A.: Control charts and minimization of costs. *J. Roy. Statist. Soc. Ser. B*, **25** (1963) 49–80.
2. Baxley, R. V., Jr.: An application of variable sampling interval control charts. *Journal of Quality Technology*, **27** (1995) 275–282.
3. Bertsekas, D. P. and Shreve, S. E. *Stochastic Optimal Control: The Discrete Time Case.* Academic Press, NY, 1978.
4. Costa, O. L. V. and Dufour, F. *Continuous Average Control of Piecewise Deterministic Markov Processes.* Springer Briefs in Mathematics. Springer, NY, 2013.
5. Davis, M. H. A. *Markov Models and Optimization.* Chapman & Hall, London, 1993.
6. DeGroot, M. H. *Optimal Statistical Decisions.* McGraw-Hill Book, NY, 1970.
7. Ferguson, T. S. *Optimal Stopping and Applications* (electronic texts). http://www.math.ucla.edu/~tom/Stopping/Contents.html.
8. Girshick, M. A. and Rubin, H.: A Bayes approach to a quality control model. *Ann. Math. Statistics*, **23** (1952) 114–125.
9. Hernández-Lerma, O. *Adaptive Markov Control Processes.* Springer-Verlag, NY, 1989.
10. Horiguchi, M. and Piunovskiy, A. B.: Optimal stopping model with unknown transition probabilities. *Control Cybernet.*, **42** (2013) 593–612.
11. Kurano, M.: Adaptive policies in Markov decision processes with uncertain transition matrices. *J. Inform. Optim. Sci.*, **4** (1983) 21–40.
12. Makis, V.: Multivariate Bayesian control chart. *Oper. Res.*, **56** (2008) 487–496.
13. Porteus, E. L. and Angelus, A.: Opportunities for improved statistical process control. *Management Sci.*, **43** (1997) 1214–1228.
14. Puterman, M. L. *Markov Decision Processes: Discrete Stochastic Dynamic Programming.* John Wiley & Sons, NY, 1994.
15. Reynolds, M. R., Jr., Arnold, J. C., Amin, R. W. and Nachlas, J. A.: $\overline{X}$ charts with variable sampling intervals. *Technometrics*, **30** (1988) 181–192.
16. Rieder, U.: Bayesian dynamic programming. *Advances in Appl. Probability*, **7** (1975) 330–348.
17. Sasaki, M., Horiguchi, M. and Kurano, M.: Adaptive methods for multivariate Bayesian control chart. *RIMS kokyuroku(In Japanese)*, **1912** (2014) 181–192.
18. Sutton, R. S. and Barto, A. G. *Reinforcement Learning: An Introduction.* MIT Press, Cambridge, MA, 1998.
19. Tagaras, G.: A dynamic programming approach to the economic design of X-charts. *IIE Trans.*, **26** (1994) 48–56.

20. Tagaras, G.: Dynamic control charts for finite production runs. *Europian J. Oper. Res.*, **91** (1998) 38–55.
21. Tagaras, G. and Nikolaidis, Y.: Comparing the effectiveness of various Bayesian $\overline{X}$ control charts. *Oper. Res.*, **50** (2002) 878–888.
22. Taylor, H. M.: Markovian sequential replacement processes. *Ann. Math. Statist.*, **36** (1965) 1677–1694.
23. Taylor, H. M.: Statistical control of a Gaussian process. *Technometrics*, **9** (1967) 29–41.
24. van Hee, K. M. *Bayesian Control of Markov Chains.* Mathematisch Centrum, Amsterdam, 1978.
25. Woodall, W. H., Lorenzen, T. J. and Vance, L. C.: Weaknesses of the economic design of control charts. *Technometrics*, **28** (1986) 408–410.

Sufficient Classes of Strategies in Continuous-Time Markov Decision Processes with Total Expected Cost

Alexey B. Piunovskiy

University of Liverpool
Dept. of Mathematical Sciences, Liverpool L69 7ZL, UK
piunov@liv.ac.uk

Abstract. In this paper we introduce a new wider class of control strategies for continuous-time Markov decision processes (MDP) with the total expected cost in such a way that the exponential semi-Markov decision process (ESMDP) is a special case. After that, we describe the sufficient classes of strategies and justify the transformation to the discrete-time MDP.
Keywords: Markov decision process, continuous-time Markov chain, occupation measure, total expected cost, constrained optimisation, convex analytic approach, linear program.
AMS 2000 subject classification: Primary 90C40, Secondary 90C39

1 Introduction

When studying controlled continuous-time Markov chains, two approaches are known. If the (randomised) action can be changed only at the jump epochs, the model is called ESMDP [5, 8, 15]. Starting from [15], another construction based on the article by Jacod [14] became more popular [6, 9, 10, 16, 19]. The latter model is currently called continuous-time MDP (CTMDP). What is important, CTMDP does not cover ESMDP in the sense that randomised, but constant during the sojourn times, actions cannot be described in the framework of traditional CTMDP. The deep connection between CTMDP and ESMDP for the total discounted cost was studied in [5], but again the author had to introduce those models separately. It looks more appropriate to build one model, but with wider class of strategies in such way that some strategies (called below 'randomised') correspond to ESMDP, and the others (called 'relaxed') correspond to the standard strategies in CTMDP. This plan was realised in [20], and the current paper provides some new insights on that model.

It is worth emphasising that the realisations of a relaxed strategy are usually impossible on practice. For a discussion, see [6, p.78]. Roughly speaking, if the decision maker wants to use two actions with non-zero probabilities at each time moment, then the trajectories of the control process are not measurable. On the opposite, randomised strategies are clearly implementable.

In difference from [5, 6], we consider the total undiscounted cost which transforms to the discounted cost in a special case. What is new and important, the simple randomised strategies (called below 'Markov standard ξ-strategies') are no more sufficient in general even if there are no constraints (see Section 5 'Example'). At the same time, the new class of randomised strategies (called below 'Poisson-related') turns to be sufficient without any restrictions in the framework of constrained optimisation. This makes it possible to prove the equivalence of the continuous-time problem with the corresponding discrete-time MDP, where transitions to the same state (loops) are allowed. Remember that in simple cases that equivalence was known long ago through the so called uniformisation technique [22] which is not directly applicable if the transition rate is unbounded. In the case of discounted cost, the transformation to the discrete-time MDP was justified in [6] without any restrictive conditions.

In Section 2, we introduce the model under study and describe the general and particular classes of strategies. Note that the transition rate may be arbitrarily unbounded and non-conservative. The process is studied up to the accumulation of jumps (if it takes place). In Sections 3 and 4, the main theoretical results regarding to occupation measures are presented. An example illustrating the theoretical issues is given in Section 5. In Section 6, we show how one can use the modern theory of discrete-time MDP for solving the underlying (constrained) continuous-time problems. The proofs are postponed to Appendix.

2 Model Description

The following notations are frequently used throughout this paper. $\mathbb{N}$ is the set of natural numbers including zero; $\delta_x(\cdot)$ is the Dirac measure concentrated at x, we call such distributions degenerate; $I\{\cdot\}$ is the indicator function. $\mathcal{B}(E)$ is the Borel σ-algebra of the Borel space E, $\mathcal{P}(E)$ is the Borel space of probability measures on E. $\mathcal{F}_1 \bigvee \mathcal{F}_2$ is the smallest σ-algebra containing the two σ-algebras $\mathcal{F}_1$ and $\mathcal{F}_2$. $\mathbb{R}_+ \triangleq (0,\infty)$, $\mathbb{R}_+^0 \triangleq [0,\infty)$, $\bar{\mathbb{R}} = [-\infty,+\infty]$, $\bar{\mathbb{R}}_+ = (0,\infty]$, $\bar{\mathbb{R}}_+^0 = [0,\infty]$. The abbreviation *w.r.t.* (resp. *a.s.*) stands for "with respect to" (resp. "almost surely"); for $b, d \in \bar{\mathbb{R}}$, $b \wedge d = \min\{b,d\}$, $b^+ \triangleq \max\{b,0\}$ and $b^- \triangleq \min\{b,0\}$. Capital letters denote random variables, and little letters are for their values.

The primitives of a continuous-time Markov decision process (CTMDP) are the following elements.

- State space: $(\mathbf{X}, \mathcal{B}(\mathbf{X}))$ (arbitrary Borel).
- Action space: $(\mathbf{A}, \mathcal{B}(\mathbf{A}))$ (arbitrary Borel), $\mathbf{A}(x) \in \mathcal{B}(\mathbf{A})$ is the non-empty space of admissible actions in state $x \in \mathbf{X}$. It is supposed that $\mathbb{K} \triangleq \{(x,a) \in \mathbf{X} \times \mathbf{A} : \ a \in \mathbf{A}(x)\} \in \mathcal{B}(\mathbf{X} \times \mathbf{A})$ and this set contains the graph of a measurable function from $\mathbf{X}$ to $\mathbf{A}$.

- Transition rate: $q(dy|x,a)$, a signed kernel on $\mathcal{B}(\mathbf{X})$ given $(x,a) \in \mathbb{K}$, taking nonnegative values on $\Gamma_{\mathbf{X}} \setminus \{x\}$ with $\Gamma_{\mathbf{X}} \in \mathcal{B}(\mathbf{X})$. We assume that $q(\mathbf{X}|x,a) \leq 0$ and $\bar{q}_x = \sup_{a \in \mathbf{A}(x)} q_x(a) < \infty$, where $q_x(a) \stackrel{\triangle}{=} -q(\{x\}|x,a)$.
- Cost rates: measurable $\bar{\mathbb{R}}$-valued functions $c_i(x,a)$ on $\mathbb{K}$, $i = 0,1,2,\ldots,N$.
- Initial distribution: $\gamma(\cdot)$, a probability measure on $(\mathbf{X}, \mathcal{B}(\mathbf{X}))$.
- Additional Borel space $(\mathbf{\Xi}, \mathcal{B}(\mathbf{\Xi}))$, the source of the control randomness.

Actually, the space $(\mathbf{\Xi}, \mathcal{B}(\mathbf{\Xi}))$ can be chosen by the decision maker (see Definition 1), but it is convenient to introduce it immediately, in order to describe the sample space. The role of the space $\mathbf{\Xi}$ will become clear after the description of control strategies.

We introduce the artificial isolated point (cemetery) Δ, put $\mathbf{X}_\Delta \stackrel{\triangle}{=} \mathbf{X} \cup \{\Delta\}$, $\mathbf{\Xi}_\Delta = \mathbf{\Xi} \cup \{\Delta\}$, and define $\mathbf{A}(\Delta) \stackrel{\triangle}{=} \mathbf{A}$, $q(\Gamma|\Delta,a) \stackrel{\triangle}{=} 0$ for all $\Gamma \in \mathcal{B}(\mathbf{X}_\Delta)$, $\alpha(x,a) \stackrel{\triangle}{=} q(\{\Delta\}|x,a) \stackrel{\triangle}{=} q_x(a) - q(\mathbf{X} \setminus \{x\}|x,a) \geq 0$ for $(x,a) \in \mathbb{K}$. The state Δ means, the process is over, i.e. escaped from the state space. We also put $c_i(\Delta,a) = 0$.

Given the above primitives, let us construct the underlying (measurable) sample space $(\Omega, \mathcal{F})$. Having firstly defined the measurable space $(\Omega^0, \mathcal{F}^0) \stackrel{\triangle}{=} (\mathbf{\Xi} \times (\mathbf{X} \times \mathbf{\Xi} \times \mathbb{R}_+)^\infty, \mathcal{B}(\mathbf{\Xi} \times (\mathbf{X} \times \mathbf{\Xi} \times \mathbb{R}_+)^\infty))$, let us adjoin all the sequences of the form

$$(\xi_0, x_0,\ \xi_1,\ \theta_1,\ x_1,\ \xi_2,\ \ldots,\ \theta_{m-1},\ x_{m-1},\ \xi_m,\ \theta_m,\ \Delta,\ \Delta,\ \infty,\ \Delta,\ \Delta,\ \ldots)$$

to Ω^0, where $m \geq 1$ is some integer, $\xi_m \in \mathbf{\Xi}$, $\theta_m \in \bar{\mathbb{R}}_+$, $\theta_l \in \mathbb{R}_+$, $x_l \in \mathbf{X}$, $\xi_l \in \mathbf{\Xi}$ for all nonnegative integers $l \leq m-1$. After the corresponding modification of the σ-algebra $\mathcal{F}^0$, we obtain the basic sample space $(\Omega, \mathcal{F})$.

Below,

$$\omega = (\xi_0,\ x_0,\ \xi_1,\ \theta_1,\ x_1,\ \xi_2,\ \theta_2,\ x_2, \ldots).$$

For $n \in \mathbb{N} \setminus \{0\}$, introduce the mapping $\Theta_n : \Omega \to \bar{\mathbb{R}}_+$ by $\Theta_n(\omega) = \theta_n$; for $n \in \mathbb{N}$, the mappings $X_n : \Omega \to \mathbf{X}_\Delta$ and $\Xi_n : \Omega \to \mathbf{\Xi}_\Delta$ are defined by $X_n(\omega) = x_n$ and $\Xi_n(\omega) = \xi_n$. As usual, the argument ω will be often omitted. The increasing sequence of random variables T_n, $n \in \mathbb{N}$ is defined by $T_n = \sum_{i=1}^n \Theta_i$; $T_\infty = \lim_{n\to\infty} T_n$. Here, Θ_n (resp. T_n, X_n) can be understood as the sojourn times (resp. the jump moments, the states of the process on the intervals $[T_n, T_{n+1})$). We do not intend to consider the process after T_∞; the isolated point Δ will be regarded as absorbing; it appears when $\theta_m = \infty$ or when $\theta_m < \infty$ and the jump $x_{m-1} \to \Delta$ is realized with intensity $\alpha(x,a)$. The meaning of the ξ_n components will be described later. Finally, for $n \in \mathbb{N}$,

$$H_n = (\Xi_0, X_0, \Xi_1, \Theta_1, X_1, \ldots, \Xi_n, \Theta_n, X_n)$$

is the n-term (random) history.

The random measure μ is a measure on $\mathbb{R}_+ \times \mathbf{\Xi} \times \mathbf{X}_\Delta$ with values in $\mathbb{N} \cup \{\infty\}$, defined by

$$\mu(\omega;\ \Gamma_{\mathbb{R}} \times \Gamma_{\mathbf{\Xi}} \times \Gamma_{\mathbf{X}}) = \sum_{n \geq 1} I\{T_n(\omega) < \infty\} \delta_{(T_n(\omega), \Xi_n(\omega), X_n(\omega))}(\Gamma_{\mathbb{R}} \times \Gamma_{\mathbf{\Xi}} \times \Gamma_{\mathbf{X}});$$

the right continuous filtration $(\mathcal{F}_t)_{t\in\mathbb{R}^0_+}$ on $(\Omega,\mathcal{F})$ is given by

$$\mathcal{F}_t = \sigma\{H_0\} \vee \sigma\{\mu(]0,s] \times B) : \ s \le t, \ B \in \mathcal{B}(\mathbf{\Xi} \times \mathbf{X}_\Delta)\}.$$

The controlled process of our interest

$$X(\omega,t) \triangleq \sum_{n\ge 0} I\{T_n \le t < T_{n+1}\}X_n + I\{T_\infty \le t\}\Delta$$

takes values in $\mathbf{X}_\Delta$ and is right continuous and adapted. The filtration $\{\mathcal{F}_t\}_{t\ge 0}$ gives rise to the predictable σ-algebra on $\Omega \times \mathbb{R}^0_+$ defined by $\mathcal{P} \triangleq \sigma\{\Gamma \times \{0\} \ (\Gamma \in \mathcal{F}_0), \Gamma \times (s,\infty) \ (\Gamma \in \mathcal{F}_{s-}, s > 0)\}$, where $\mathcal{F}_{s-} \triangleq \bigvee_{t<s} \mathcal{F}_t$.

Definition 1. *A control strategy is defined as follows*

$$S = \{\mathbf{\Xi}, p_0, \langle p_n, \pi_n\rangle, \ n = 1,2,\ldots\},$$

where $p_0(d\xi_0)$ is a probability distribution on $\mathbf{\Xi}$; for $x_{n-1} \in \mathbf{X}$, $p_n(d\xi_n|h_{n-1})$ is a stochastic kernel on $\mathbf{\Xi}$ given $\mathbf{H}_{n-1}$ (the space of $(n-1)$-component histories); $\pi_n(da|h_{n-1},\xi_n,s)$ is a stochastic kernel on $\mathbf{A}(x_{n-1})$ given $\mathbf{H}_{n-1} \times \mathbf{\Xi} \times \mathbb{R}_+$. If $x_{n-1} = \Delta$, then we assume that $p_n(d\xi_n|h_{n-1}) = \delta_\Delta(d\xi_n)$ and $\pi_n(da|h_{n-1},\Delta,s) = \delta_\Delta(da)$.

The p_n components mean the randomizations of controls; the π_n components mean relaxations.

If the randomizations are absent, that is, the kernels π_n do not depend on the ξ-components, then we deal with a relaxed strategy. One can omit the ξ_n components; as a result we obtain the standard control strategy $\{\pi_n, \ n = 1,2,\ldots\}$. Such models were built and investigated by many authors [5, 6, 16, 19].

On the other hand, if the relaxations are absent, that is, all kernels π_n are degenerate and concentrated at singletons

$$\varphi_n(\xi_0, x_0, \theta_1, \ldots, x_{n-1}, \xi_n, s) \in \mathbf{A}(x_{n-1}), \tag{1}$$

then the control process $A(t)$ can be defined like follows

$$A(\omega,t) = \sum_{n\ge 1} I\{T_{n-1} < t \le T_n\}\varphi_n(\Xi_0, X_0, \Xi_1, \Theta_1, \ldots, X_{n-1}, \Xi_n, t - T_{n-1})$$
$$+ I\{T_\infty \le t\}\Delta. \tag{2}$$

Below, we call such (purely randomized) strategies as ξ-strategies; they are defined by sequences $\{\mathbf{\Xi}, p_0, \langle p_n, \varphi_n\rangle, \ n = 1,2,\ldots\}$. According to (2), after the history H_{n-1} is realized, the decision maker flips a coin resulting in the value of Ξ_n having the distribution p_n. Afterwards, up to the next jump epoch T_n, the control $A(t)$ is just a (deterministic measurable) function φ_n.

Definition 2. *ξ-strategies were defined just above. Purely relaxed strategies introduced earlier will be called π-strategies. General strategies S can be called π-ξ-strategies. If $\pi_n(da|x_0,\theta_1,x_1,\theta_2,\ldots,x_{n-1},s) = \pi^M_n(da|x_{n-1},s)$ for all $n = 1,2,\ldots$ then the π-strategy is called Markov.*

Suppose a π-ξ-strategy S is fixed. The dynamics of the controlled process can be described like follows. First of all, $\Xi_0 = \xi_0$ is realized based on the chosen distribution $p_0(d\xi_0)$. If p_0 is a combination of two Dirac measures, then in the future this or that control will be applied: p_0 is responsible for the mixtures of simpler control strategies. After that, the initial state X_0, having the distribution $\gamma(dx)$, is realized. Later, when the realized state $x_{n-1} \in \mathbf{X}$ becomes known at the realized jump epoch t_{n-1} ($n = 1, 2, \ldots$), the dynamics is controlled in the following way. The decision maker flips a coin resulting in the $\Xi_n = \xi_n$ component having distribution $p_n(d\xi_n|h_{n-1})$; after that the stochastic kernel $\pi_n(da|h_{n-1}, \xi_n, s)$ gives rise to the jumps intensity $\lambda_n(\Gamma|h_{n-1}, s)$ from the current state x_{n-1} to $\Gamma \in \mathcal{B}(\mathbf{X}_\Delta)$, where

$$\lambda_n(\Gamma|h_{n-1}, \xi_n, s) = \int_{\mathbf{A}} \pi_n(da|h_{n-1}, \xi_n, s) q(\Gamma \setminus \{x_{n-1}\}|x_{n-1}, a); \tag{3}$$

parameter $s > 0$ is the time interval passed after the jump epoch t_{n-1}. After the corresponding interval θ_n, the new state $x_n \in \mathbf{X}_\Delta$ of the process $X(t)$ is realized at the jump epoch $t_n = t_{n-1} + \theta_n$. The joint distribution of (Θ_n, X_n) is given below. And so on. If $\theta_n = \infty$ then $x_n = \Delta$ and actually the process is over: the triples $(\theta = \infty, \Delta, \Delta)$ will be repeated endlessly. The same happens if $\theta_n < \infty$ and $x_n = \Delta$. Along with the intensity λ_n, we need the following integral

$$\Lambda_n(\Gamma, h_{n-1}, \xi_n, t) = \int_{(0,t]} \lambda_n(\Gamma|h_{n-1}, \xi_n, s) ds. \tag{4}$$

Note that, in case $q_x(a) \geq \varepsilon > 0$, $\Lambda_n(\mathbf{X}_\Delta|h_{n-1}, \xi_n, \infty) = \infty$ if $x_{n-1} \neq \Delta$.

Now, the distribution of $H_0 = (\Xi_0, X_0)$ is given by $p_0(d\xi_0) \cdot \gamma(dx_0)$ and, for any $n \in \mathbb{N} \setminus \{0\}$, the stochastic kernel G_n on $\bar{\mathbb{R}}_+ \times \mathbf{\Xi}_\Delta \times \mathbf{X}_\Delta$ given $\mathbf{H}_{n-1}$ is defined by formulae

$$\begin{aligned}
G_n(\{\infty\} \times \{\Delta\} \times \{\Delta\}|h_{n-1}) &= \delta_{x_{n-1}}(\{\Delta\}); \\
G_n(\{\infty\} \times \Gamma_{\mathbf{\Xi}} \times \{\Delta\}|h_{n-1}) &= \delta_{x_{n-1}}(\mathbf{X}) \int_{\Gamma_{\mathbf{\Xi}}} e^{-\Lambda(\mathbf{X}_\Delta, h_{n-1}, \xi_n, \infty)} p_n(d\xi_n|h_{n-1}); \\
G_n(\Gamma_{\mathbb{R}} \times \Gamma_{\mathbf{\Xi}} \times \Gamma_{\mathbf{X}}|h_{n-1}) &= \delta_{x_{n-1}}(\mathbf{X}) \int_{\Gamma_{\mathbf{\Xi}}} \int_{\Gamma_{\mathbb{R}}} \lambda_n(\Gamma_{\mathbf{X}}|h_{n-1}, \xi_n, t) \\
&\quad \times e^{-\Lambda_n(\mathbf{X}_\Delta, h_{n-1}, \xi_n, t)} dt\, p_n(d\xi_n|h_{n-1}); \\
G_n(\{\infty\} \times \mathbf{\Xi}_\Delta \times \mathbf{X}|h_{n-1}) &= G_n(\mathbb{R}_+ \times \{\Delta\} \times \mathbf{X}_\Delta|h_{n-1}) = 0.
\end{aligned} \tag{5}$$

Here $\Gamma_{\mathbb{R}} \in \mathcal{B}(\mathbb{R}_+)$, $\Gamma_{\mathbf{\Xi}} \in \mathcal{B}(\mathbf{\Xi})$, $\Gamma_{\mathbf{X}} \in \mathcal{B}(\mathbf{X}_\Delta)$.

It remains to apply the induction and Ionescu-Tulcea's theorem [2, Prop.7.28] or [17, p.294] to obtain the probability measure P_γ^S on $(\Omega, \mathcal{F})$ called strategic measure. A more detailed discussion and connection to the martingales, compensator etc [14] can be found in [20].

Below, when $\gamma(\cdot)$ is a Dirac measure concentrated at $x \in \mathbf{X}$, we use the 'degenerated' notation P_x^S. Expectations with respect to P_γ^S and P_x^S are denoted as E_γ^S and E_x^S, respectively. The set of all π-ξ-strategies S will be denoted as

Π_S; the collections of all π- and ξ-strategies will be denoted as Π_π and Π_ξ correspondingly.

We aim to study several classes of control strategies and the associated measures. That is important for stochastic optimal control. For example, one can consider the following problem:

$$\begin{aligned} W_0(S) &= E_\gamma^S\left[\sum_{n=1}^{\infty}\int_{(T_{n-1},T_n]}\int_{\mathbf{A}}\pi_n(da|H_{n-1},\Xi_n,t-T_{n-1})c_0^+(X_{n-1},a)dt\right] \\ &\quad +E_\gamma^S\left[\sum_{n=1}^{\infty}\int_{(T_{n-1},T_n]}\int_{\mathbf{A}}\pi_n(da|H_{n-1},\Xi_n,t-T_{n-1})c_0^-(X_{n-1},a)dt\right] \\ &= E_\gamma^S\left[\int_{(0,T_\infty)}\int_{\mathbf{A}}\pi(da|t)c_0(X(t),a)\,dt\right]\to\inf_{S\in\Pi_S} \qquad (6) \end{aligned}$$

subject to

$$W_i(S)\le d_i,\quad i=1,2,\ldots,N,$$

where all the objectives $W_i(S)$ have the form similar to $W_0(S)$ with function c_0 being replaced with other given cost rates c_i; d_i are given numbers. Here and below, $\infty-\infty\triangleq+\infty$ and

$$\pi(da|t)=\sum_{n=1}^{\infty}I\{T_{n-1}<t\le T_n\}\pi_n(da|H_{n-1},\Xi_n,t-T_{n-1})$$

is the $\mathcal{P}(\mathbf{A})$-valued random process. The notions of optimal and ε-optimal strategies are conventional.

Remark 1. Suppose a strategy S is such that, for some $m\ge 0$, all kernels $\{\pi_n\}_{n=1}^{\infty}$ for $x_{n-1}\ne\Delta$ do not depend on the ξ_m-component. Then one can omit $\xi_m\in\mathbf{\Xi}_\Delta$ and $\Xi_m\in\mathbf{\Xi}_\Delta$ from the consideration. In this case, instead of the strategic measure $P_\gamma^S(d\omega)$, we can everywhere use the marginal $\tilde{P}_\gamma^S(d\tilde{\omega})=P_\gamma^S(d\tilde{\omega}\times\mathbf{\Xi})$. Here

$$\tilde{\omega}=(\xi_0,x_0,\xi_1,\theta_1,\ldots,x_{m-1},\theta_m,x_m,\xi_{m+1},\theta_{m+1},\ldots)$$

and $\tilde{\omega}\times\mathbf{\Xi}=(\xi_0,x_0,\xi_1,\theta_1,\ldots,x_{m-1},\mathbf{\Xi},\theta_m,x_m,\xi_{m+1},\theta_{m+1},\ldots)$. Below, we omit the tilde and hope this will not lead to a confusion.

For example, for a purely relaxed strategy $S\in\Pi_\pi$, the strategic measure is defined on the space of sequences

$$\omega=(x_0,\theta_1,x_1,\ldots),$$

and that is standard for CTMDP [5, 6, 16, 19].

As was mentioned, the space $\mathbf{\Xi}$ can be chosen by the decision maker. Let us look at several possibilities.

Definition 3. *Suppose $\mathbf{\Xi} = \mathbf{A}$, the relaxations are absent, and the functions φ_n in (2) have the form $\varphi_n(h_{n-1}, \xi_n, s) = \xi_n$, so that the argument ξ_0 never appears and thus can be omitted. Then such a strategy will be called a standard ξ-strategy. It will be denoted as $S = \{\mathbf{A},\ p_n,\ n = 1, 2, \ldots\}$ and below we usually write A_n (or a_n) instead of Ξ_n (or ξ_n), $n = 1, 2, \ldots$. If we consider only such strategies then we deal with the so called exponential semi-Markov decision process [5, p.498]. In case $p_n(d\xi_n|h_{n-1}) = p_n(da_n|h_{n-1}) = p_n^M(da_n|x_{n-1})$ $(n = 1, 2, \ldots)$, the standard ξ-strategy will be called Markov. The collection of all Markov standard ξ-strategies will be denoted as Π_ξ^M, they are often denoted as p^m.*

According to Remark 1, slightly modified sample spaces are associated with different types of strategies which are again denoted in different ways. For the reader's convenience, we summarize the main notations in the following table.

Strategy	Sample space
General (π-ξ-strategy) $S = \{\mathbf{\Xi},\ p_0,\ \langle p_n, \pi_n\rangle,\ n = 1, 2, \ldots\} \in \Pi_S$	$\Omega = \{(\xi_0, x_0, \xi_1, \theta_1, x_1, \xi_2, \theta_2, \ldots)\}$
Purely randomized (ξ-strategy) $S = \{\mathbf{\Xi},\ p_0,\ \langle p_n, \varphi_n\rangle,\ n = 1, 2, \ldots\} \in \Pi_\xi$	$\Omega = \{(\xi_0, x_0, \xi_1, \theta_1, x_1, \xi_2, \theta_2, \ldots)\}$
Purely relaxed (π-strategy) $S = \{\pi_n,\ n = 1, 2, \ldots\} \in \Pi_\pi$	$\Omega = \{(x_0, \theta_1, x_1, \theta_2, \ldots)\}$
Markov standard ξ-strategy $S = \{\mathbf{A},\ p_n^M(da_n|x_{n-1}),\ n = 1, 2, \ldots\}$ $= p^M \in \Pi_\xi^M$	$\Omega = \{(x_0, a_1, \theta_1, x_1, a_2, \theta_2, \ldots)\}$

We introduced the new, richer set of strategies Π_S, and one of the targets is to establish the sufficiency of smaller classes Π_π and Π_ξ. More about the model in [20].

3 Occupation Measures and Sufficient Classes of Strategies

Definition 4. *Following [5, 6], for a fixed strategy $S \in \Pi_S$, we introduce the occupation measures for $n = 1, 2, \ldots$:*

$$\eta_n^S(\Gamma_{\mathbf{X}} \times \Gamma_{\mathbf{A}}) = E_\gamma^S\left[\int_{(T_{n-1}, T_n] \cap \mathbb{R}_+} I\{X_{n-1} \in \Gamma_{\mathbf{X}}\}\pi_n(\Gamma_{\mathbf{A}}|H_{n-1}, \Xi_n, t - T_{n-1})dt\right],$$

where $\Gamma_{\mathbf{X}} \in \mathcal{B}(\mathbf{X}), \Gamma_{\mathbf{A}} \in \mathcal{B}(\mathbf{A})$.

Remark 2. If S is a standard ξ-strategy then, for $n = 1, 2, \ldots$

$$\eta_n^S(\Gamma_{\mathbf{X}} \times \Gamma_{\mathbf{A}}) = E_\gamma^S\left[I\{X_{n-1} \in \Gamma_{\mathbf{X}}\}I\{A_n \in \Gamma_{\mathbf{A}}\}\Theta_n\right] = E_\gamma^S[\delta_{X_{n-1}}(\Gamma_{\mathbf{X}})\delta_{A_n}(\Gamma_{\mathbf{A}})\Theta_n],$$

and

$$\int_{\Gamma_{\mathbf{X}}}\int_{\Gamma_{\mathbf{A}}} q_x(a)\eta_n^S(dx,da)$$
$$= E_\gamma^S\left[I\{X_{n-1}\in\Gamma_{\mathbf{X}}\}I\{A_n\in\Gamma_{\mathbf{A}}\}I\{q_{X_{n-1}}(A_n)>0\}E_\gamma^S[\Theta_n|X_{n-1},A_n]\right]$$
$$= E_\gamma^S\left[I\{X_{n-1}\in\Gamma_{\mathbf{X}}\}I\{A_n\in\Gamma_{\mathbf{A}}\}\right]$$

confirming that, e.g., if S is Markov standard then $\sum_{n=1}^\infty q_x(a)\eta_n^S$ coincides with the (total) occupation measure on $\mathbb{K}$ in the discrete-time MDP with the same state and action spaces $\mathbf{X}_\Delta$ and $\mathbf{A}$ and transition probability

$$Q(\Gamma_{\mathbf{X}}|x,a) = I\{q_x(a)=0\}I\{\Gamma_{\mathbf{X}}\ni\Delta\} + I\{q_x(a)>0\}\frac{q(\Gamma_{\mathbf{X}}\setminus\{x\}|x,a)}{q_x(a)},$$

under the control strategy $p_n^M(da|x_{n-1})$. We discuss the relations to the discrete-time MDP in Section 6.

For any non-negative function r, for any $S\in\Pi_S$,

$$E_\gamma^S\left[\sum_{n=1}^\infty\int_{(T_{n-1},T_n]\cap\mathbb{R}_+}\int_{\mathbf{A}}\pi_n(da|H_{n-1},\Xi_n,t-T_{n-1})r(X_{n-1},a)dt\right] \tag{7}$$
$$= \sum_{n=1}^\infty\int_{\mathbf{X}\times\mathbf{A}} r(x,a)\eta_n^S(dx,da).$$

Now, after we introduce the sets
$\mathcal{D}_S = \{\ \{\eta_n^S\}_{n=1}^\infty,\ S\in\Pi_S\}$,
$\mathcal{D}_\pi = \{\ \{\eta_n^S\}_{n=1}^\infty,\ S\in\Pi_\pi,\ S$ is Markov$\}$ and
$\mathcal{D}_\xi = \{\ \{\eta_n^S\}_{n=1}^\infty,\ S\in\Pi_\xi$ with $\mathbf{\Xi}=\mathbf{A}$, ξ-strategy S is Markov standard$\}$,
the problem (6) can be reformulated as

$$\left.\begin{array}{l} \displaystyle\sum_{n=1}^\infty\int_{\mathbf{X}\times\mathbf{A}} c_0(x,a)\eta_n(dx,da) \to \inf_{\{\eta_n\}_{n=1}^\infty\in\mathcal{D}_S} \\ \text{subject to} \\ \displaystyle\sum_{n=1}^\infty\int_{\mathbf{X}\times\mathbf{A}} c_i(x,a)\eta_n(dx,da) \le d_i, \quad i=1,2,\ldots,N. \end{array}\right\}$$

Condition 1. *(a)* $q_x(a)>0$ *for all* $(x,a)\in\mathbb{K}$. *(b)* $\exists\varepsilon>0:\ \forall x\in\mathbf{X}$ $\inf_{a\in\mathbf{A}(x)} q_x(a)\ge\varepsilon$.

It is clear that the possible gap

$$\alpha(x,a) \triangleq q_x(a) - q(\mathbf{X}\setminus\{x\}|x,a) = q(\{\Delta\}|x,a) \ge 0$$

can be understood as the discount factor depending on the current state and action. More about this in [20]. If $\alpha>0$ is a constant then we deal with the classical discounted model [5, 6, 10, 19] satisfying the requirement 1-(b). Certainly, if $q_x(a) = 0$ for some $(x, a) \in \mathbb{K}$, and that state x cannot be reached under any control strategy S, then one can consider the state space $\mathbf{X} \setminus \{x\}$. Similarly, if $q_x(a) \equiv 0$ for all $a \in \mathbf{A}(x)$ and $\forall i = 0, 1, 2, \ldots, N$, $\forall n = 1, 2, \ldots$ $c_i(x, a) \equiv 0$ for all $a \in \mathbf{A}(x)$, then one can denote that state x as Δ (meaning, the process escaped from the state space $\mathbf{X}$). The situation, when $q_x(a) = 0$ and $c_i(x, a) \neq 0$ for a reachable state x and for some i and $a \in \mathbf{A}(x)$, is more delicate.

Theorem 1. *Suppose Condition 1-(a) is satisfied. Then, for any π-ξ-strategy S, there is a Markov standard ξ-strategy S_ξ such that $\eta_n^{S_\xi} \geq \eta_n^S$ for all $n = 1, 2, \ldots$. Hence, Markov standard ξ-strategies are sufficient for solving optimization problem (6) with negative costs c_i.*

If Condition 1-(b) is satisfied, then $\mathcal{D}_S = \mathcal{D}_\xi$. Hence, Markov standard ξ-strategies are sufficient in the problem (6).

Theorem 2. $\mathcal{D}_S = \mathcal{D}_\pi$. *Thus, Markov π-strategies are sufficient in the problem (6).*

The proofs will appear in [20].

If $\eta_n^{S_1} = \eta_n^{S_2}$ for all $n = 1, 2, \ldots$ then, for any cost rate c_0, the expected total costs $W_0(S_1) = W_0(S_2)$ are the same. But other important objectives (e.g. the variances) may be different. Consider the following simple example: $\mathbf{X} = \{1\}$, $\mathbf{A} = \mathbf{A}(1) = \{a_1, a_2\}$, $\gamma(1) = 1$, $q_1(a_1) = \lambda_1$, $q_1(a_2) = \lambda_2$, $N = 0$. Note that $q(\mathbf{X} \setminus \{1\}|1, a) = 0$ and $q(\mathbf{X}|1, a) = -q_1(a) < 0$. After introducing the cemetery Δ with $\alpha(1, a) = q(\{\Delta\}|1, a) = q_1(a)$, we obtain the standard conservative transition rate q. In this model, we have a single sojourn time $\Theta = T$, so that the n index is omitted. Suppose the cost rate $c_0(1, a) = c^a$ is given. Let S_1 be the stationary π-strategy with the values $\pi^s(a_1|x) = \beta = 1 - \pi^s(a_2|1)$. Then $\eta^{S_1} = \eta^{S_2}$ for the Markov standard ξ-strategy S_2 defined by $p(a_1|1) = \frac{\beta\lambda_1}{\beta\lambda_1 + (1-\beta)\lambda_2} = 1 - p(a_2|1)$: see the proof of Theorem 1 in [20]. For S_1, the sample space contains only the pairs $(x_0 = 1, \theta)$, the (random) total cost is $C = [\beta c^{a_1} + (1 - \beta)c^{a_2}]\theta$,

$$E_\gamma^{S_1}[C] = \frac{\beta c^{a_1} + (1 - \beta)c^{a_2}}{\beta\lambda_1 + (1 - \beta)\lambda_2} \text{ and } E_\gamma^{S_1}[C^2] = \frac{2[\beta c^{a_1} + (1 - \beta)c^{a_2}]^2}{[\beta\lambda_1 + (1 - \beta)\lambda_2]^2}.$$

For S_2, the sample space contains the triplets $(x_0 = 1, \xi = a, \theta)$, the total (random) cost is $C = I\{a = a_1\}c^{a_1}\theta + I\{a = a_2\}c^{a_2}\theta$,

$$E_\gamma^{S_2}[C] = \frac{\beta\lambda_1}{\beta\lambda_1 + (1 - \beta)\lambda_2} \cdot \frac{c^{a_1}}{\lambda_1} + \frac{(1 - \beta)\lambda_2}{\beta\lambda_1 + (1 - \beta)\lambda_2} \cdot \frac{c^{a_2}}{\lambda_2} = E_\gamma^{S_1}[C],$$

but

$$E_\gamma^{S_2}[C^2] = \frac{\beta\lambda_1}{\beta\lambda_1 + (1 - \beta)\lambda_2} \cdot \frac{2(c^{a_1})^2}{\lambda_1^2} + \frac{(1 - \beta)\lambda_2}{\beta\lambda_1 + (1 - \beta)\lambda_2} \cdot \frac{2(c^{a_2})^2}{\lambda_2^2}.$$

The difference

$$E_\gamma^{S_2}[C^2] - E_\gamma^{S_2}[C^2] = \frac{2}{\beta\lambda_1 + (1 - \beta)\lambda_2} \left\{ \frac{\beta(c^{a_1})^2}{\lambda_1} + \frac{(1 - \beta)(c^{a_2})^2}{\lambda_2} \right.$$

$$-\frac{[\beta c^{a_1}+(1-\beta)c^{a_2}]^2}{\beta\lambda_1+(1-\beta)\lambda_2}\Bigg\} = \frac{2\beta(1-\beta)[\lambda_2 c^{a_1}-\lambda_1 c^{a_2}]^2}{\lambda_1\lambda_2[\beta\lambda_1+(1-\beta)\lambda_2]^2}$$

is non-negative, so that the variance of the total cost is bigger for the ξ-strategy S_2.

4 Sufficiency of ξ-strategies, General Case

Example presented in Section 5 shows that, if Condition 1 is not satisfied, then it can happen that, for a π-strategy S, there is no equivalent Markov standard ξ-strategy having the same occupation measures. Below, we describe a more general class of ξ-strategies which turns to be sufficient in the general case.

Definition 5. *A Poisson-related ξ-strategy*

$$S = \{\mathbf{\Xi}, \varepsilon, \tilde{p}_{n,k}(da|x_{n-1}),\ n=1,2,\dots,\ k=1,2,\dots\}$$

is defined by a constant $\varepsilon > 0$ and a sequence of stochastic kernels $\tilde{p}_{n,k}(da|x)$ from $\mathbf{X}_\Delta$ to $\mathbf{A}$ with $\tilde{p}_{n,k}(\mathbf{A}(x)|x) = 1$. Here $\mathbf{\Xi} = (\mathbf{A}\times\mathbb{R})^\infty = \{(\alpha_1,\tau_1,\alpha_2,\tau_2,\dots)\}$, and for $n = 1,2,\dots$ the distribution p_n of $\Xi_n = (A_1^n, T_1^n, A_2^n,\dots)$ given $\mathbf{H}_{n-1}$ is defined as follows:

- *for all $k \ge 1$, $p_n(A_k^n \in \Gamma_\mathbf{A}|h_{n-1}) = \tilde{p}_{n,k}(\Gamma_\mathbf{A}|x_{n-1})$;*
- *for all $k \ge 1$, $p_n(T_k^n \le t|h_{n-1}) = 1-e^{-\varepsilon t}$; random variables T_k^n are mutually independent and also independent of $\mathcal{F}_{T_{n-1}} = \mathcal{B}(\mathbf{H}_{n-1})$;*
- *finally,*

$$\varphi_n(\xi_0,x_0,\xi_1,\theta_1,\dots,x_{n-1},\xi_n,s) = \sum_{k=1}^{\infty} I\{\tau_1^n+\dots+\tau_{k-1}^n < s \le \tau_1^n+\dots+\tau_k^n\}\alpha_k^n,$$

and the mapping φ_n in fact depends only on ξ_n.

The Ξ_0 component plays no role and is omitted.

Such a strategy means that, after any jump of the controlled process $X(t)$, we simulate a Poisson process and apply different randomized controls during the different sojourn times of that Poisson process.

Theorem 3. *For any control strategy S, there is a Poisson-related ξ-strategy S^P such that $\{\eta_n^S\}_{n=1}^\infty = \{\eta_n^{S^P}\}_{n=1}^\infty$. The value of $\varepsilon > 0$ can be chosen arbitrarily.*

The proof can be found in [20]. The explicit form of the S^P strategy is given by the following expressions. Suppose $S = \{\mathbf{\Xi}, p_0, \langle p_n,\pi_n\rangle, n=1,2,\dots\} \in \Pi_S$ is a given control strategy and fix an arbitrary $\varepsilon > 0$. The (standard) space $\widetilde{\mathbf{\Xi}} = (\mathbf{A}\times\mathbb{R})^\infty$ which appears in the definition of a Poisson-related strategy, is equipped with tilde. It has no concern to the calculations. For a fixed $n \ge 1$, we introduce random functions $Q_k(w)$ depending on $\omega \in \Omega$:

$$Q_k(w) \triangleq \frac{\varepsilon(\varepsilon w)^{k-1}}{(k-1)!}e^{-\varepsilon w - \Lambda_n(\mathbf{X}_\Delta, H_{n-1}, \Xi_n, w)},\quad k=1,2,\dots,\ w\in\mathbb{R}_+^0$$

and (random) function $f_w(t)$:

$$f_w(t) \stackrel{\triangle}{=} [\lambda_n(\mathbf{X}_\Delta|H_{n-1},\Xi_n,w+t)+\varepsilon]e^{-\Lambda_n(\mathbf{X}_\Delta,H_{n-1},\Xi_n,w+t)+\Lambda_n(\mathbf{X}_\Delta,H_{n-1},\Xi_n,w)-\varepsilon t},$$

$w,t \in \mathbb{R}^0_+$.

Now, the Poisson-related ξ-strategy S^P of our interest is defined by

$$\tilde{p}_{n,1}(\Gamma_\mathbf{A}|x_{n-1}) \stackrel{\triangle}{=} E^S_\gamma\left[\int_{(0,\infty)} f_0(t)\int_{(0,t]}\int_{\Gamma_\mathbf{A}}\pi_n(da|H_{n-1},\Xi_n,u)\right.$$
$$\left.\times\ [q_{X_{n-1}}(a)+\varepsilon]du\ dt|X_{n-1}=x_{n-1}\right];$$

$$\tilde{p}_{n,k}(\Gamma_\mathbf{A}|x_{n-1}) \stackrel{\triangle}{=} \frac{1}{E^S_\gamma\left[\int\limits_{(0,\infty)} Q_{k-1}(w)dw|X_{n-1}=x_{n-1}\right]}$$
$$\times E^S_\gamma\left[\int\limits_{(0,\infty)} Q_{k-1}(w)\int\limits_{(0,\infty)} f_w(t)\int\limits_{(0,t]}\int\limits_{\Gamma_\mathbf{A}}\pi_n(da|H_{n-1},\Xi_n,w+u)\right.$$
$$\left.\times\ [q_{X_{n-1}}(a)+\varepsilon]du\ dt\ dw|X_{n-1}=x_{n-1}\right],$$

for $k\geq 2$.

By the way, the normalizing denominator $E^S_\gamma\left[\int\limits_{(0,\infty)} Q_{k-1}(w)dw|X_{n-1}=x_{n-1}\right]$ equals the P^S_γ-probability and also the $P^{S^P}_\gamma$-probability that Θ_n is bigger than the $Erlang(\varepsilon,k-1)$ RV, i.e. that the action A_k is actually applied when using the S^P strategy.

5 Example

This example illustrates that Markov standard strategies (as well as stationary standard ξ-strategies and stationary π-strategies) are not sufficient in optimization problems.

Consider the following CTMDP, very similar to the one described in [9, Ex.3.1]. $\mathbf{X}=\{1\}$, $\mathbf{A}=\mathbf{A}(1)=(0,1]$, $\gamma(\{1\})=1$, $q_1(a)=a$, $c_0(x,a)=a$, $N=0$. Note that $q(\mathbf{X}\setminus\{1\}|1,a)=0$ and $q(\mathbf{X}|1,a)=-q_1(a)=-a<0$. After introducing the cemetery Δ with $\alpha(1,a)=q(\{\Delta\}|1,a)=q_1(a)$, we obtain the standard conservative transition rate q. In this model, we have a single sojourn time $\Theta=T$, so that the n index is omitted.

It is obvious that, for any Markov standard ξ-strategy p^M (which is also stationary),

$$\eta^{p^M}(\{1\}\times\Gamma_\mathbf{A}) = E^{p^M}_\gamma\left[\int_{(0,T]\cap\mathbb{R}_+} I\{A(t)\in\Gamma_\mathbf{A}\}dt\right] = \int_{\Gamma_\mathbf{A}} p^M(da|1)\cdot\frac{1}{a}$$

and

$$W_0(p^M) = E_\gamma^{p^M}\left[\int_{(0,T]\cap\mathbb{R}_+} A(t)dt\right] = \int_{\mathbf{A}} a\,\eta^{p^M}(\{1\}\times da) = \int_{\mathbf{A}} a\frac{1}{a}p^M(da|1) = 1.$$

For an arbitrary stationary π-strategy S_π, we similarly obtain

$$\eta^{S_\pi}(\{1\}\times\Gamma_{\mathbf{A}}) = \pi(\Gamma_{\mathbf{A}}) \Big/ \int_{\mathbf{A}} a\ \pi(da)$$

and

$$W_0(S_\pi) = \int_{\mathbf{A}} a\ \eta^{S_\pi}(\{1\}\times da) = 1.$$

On the other hand, under an arbitrarily fixed $\kappa > 0$, for the purely deterministic strategy $\varphi(1,s) = e^{-\kappa s}$, the (first) sojourn time $\Theta = T$ has the cumulative distribution function (CDF) $1 - e^{\frac{-1+e^{-\kappa\theta}}{\kappa}}$, so that $P_\gamma^\varphi(\Theta = \infty) = e^{-\frac{1}{\kappa}}$. Under an arbitrarily fixed $U \in (0,1]$ we have

$$\eta^\varphi(\{1\}\times(U,1]) = \int_U^1 \frac{e^{\frac{-1+a}{\kappa}}}{\kappa a}da. \tag{8}$$

The detailed calculation is given in [20]. The measure $\eta^\varphi(\{1\}\times da)$ is absolutely continuous w.r.t. the Lebesgue measure, the density being $\frac{e^{\frac{-1+a}{\kappa}}}{\kappa a}$ and

$$W_0(\varphi) = \int_{\mathbf{A}} a\ \eta^\varphi(\{1\}\times da) = 1 - e^{-\frac{1}{\kappa}}. \tag{9}$$

It is clear that $\inf_{S\in\Pi_S} W_0(S) = 0$: see (9) with $\kappa \to \infty$, but the optimal strategy does not exist because $\Theta > 0$ and $c_0(x,a) > 0$. Note also that, if we extend the action space to $[0,1]$ and keep q_1 and c_0 continuous, i.e., $q_1(0) = c_0(0) = 0$, then stationary deterministic strategy $\varphi^*(x) = 0$ is optimal with $W_0(\varphi^*) = 0$.

According to Theorem 1, there is a Markov standard ξ-strategy S_ξ such that $\eta^{S_\xi} \ge \eta^\varphi$. It is given by the following formula:

$$P^M((U,1]|1) = \frac{E_\gamma^\varphi\left[\int_{(0,\Theta]} I\{e^{-\kappa t}\in(U,1]\}e^{-\kappa t}dt\right]}{E_\gamma^\varphi\left[\int_{(0,\Theta]} e^{-\kappa t}dt\right]}.$$

After the change of variables $y = e^{-\kappa t}$, the numerator becomes

$$E_\gamma^\varphi\left[\int_{[e^{-\kappa\Theta},1)} I\{y\in(U,1]\}\frac{dy}{\kappa}\right] = 1 - e^{\frac{U-1}{\kappa}}$$

and

$$P^M((U,1]|1) = \frac{1-e^{\frac{U-1}{\kappa}}}{1-e^{-\frac{1}{\kappa}}} = \int_U^1 \frac{\frac{1}{\kappa}e^{\frac{a-1}{\kappa}}}{1-e^{-\frac{1}{\kappa}}}da.$$

Now, since for any $a \in \mathbf{A}$ the expectation of Θ is $\frac{1}{a}$,

$$\eta^{S_\varepsilon}(\{1\} \times \Gamma_{\mathbf{A}}) = \int_{\Gamma_{\mathbf{A}}} \frac{\frac{1}{\kappa a} e^{\frac{a-1}{\kappa}}}{1 - e^{-\frac{1}{\kappa}}} da \geq \int_{\Gamma_{\mathbf{A}}} e^{\frac{a-1}{\kappa}} \frac{1}{\kappa a} da = \eta^{\varphi}(\{1\} \times \Gamma_{\mathbf{A}}).$$

Let us construct the Poisson-related ξ-strategy S^P such that $\eta^{S^P} = \eta^{\varphi}$, using the expressions given at the end of Section 4.

As usual, we omit index $n = 1$. Now $\Lambda(\mathbf{X}_\Delta, h_0, \xi_0, t) = \int_0^t e^{-\kappa s} ds = \frac{1-e^{-\kappa t}}{\kappa}$ and, using the formulae for $Q_k(w)$ and $f_w(t)$, we obtain

$$\begin{aligned}
\tilde{p}_1((U,1]|1) &= \int_0^\infty \left[\int_0^t I\{e^{-\kappa s} \in (U,1]\}[e^{-\kappa s} + \varepsilon] ds\right] (e^{-\kappa t} + \varepsilon) e^{\frac{-1+e^{-\kappa t}}{\kappa} - \varepsilon t} dt \\
&= \int_0^{-\frac{\ln U}{\kappa}} \left[\frac{1}{\kappa}(1 - e^{-\kappa t}) + \varepsilon t\right] (e^{-\kappa t} + \varepsilon) e^{\frac{-1+e^{-\kappa t}}{\kappa} - \varepsilon t} dt \\
&\quad + \frac{1}{\kappa} \int_{-\frac{\ln U}{\kappa}}^\infty [1 - U - \varepsilon \ln U](e^{-\kappa t} + \varepsilon) e^{\frac{-1+e^{-\kappa t}}{\kappa} - \varepsilon t} dt,
\end{aligned}$$

and the density of the $\tilde{p}_1$ distribution is given by

$$-\frac{d\tilde{p}_1((u,1]|1)}{du} = \frac{1}{\kappa}\left(1 + \frac{\varepsilon}{u}\right) u^{\frac{\varepsilon}{\kappa}} e^{\frac{-1+u}{\kappa}}.$$

The starting point for the description of the desired Poisson-related strategy S^P is as follows.

- On the interval $(0, T_1]$ one should choose the action A_1 using the CDF $a^{\frac{\varepsilon}{\kappa}} e^{\frac{-1+a}{\kappa}}$, $a \in (0, 1] = \mathbf{A}$.
- The expected cost on the interval $(0, T_1 \wedge \Theta]$ equals

$$\frac{1}{\kappa} \int_0^1 \left(1 + \frac{\varepsilon}{a}\right) a^{\frac{\varepsilon}{\kappa}} e^{\frac{a-1}{\kappa}} \cdot \frac{a}{a + \varepsilon} da = \frac{1}{\kappa} \int_0^1 a^{\frac{\varepsilon}{\kappa}} e^{\frac{a-1}{\kappa}} da.$$

For $k \geq 2$, we have

$$\tilde{p}_k((U,1]|1) = \frac{\int_0^{\frac{-\ln U}{\kappa}} \frac{\varepsilon(\varepsilon w)^{k-2}}{(k-2)!} e^{-\varepsilon w} e^{\frac{-1+e^{-\kappa w}}{\kappa}} [1 - e^{\frac{U - e^{-\kappa w}}{\kappa} + \varepsilon w + \frac{\varepsilon}{\kappa} \ln U}] dw}{\int_0^\infty \frac{\varepsilon(\varepsilon w)^{k-2}}{(k-2)!} e^{-\varepsilon w} e^{\frac{-1+e^{-\kappa w}}{\kappa}} dw},$$

and the desired Poisson-related strategy S^P is as follows.

- On the interval $\left(\sum_{i=1}^{k-1} T_i, \sum_{i=1}^k T_i\right]$ one should choose the action A_k using probability density

$$\begin{aligned}
-\frac{d\tilde{p}_k((a,1]|1)}{da} &= \frac{\varepsilon \int_0^{-\frac{\ln a}{\kappa}} \frac{(\varepsilon w)^{k-2}}{(k-2)!} e^{-\frac{1}{\kappa}} \left(\frac{1}{\kappa} + \frac{\varepsilon}{\kappa a}\right) e^{\frac{a}{\kappa} + \frac{\varepsilon}{\kappa} \ln a} dw}{\int_0^\infty \frac{\varepsilon(\varepsilon w)^{k-2}}{(k-2)!} e^{-\varepsilon w} e^{\frac{-1+e^{-\kappa w}}{\kappa}} dw} \\
&= \frac{\frac{a+\varepsilon}{a\kappa(k-1)!} \left(\frac{-\varepsilon \ln a}{\kappa}\right)^{k-1} e^{\frac{a}{\kappa} + \frac{\varepsilon \ln a}{\kappa} - \frac{1}{\kappa}}}{\int_0^\infty \frac{\varepsilon(\varepsilon w)^{k-2}}{(k-2)!} e^{-\varepsilon w} e^{\frac{-1+e^{-\kappa w}}{\kappa}} dw}.
\end{aligned}$$

- If the action A_k is actually applied then the duration $\tilde{T}_k$ is the smallest RV between the sojourn time (in state 1 under the action A_k) and the independent $\exp(\varepsilon)$ random variable T_k. If A_k is not applied, we put $\tilde{T}_k = 0$. The expected length of that interval $\tilde{T}_k$ (if positive, that is, with probability $\int_0^\infty \frac{\varepsilon(\varepsilon w)^{k-2}}{(k-2)!} e^{-\varepsilon w} e^{\frac{-1+e^{-\kappa w}}{\kappa}} dw$: see the proof of Th.5 in [20]) equals $\frac{1}{A_k+\varepsilon}$ and

$$E_\gamma^{S^P}\left[\int_{(0,\tilde{T}_k]} I\{A_k \in (U,1]\}dt\right] = \int_U^1 \frac{1}{\kappa a} e^{\frac{-1+\varepsilon \ln a + a}{\kappa}} \frac{\left(\frac{-\varepsilon \ln a}{\kappa}\right)^{k-1}}{(k-1)!} da.$$

- The expected cost on that interval equals $\int_0^1 \frac{1}{\kappa} e^{\frac{-1+\varepsilon \ln a + a}{\kappa}} \frac{\left(\frac{-\varepsilon \ln a}{\kappa}\right)^{k-1}}{(k-1)!} da.$

One can easily compute

$$\begin{aligned}
\eta^{S^P}(\{1\} \times (U,1]) &= \sum_{k=1}^\infty E_\gamma^{S^P}\left[\int_{(0,\tilde{T}_k]} I\{A_k \in (U,1]\}dt\right] \\
&= \int_U^1 \frac{1}{\kappa} a^{\frac{\varepsilon}{\kappa}-1} e^{\frac{a-1}{\kappa}} da + \int_U^1 \frac{1}{\kappa} a^{\frac{\varepsilon}{\kappa}-1} e^{\frac{a-1}{\kappa}} \left(a^{-\frac{\varepsilon}{\kappa}} - 1\right) da \\
&= \int_U^1 \frac{e^{\frac{a-1}{\kappa}}}{\kappa a} da = \eta^\varphi(\{1\} \times (U,1]) :
\end{aligned}$$

see (8).

Similarly,

$$\begin{aligned}
W_0(S^P) &= \frac{1}{\kappa} \int_0^1 a^{\frac{\varepsilon}{\kappa}} e^{\frac{a-1}{\kappa}} da + \int_0^1 \frac{1}{\kappa} a^{\frac{\varepsilon}{\kappa}} e^{\frac{a-1}{\kappa}} \left(a^{-\frac{\varepsilon}{\kappa}} - 1\right) da \\
&= \int_0^1 \frac{1}{\kappa} e^{\frac{a-1}{\kappa}} da = 1 - e^{-\frac{1}{\kappa}} = W_0(\varphi) :
\end{aligned}$$

see (9).

6 Continuous and Discrete-Time MDP

6.1 Non-Zero Jumps Intensity

Suppose Condition 1-(b) is satisfied (or Condition 1-(a) if the cost rates c_i are negative). Then, according to Theorem 1, Markov standard ξ-strategies are sufficient in problem (6). Formula (6) takes the form

$$\begin{aligned}
W_0(S) &= \sum_{n=1}^\infty E_\gamma^S \left[I\{X_{n-1} \neq \Delta\} E_\gamma^S \left[c_0(X_{n-1}, A_n)\Theta_n | \mathcal{F}_{T_{n-1}}\right]\right] \qquad (10) \\
&= \sum_{n=1}^\infty E_\gamma^S \left[I\{X_{n-1} \neq \Delta\} \int_{\mathbf{A}} \frac{c_0(X_{n-1}, a)}{q_{X_{n-1}}(a)} p_n^M(da|X_{n-1})\right] \to \inf_{S \in \Pi_\xi^M} .
\end{aligned}$$

It remains to notice that $\forall \Gamma_{\mathbf{X}} \in \mathcal{B}(\mathbf{X}_\Delta)$

$$P_\gamma^S(X_n \in \Gamma_{\mathbf{X}} | \mathcal{F}_{T_{n-1}}) = \int_{\mathbf{A}} \frac{q(\Gamma_{\mathbf{X}} \setminus \{X_{n-1}\} | X_{n-1}, a)}{q_{X_{n-1}}(a)} p_n^M(da|X_{n-1}) \tag{11}$$

to deduce that actually we deal with a discrete-time MDP in the class of randomized Markov control strategies p^M. Indeed, at any one time moment n, having the current state $x_{n-1} \in \mathbf{X}$ and choosing action $a \in \mathbf{A}(x_{n-1})$, we face the one-step cost $c_0(x_{n-1}, a)/q_{x_{n-1}}(a)$, and the process moves to a state $x_n \in \Gamma_{\mathbf{X}} \in \mathcal{B}(\mathbf{X}_\Delta)$ with probability $q(\Gamma_{\mathbf{X}} \setminus \{x_{n-1}\} | x_{n-1}, a)/q_{x_{n-1}}(a)$. State Δ is absorbing with zero one-step cost. It is known that randomized Markov control strategies are sufficient for solving discrete-time problems with the total expected cost [17, Lemma 2].

The optimality equation looks as follows

$$\inf_{a \in \mathbf{A}(x)} \left\{ c_0(x,a)/q_x(a) + \int_{\mathbf{X}\setminus\{x\}} v(y) q(dy|x,a)/q_x(a) - v(x) \right\} = 0, \quad x \in \mathbf{X}_\Delta, \tag{12}$$

and all the theory of discrete-time MDP is applicable.

Remark 3. If $c_0(x,a) \geq 0$ and the model is semi-continuous (see [2, Def.8.7], or [3, Ass.2.1], or [17, Con.5]) then the Bellman function $v(x) = \inf_{p^M \in \Pi_\xi^M} W_0(p^M)$, where x is the initial state (i.e. $\gamma(dy) = \delta_x(dy)$), is the minimal non-negative solution to (12). Moreover, there exists a stationary deterministic uniformly (or persistently) optimal strategy φ^*, that is, a strategy satisfying $v(x) = W_0(\varphi^*)$ for all initial states $x \in \mathbf{X}$. The (measurable) mapping $\varphi^* : \mathbf{X}_\Delta \to \mathbf{A}$ provides the infimum in (12) [2, Prop.9.12, Cor.9.17.2]. One can find more about the total-cost MDP in [1,2,7,13] and other monographs and articles.

Note that MDP with total (undiscounted) expected cost is a challenging area, full of unexpected: strategy and value iterations may be unsuccessful, a conserving strategy (providing the infimum in (12)) may be not optimal, and so on: see the corresponding counter-examples in [18, Ch.2]. At the same time, particular cases, like transient and discounted models are well studied [1,2,13, 17]. For instance, in the standard discounted case, if $\alpha(x,a) = \alpha > 0$, the model is semi-continuous, and the cost rate c_0 is bounded, then equation (12) has a single bounded solution on $\mathbf{X}$ with $v(\Delta) = 0$, and the stationary conserving strategy exists and is optimal. By the way, here equation (12) takes the form

$$\inf_{a \in \mathbf{A}(x)} \left\{ c_0(x,a) + \int_{\mathbf{X}\setminus\{x\}} v(y) q(dy|x,a) - q(\mathbf{X} \setminus \{x\}|x,a) v(x) - \alpha v(x) \right\} = 0, \quad x \in \mathbf{X},$$

coincident with the Bellman equation investigated in many works on CTMDP [10, 19, 21]. Note that the discount factor was state -dependent in [23]. One can investigate also the case when the cost rate c_0 is not necessarily bounded, working in the spaces with 'weighted' norms. Similar approach was demonstrated in [10, 19, 21] for CTMDP and in [1, 13], [22, §6.10] for discrete-time MDP.

In the cited works on CTMDP, many efforts were made to ensure that the controlled process is non-explosive, that is, $P_\gamma^S(T_\infty = \infty) = 1$ for all strategies S. We underline here that explosions are not excluded in the current article: we simply consider the $X(t)$ process up to the moment T_∞ which may be finite.

Let us apply the recent results on constrained discrete-time MDP with total expected cost [4] to the problem (6).

Condition 2. *(a) There exists a dominating probability measure m on $\mathbf{X}$: $\forall (x,a) \in \mathbb{K}$ $q(\cdot|x,a) \ll m$. (Here the measure $q(\cdot|x,a)$ is considered to be defined on $\mathbf{X} \setminus \{x\}$ for $(x,a) \in \mathbb{K}$.)*

(b) $\mathbf{A}$ is compact, $\forall x \in \mathbf{X}$ $\mathbf{A}(x) = \mathbf{A}$; for any $\Gamma_\mathbf{X} \in \mathcal{B}(\mathbf{X}_\Delta)$ and $x \in \mathbf{X}$ function $q(\Gamma_\mathbf{X} \setminus \{x\}|x,\cdot)/q_x(\cdot)$ is continuous on $\mathbf{A}$; functions $c_i(x,\cdot)$, $i = 0,1,2,\ldots,N$ are continuous on $\mathbf{A}$ for any $x \in \mathbf{X}$.

The linear program $\mathbb{LP}$ associated with the constrained problem (6) looks as follows:

$$W_0 = \int_{\mathbf{X}\times\mathbf{A}} \frac{c_0(x,a)}{q_x(a)} \eta(dx,da) \to \inf \tag{13}$$

subject to $\eta \in \mathbb{L}_C$, where $\mathbb{L}_C$ is the space of (possibly infinite-valued) feasible measures, that is, satisfying equation

$$\eta(\Gamma_\mathbf{X} \times \mathbf{A}) = \gamma(\Gamma_\mathbf{X}) + \int_{\mathbf{X}\times\mathbf{A}} \frac{q(\Gamma_\mathbf{X} \setminus \{x\}|x,a)}{q_x(a)} \eta(dx,da), \tag{14}$$

and such that, for any $i = 0,1,2,\ldots,N$, the integral

$$W_i = \int_{\mathbf{X}\times\mathbf{A}} \frac{c_i(x,a)}{q_x(a)} \eta(dx,da) \tag{15}$$

is well defined and satisfies the constraints

$$W_i \le d_i, \quad i = 1,2,\ldots,N. \tag{16}$$

Note that, for any Markov standard ξ-strategy p^M, the total sum of (slightly modified) occupation measures $\sum_{n=1}^{\infty} q_x(a)\eta_n^{p^M}(dx,da)$ satisfies equation (14).

We also need auxiliary linear programs $\mathbb{LP}_i$, $i = 0,1,2,\ldots,N$

$$W_i \to \inf$$

subject to $\eta \in \mathbb{L}_i$, where $\mathbb{L}_i$ is the space of measures satisfying (14) and such that the integral (15) is well defined.

Proposition 1. *Let Condition 2 be satisfied. Suppose, for any $i = 0,1,2,\ldots,N$, the minimal value of the linear program $\mathbb{LP}_i$ is finite and let η^* be the optimal solution of the constrained linear program $\mathbb{LP}$ (13),(14),(15),(16). Then η^* gives rise to the so called 'induced' stochastic kernel $p^*(da|x)$ which defines the stationary standard ξ-strategy solving problem (6).*

The proof follows from [4, Th.5.2]. Generally speaking, after the measure η^* is obtained, the state space $\mathbf{X}$ is split into two disjoint parts $\mathbf{X} = V \cup V^c$. The subset V^c is the largest (in some sense) such that the measure $\eta^*(dx, da)$ is σ-finite on it and hence can be disintegrated: $\eta^*(dx, da) = p^*(da|x)\eta^*(dx \times \mathbf{A})$. On the set V, $p^*(da|x) = \delta_{f(x)}(da)$, where $f(x)$ is a specially constructed function: see Lemma 5.1, Prop.5.1 and Def.5.1 in [4]. Easier constructions can be found in [4, §4], where the set $\mathbf{A}$ is finite.

Note that in [4] the number of constraints N was not necessarily finite.

If all cost rates $c_i \geq 0$ are non-negative, then a stronger version of Proposition 1 is valid (see [3]): if the model is semi-continuous and the objective (13) is finite for some feasible measure η, then there is a stationary standard ξ-strategy solving problem (6).

6.2 General Case

Now we investigate the general case when Condition 1 is not necessarily fulfilled. For an arbitrarily fixed $\varepsilon > 0$, consider the discrete-time MDP $\mathcal{M}$ with the same state and action spaces $\mathbf{X}_\Delta$ and $\mathbf{A}$ and the same set $\mathbb{K}$ of admissible state-action pairs. Transition probability on $\mathbf{X}_\Delta$ is defined by

$$Q(\Gamma_\mathbf{X}|y, b) = \frac{q(\Gamma_\mathbf{X} \setminus \{y\}|y, b) + \varepsilon I\{\Gamma_\mathbf{X} \ni y\}}{q_y(b) + \varepsilon};$$

the initial distribution is γ. Here and below, it is convenient to denote the states and actions in the $\mathcal{M}$ model as y and b. The notions of a control strategy p and the corresponding strategic measure ${}^{\mathcal{M}}P^p_\gamma$ in $\mathcal{M}$ are conventional [12, 17]. The (total) occupation measure ${}^{\mathcal{M}}\eta^p$ on $\mathbb{K}$ is defined by the standard formula (see [13, §9.4]):

$${}^{\mathcal{M}}\eta^p(\Gamma_\mathbf{X} \times \Gamma_\mathbf{A}) = \sum_{m=1}^{\infty} {}^{\mathcal{M}}\eta^p_m(\Gamma_\mathbf{X} \times \Gamma_\mathbf{A}) = \sum_{m=1}^{\infty} {}^{\mathcal{M}}E^p_\gamma[I\{Y_{m-1} \in \Gamma_\mathbf{X},\ B_m \in \Gamma_\mathbf{A}\}]. \tag{17}$$

Let $\mathcal{D}$ be the full collection of such occupation measures under different Markov strategies p. Other strategies do not extend the set $\mathcal{D}$ [17, Lemma 2].

Lemma 1. *$\mathcal{D}$ coincides with the space of all (total) occupation measures $\sum_{n=1}^{\infty}(q_x(a) + \varepsilon)\eta^S_n$ under different strategies $S \in \Pi_S$ in the original continuous-time model. (See Section 3.)*

Now it is clear that solving the original constrained problem (6) is equivalent to solving the corresponding discrete-time MDP with one-step costs $c_i(y, b)/[q_y(b)+\varepsilon]$. By the way, in the unconstrained case, the optimality equation takes the form

$$\inf_{b\in\mathbf{A}(y)}\left\{c_0(y,b)/[q_x(a)+\varepsilon]+\int_{\mathbf{X}\setminus\{y\}}v(z)q(dz|y,b)/[q_y(b)+\varepsilon]\right.$$
$$\left.+\,\varepsilon v(y)/[q_y(b)+\varepsilon]-v(y)\right\}=0,\qquad y\in\mathbf{X}_\Delta$$

yielding

$$\inf_{b\in\mathbf{A}(y)}\left\{c_0(y,b)+\int_{\mathbf{X}\setminus\{y\}}v(z)q(dz|y,b)-q_y(b)v(y)\right\}=0 \tag{18}$$

for such $y\in\mathbf{X}_\Delta$ that $v(y)\in\mathbb{R}$. The last equation is well known for the problems with total (undiscounted) cost [11].

All the assertions in Remark 3 hold true. Note also that, for any stationary deterministic strategy φ^s ${}^{\mathcal{M}}\eta^{\varphi^s}=\sum_{n=1}^{\infty}(q_x(a)+\varepsilon)\eta_n^{\varphi^s}$: see the proof of Lemma 1.

Under Condition 2 one can investigate the linear program similar to (13), (14),(15),(16) and apply Proposition 1. If, for instance, the problem is unconstrained ($N=0$) and the value of $\mathbb{LP}$

$$\int_{\mathbf{X}\times\mathbf{A}}\frac{c_0(x,a)}{q_x(a)+\varepsilon}\eta(dx,da)\to\inf_\eta \tag{19}$$

subject to

$$\hat\eta(\Gamma_\mathbf{X})=\eta(\Gamma_\mathbf{X}\times\mathbf{A})=\gamma(\Gamma_\mathbf{X})+\int_{\mathbf{X}\times\mathbf{A}}\left[\frac{q(\Gamma_\mathbf{X}\setminus\{x\}|x,a)+\varepsilon I\{x\in\Gamma_\mathbf{X}\}}{q_x(a)+\varepsilon}\right]\eta(dx,da)$$

is finite, then, in case the optimal marginal $\hat\eta^*(\cdot)$ is σ-finite, one can disintegrate the optimal measure $\eta^*(dx,da)=\hat\eta^*(dx)p^*(da|x)$ and obtain an optimal stationary Poisson-related ξ-strategy with (n,k)-independent stochastic kernels $\tilde p_{n,k}(da|x)=p^*(da|x)$. Here we assumed that the $\mathbb{LP}$ (19) has an optimal solution η^*.

In the example presented in Section 5 all the conditions 2 were satisfied except for the compactness of the action space $\mathbf{A}$. Remember, the set of Markov standard ξ-strategies was not sufficient there in the problem $W_0(S)\to\inf_{S\in\Pi_S}$. In the corresponding discrete-time MDP, there is no optimal stationary strategy. The $\mathbb{LP}$ looks as follows:

$$\int_{(0,1]}\frac{a}{a+\varepsilon}\eta(\{1\}\times da)\to\inf_\eta \tag{20}$$

subject to

$$\hat\eta(\{1\})=\eta(\{1\}\times(0,1])=1+\int_{(0,1]}\frac{\varepsilon}{a+\varepsilon}\eta(\{1\}\times da). \tag{21}$$

If $\hat{\eta}(\{1\}) < \infty$ (note that $\hat{\eta}^p(\{1\}) < \infty$ for any stationary strategy p), one can write η in the form $\eta(\{1\}\times da) = \hat{\eta}(\{1\})\times p(da)$ and, for any probability measure p, we have

$$\hat{\eta}(\{1\}) = \left[1 - \int_{(0,1]} \frac{\varepsilon}{a+\varepsilon} p(da)\right]^{-1} = \left[\int_{(0,1]} \frac{a}{a+\varepsilon} p(da)\right]^{-1}$$

$$\Longrightarrow \text{Total cost equals} \int_{(0,1]} \frac{a}{a+\varepsilon} p(da) \left[1 - \int_{(0,1]} \frac{\varepsilon}{a+\varepsilon} p(da)\right]^{-1} = 1$$

as expected. Proposition 1 does not help because the solution to $\mathbb{LP}$ (20) does not exist. The infimum equals zero because, e.g. for the measure $\eta(\{1\} \times da) = (a+\varepsilon)\frac{e^{\frac{a-1}{\kappa}}}{\kappa a} da$ with $\kappa > 0$ we have that $\hat{\eta}((\{1\}) = \infty$ and $\int_{(0,1]} \frac{a}{a+\varepsilon} \eta(\{1\}\times da) = 1 - e^{-\frac{1}{\kappa}}$. On the other hand, to obtain zero in (20) we must have $\eta(\{1\} \times da) = 0$ which violates the requirement (21).

Let us show that, for any $\delta > 0$, there is a non-stationary Markov strategy in the assiciated discrete-time MDP $\mathcal{M}$ with the total expected cost smaller than δ. Take a_1 such that $\frac{a_1}{a_1+\varepsilon} < \frac{\delta}{2}$, take a_2 such that $\frac{a_2}{a_2+\varepsilon} < \frac{\delta}{4}$ and so on: take a_m such that $\frac{a_m}{a_m+\varepsilon} < \frac{\delta}{2^m}$. Then, for this Markov deterministic non-stationary strategy $p^*_m(da|x) = \delta_{a_m}(da)$, we have

$$^{\mathcal{M}}E^{p^*}_{\gamma}\left[\sum_{m=1}^{\infty} \frac{c_0(X_{m-1}, A_m)}{q_{X_{m-1}}(A_m) + \varepsilon}\right] = \frac{a_1}{a_1+\varepsilon} + \frac{\varepsilon}{a_1+\varepsilon}\left[\frac{a_2}{a_2+\varepsilon} + \frac{\varepsilon}{a_2+\varepsilon} \times \left[\frac{a_3}{a_3+\varepsilon} + \ldots\right]\right] < \frac{\delta}{2} + \frac{\delta}{4} + \frac{\delta}{8} + \ldots = \delta.$$

This strategy p^* gives rise to the corresponding Poisson-related δ-optimal strategy in the original CTMDP, with degenerate probabilities $\tilde{p}_k$: $\tilde{p}_k(da|1) = \delta_{a_k}(da)$. More examples of discrete-time MDP, where only non-stationary strategies can be δ-optimal, in [18, §2.2.11].

Remark 4. In the case of unconstrained problem with $N = 0$ and $c_0 \geq 0$, in the associated discrete-time MDP $\mathcal{M}$, there is a δ-optimal non-randomized Markov strategy for any $\delta > 0$ [2, Prop.9.19]. That strategy gives rise to the δ-optimal Poisson-related strategy with degenerate probabilities $\tilde{p}_{n,k}$. Many other known statements from the discrete-time theory can be directly applied to the Poisson-related strategies in the framework of CTMDP. See for example the transient and absorbing MDPs in [1].

7 Acknowledgement

The author is thankful to Dr. Y.Zhang for fruitful discussions and careful reading of the draft of this article.

8 Appendix

Proof of Lemma 1 (sketch). Assume $\varepsilon > 0$ is fixed.

1. For the proof of inclusion $\{\sum_{n=1}^{\infty}(q_x(a)+\varepsilon)\eta_n^S,\ S \in \Pi_S\} \subset \mathcal{D}$, according to Theorem 3, it is sufficient to consider only Poisson-related strategies S^P. Suppose such a strategy $S^P = \{\Xi, \varepsilon, \tilde{p}_{n,k}(da|x_{n-1}),\ n,k = 1,2,\ldots\}$ is given. The elements of $\xi_n \in \Xi$ are denoted as $\xi_n = (\alpha_1^n, \tau_1^n, \alpha_2^n, \tau_2^n, \ldots)$. We intend to build a control strategy $p = \{p_{m+1}(da|\ ^{\mathcal{M}}h_m)\}_{m=0}^{\infty}$ in the $\mathcal{M}$ model such that

$$^{\mathcal{M}}\eta^p(dx, da) = \sum_{n=1}^{\infty}(q_x(a)+\varepsilon)\eta_n^{S^P}(dx, da). \tag{22}$$

The elements relevant to the $\mathcal{M}$ model are equipped with the left upper index $\mathcal{M}$. It will be convenient to denote trajectories in $\mathcal{M}$ as $^{\mathcal{M}}\omega = (y_0, b_1, y_1, \ldots)$.

For a given history $^{\mathcal{M}}h_m = (y_0, b_1, y_1, \ldots, b_m, y_m)$ with $y_m \neq \Delta$ we define $l_1(^{\mathcal{M}}h_m) = \min\{l \geq 1:\ l \leq m;\ y_l \neq y_{l-1}\} \wedge (m+1)$.

For $k \geq 1$, if $l_k(^{\mathcal{M}}h_m) = m+1-\sum_{i=1}^{k-1} l_i(^{\mathcal{M}}h_m)$ then $n(^{\mathcal{M}}h_m) = k$; otherwise

$$l_{k+1}(^{\mathcal{M}}h_m) = \min\left\{l \geq 1:\ l \leq m - \sum_{i=1}^{k} l_i(^{\mathcal{M}}h_m);\right.$$
$$\left. y_{\sum_{i=1}^{k} l_i(^{\mathcal{M}}h_m)+l} \neq y_{\sum_{i=1}^{k} l_i(^{\mathcal{M}}h_m)+l-1}\right\} \wedge \left(m+1-\sum_{i=1}^{k} l_i(^{\mathcal{M}}h_m)\right).$$

After that, $\sum_{i=1}^{n(^{\mathcal{M}}h_m)} l_i(^{\mathcal{M}}h_m) = m+1$, and we put $k(^{\mathcal{M}}h_m) \triangleq l_{n(^{\mathcal{M}}h_m)}(^{\mathcal{M}}h_m)$ and apply the randomized action according to the distribution

$$p_{m+1}(da|^{\mathcal{M}}h_m) = \tilde{p}_{n(^{\mathcal{M}}h_m),k(^{\mathcal{M}}h_m)}(da|y_m).$$

This past-dependent randomized strategy p in $\mathcal{M}$ is the desired one. Figure 1 illustrates this construction and the connection between (random) histories $^{\mathcal{M}}H_m$ and trajectories of the original control process $X(t)$.

For an (infinite) trajectory $^{\mathcal{M}}\omega$, the values $l_i(^{\mathcal{M}}\omega) \in \mathbb{N} \cup \{\infty\}$, $i = 1,2,\ldots$ are defined in the similar way: $l_1(^{\mathcal{M}}\omega) = \min\{l \geq 1:\ y_l \neq y_{l-1}\}$ and for $k \geq 1$ such that $l_k(^{\mathcal{M}}\omega) < \infty$ and $y_{\sum_{i=1}^{k} l_i(^{\mathcal{M}}\omega)+l-1} \neq \Delta$, we put

$$l_{k+1}(^{\mathcal{M}}\omega) = \min\left\{l \geq 1:\ y_{\sum_{i=1}^{k} l_i(^{\mathcal{M}}\omega)+l} \neq y_{\sum_{i=1}^{k} l_i(^{\mathcal{M}}\omega)+l-1}\right\}.$$

Below, these functions on the sample space of the $\mathcal{M}$ model are, as usual, denoted by capital letters L_k (random variables).

For any $n = 1,2,\ldots$ for arbitrary $\Gamma_{\mathbf{X}} \in \mathcal{B}(\mathbf{X})$, $\Gamma_{\mathbf{A}} \in \mathcal{B}(\mathbf{A})$

$$\int_{\Gamma_{\mathbf{X}}}\int_{\Gamma_{\mathbf{A}}}(q_x(a)+\varepsilon)\eta_n^{S^P}(dx, da) = {}^{\mathcal{M}}E_\gamma^p\left[\sum_{m=\sum_{i=1}^{n-1} L_i+1}^{\sum_{i=1}^{n} L_i} I\{Y_{m-1} \in \Gamma_{\mathbf{X}}\}I\{B_m \in \Gamma_{\mathbf{A}}\}\right].$$

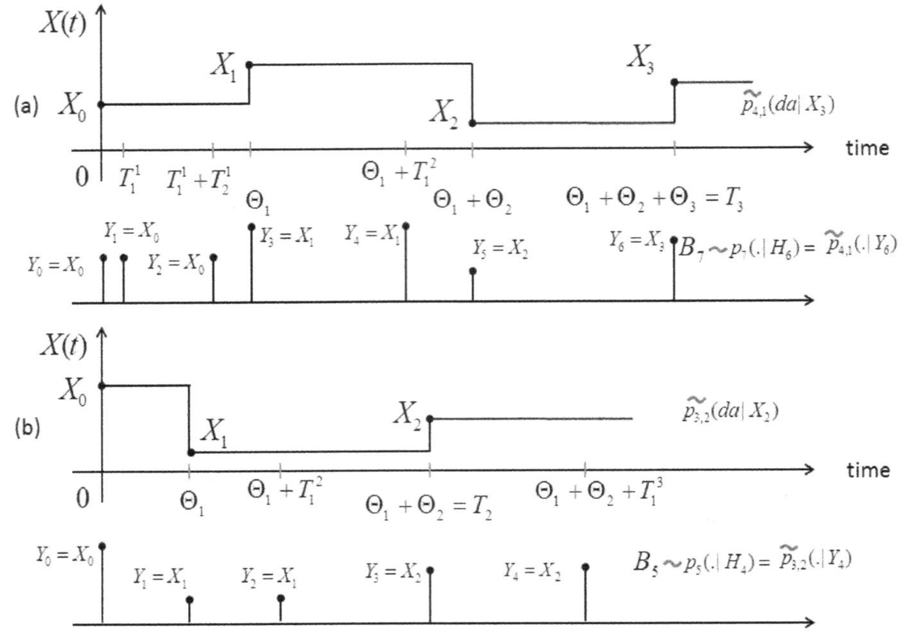

Fig. 1. Two scenarios illustrating the construction of the $\mathcal{M}$ model:
(a) ${}^{\mathcal{M}}H_6 = (Y_0, B_1, Y_1, \dots, B_6, Y_6)$; $l_1({}^{\mathcal{M}}H_6) = 3$, $l_2({}^{\mathcal{M}}H_6) = 2$, $l_3({}^{\mathcal{M}}H_6) = 1$, $l_4({}^{\mathcal{M}}H_6) = 1$, $n({}^{\mathcal{M}}H_6) = 4$, $k({}^{\mathcal{M}}H_6) = 1$;
(b) ${}^{\mathcal{M}}H_4 = (Y_0, B_1, Y_1, \dots, B_4, Y_4)$; $l_1({}^{\mathcal{M}}H_4) = 1$, $l_2({}^{\mathcal{M}}H_4) = 2$, $l_3({}^{\mathcal{M}}H_4) = 2$, $n({}^{\mathcal{M}}H_4) = 3$, $k({}^{\mathcal{M}}H_4) = 2$.

This equality is based on the formulae

$$E_\gamma^{S^P}\left[I\{X_{n-1} \neq \Delta\} \int_{(\sum_{i=1}^k T_i^n, \sum_{i=1}^{k+1} T_i^n \wedge T_n]} (q_{X_{n-1}}(A_{k+1}^n) + \varepsilon) dt \,\middle|\, \sum_{i=1}^k T_i^n < T_n, X_{n-1}, A_{k+1}^n\right] = E_\gamma^{S^P}\left[I\{X_{n-1} \neq \Delta\}\right];$$

$$P_\gamma^{S^P}(T_n < \infty,\ X_n \in \Gamma_{\mathbf{X}}) = {}^{\mathcal{M}}P_\gamma^p\left(\sum_{i=1}^n L_i < \infty,\ Y_{\sum_{i=1}^n L_i} \in \Gamma_{\mathbf{X}}\right)$$

valid for all $n = 1, 2, \dots$; $k = 0, 1, 2, \dots$. Therefore, (22) follows.

2. For the inverse inclusion $\{\sum_{n=1}^\infty (q_x(a) + \varepsilon)\eta_n^S,\ S \in \Pi_S\} \supset \mathcal{D}$, suppose a Markov strategy $p_m(da|x)$ in $\mathcal{M}$ is fixed and construct a past-dependent version

S of a Poisson-related strategy such that

$$\sum_{n=1}^{\infty}(q_x(a)+\varepsilon)\eta_n^S(dx,da) = {}^{\mathcal{M}}\eta^p(dx,da). \qquad (23)$$

Past-dependent means that the stochastic kernels $\tilde{p}_{n,k}$ will depend on the histories h_{n-1} rather than on the current states x_{n-1}.

Let $\tilde{p}_{1,k}(da|x_0) = p_k(da|x_0)$. For any history h_n with $x_n \neq \Delta$ we compute

$$k_n(h_n) \triangleq \min\{k \geq 1 : \sum_{i=1}^{k} \tau_i^n \geq \theta_n\}$$

and, in case $k_n(h_n) < \infty$, we put

$$\tilde{p}_{n+1,k}(da|h_n) = p_{\sum_{i=1}^{n} k_i(h_n)+k}(da|x_n).$$

If $k_n(h_n) = \infty$, the stochastic kernels $\tilde{p}_{n+1,k}$ can be defined arbitrarily.

For this strategy S, similarly to the ideas described above, one can prove equality

$$\int_{\Gamma_{\mathbf{X}}}\int_{\Gamma_{\mathbf{A}}}(q_x(a)+\varepsilon)\eta_n^S(dx,da) = {}^{\mathcal{M}}E_\gamma^p\left[\sum_{m=\sum_{i=1}^{n-1}L_i+1}^{\sum_{i=1}^{n}L_i} I\{Y_{m-1}\in\Gamma_{\mathbf{X}}\}I\{B_m\in\Gamma_{\mathbf{A}}\}\right]$$

for all $n = 1, 2, \ldots$, $\Gamma_{\mathbf{X}} \in \mathcal{B}(\mathbf{X})$, $\Gamma_{\mathbf{A}} \in \mathcal{B}(\mathbf{A})$.

After that, equality (23) is obvious. ■

References

1. Altman, E. *Constrained Markov Decision Processes.* Chapman and Hall/CRC, Boca Raton, 1999.
2. Bertsekas, D. and Shreve, S. *Stochastic Optimal Control.* Academic Press, NY, 1978.
3. Dufour, F., Horiguchi, M. and Piunovskiy, A.: The expected total cost criterion for Markov decision processes under constraints: a convex analytic approach. *Adv. Appl. Prob.* **44** (2012) 774-793.
4. Dufour, F. and Piunovskiy, A.: The expected total cost criterion for Markov decision processes under constraints. *Adv. Appl. Prob.* **45** (2013) 837-859.
5. Feinberg, E.: Continuous time discounted jump Markov decision processes: a discrete-event approach. *Math. Oper. Res.* **29** (2004) 492-524.
6. Feinberg, E.: Reduction of discounted continuous-time MDPs with unbounded jump and reward rates to discrete-time total-reward MDPs. In *Optimization, Control, and Applications of Stochastic Systems* (D.Hernandez-Hernandez and J.A.Minjares-Sosa ed.), Birkhauser, 2012, 77-97.
7. Feinberg, E.: Total reward criteria. In *Handbook of Markov Decision Processes.* (E.Feinberg and A.Shwartz ed.), Kluwer, Boston/Dordrecht/London, 2002, 173-207.

8. Ghosh, M. and Saha, S.: Non-stationary semi-Markov decision processes on a finite horizon. *Stoch. Anal. Appl.* **31** (2013) 183-190.
9. Guo, X. and Zhang, Y.: Constrained total undiscounted continuous-time Markov decision processes. *Bernoulli*, accepted; http://arxiv.org/pdf/1304.3314v5.pdf
10. Guo, X. and Piunovskiy, A.: Discounted continuous-time Markov decision processes with constraints: unbounded transition and loss rates. *Math. Oper. Res.* **36** (2011) 105-132.
11. Guo, X., Vykertas, M., Zhang, Y.: Absorbing continuous-time Markov decision processes with total cost criteria. *Adv. Appl. Prob.* **45** (2013) 490-519.
12. Hernández-Lerma, O. and Lasserre, J.B. *Discrete-Time Markov Control Processes.* Springer-Verlag, NY, 1996.
13. Hernández-Lerma, O. and Lasserre, J.B. *Further Topics on Discrete-Time Markov Control Processes.* Springer-Verlag, NY, 1999.
14. Jacod, J.: Multivariate point processes: predictable projection, Radon-Nykodym derivatives, representation of martingales. *Z. Wahrscheinlichkeitstheorie verw. Gebite.* **31** (1975) 235-253.
15. Kitayev, M.: Semi-Markov and jump Markov controlled models: average cost criterion. *Theory. Probab. Appl.* **30** (1986) 272-288.
16. Kitaev, M and Rykov, V. *Controlled Queueing Systems.* CRC Press, Boca Raton, 1995.
17. Piunovskiy, A. *Optimal Control of Random Sequences in Problems with Constraints.* Kluwer, Dordrecht, 1997. 51-71.
18. Piunovskiy, A. *Examples in Markov Decision Processes.* Imperial College Press, London, 2013.
19. Piunovskiy, A. and Zhang, Y.: Discounted continuous-time Markov decision processes with unbounded rates: the convex analytic approach. *SIAM J. Control Optim.*, **49** (2011) 2032-2061.
20. Piunovskiy, A.: Randomized and relaxed strategies in continuous-time Markov decision processes. *SIAM J. Control Optim.*, in press.
21. Prieto-Rumeau, T. and Hernandez-Lerma, O. *Selected Topics on Continuous-Time Controlled Markov Chains and Markov Games.* Imperial College Press, London, 2012.
22. Puterman, M. *Markov Decision Processes: Discrete Stochastic Dynamic Programming.* Wiley, NY, 1994. 215-235.
23. Zhang, Y.: Convex analytic approach to constrained discounted Markov decision processes with non-constant discount factors. *TOP*, **21** (2013) 378-408.

Ergodic Control of Pollution with Cost Constraints

Armando F. Mendoza-Pérez[1], Héctor Jasso-Fuentes*[2] and Onésimo Hernández-Lerma[2]

[1] CEFyMAP-UNACH
Cuarta Oriente Norte 1428, C.P. 29040, Tuxtla Gutiérrez , Chiapas , México
mepa680127@hotmail.com

[2] Departamento de Matemáticas
Cinvestav-IPN. A. Postal 14-740, México , D.F. 07000 , México
hjasso@math.cinvestav.mx; ohernand@math.cinvestav.mx

Abstract. In this chapter we study the ergodic (a.k.a. long-run average) control problem for a pollution accumulation model with cost constraints governed by a diffusion process. Our problem is to maximize the ergodic social welfare subject to a constraint on a certain ergodic cost. This constraint might be applied, for example, to the costs of environmental cleaning ensuring that these costs do not exceed some given value. To obtain optimality results, we follow the Lagrange multipliers approach that relates our constrained problem to some unconstrained one that is easier to handle. Explicit optimal solutions are provided for the cases when both the social welfare and the cost constraint are linear functions.
Keywords: pollution accumulation, controlled diffusions, ergodic control, constrained control problems.
AMS 2010 subject classification: Primary 93E20, Secondary 60J60

1 Introduction

In this chapter, we study an ergodic (or long-run average) control problem of a pollution accumulation model similar to [8] but imposing some cost constraints which naturally appear as part of the control model. The optimal control of pollution is an important challenge for sustainable development policies. The idea is to manage the quantity of some good to be consumed by some economy, which, as a result of this consumption, it generates pollution. In most of the models, the pollution stock is assumed to be gradually degraded and incorporation of random shocks of pollution depend of some instantaneous growth rate. As a reponse to both the consumption and the pollution, the economy has a utility (associated to the consumption) and a disutility (concerned with the pollution). The subtraction of these two components is known as social welfare (see, for instance, [7, 8]).

* Corresponding author

On the other hand, the restoration of a polluted environment will require an initial investment, which implies a social cost to compensate for the rising disutility of pollution. Assuming that the pollution evolves along the time as a (controlled) diffusion process, our objective is to find a policy of consumption —or quantity to be sold to the consumers— that maximizes the expected social welfare for the society, subject to a constraint on a general expected cost that may include the disutility of pollution; for example, this constraint might be applied to the cost of environmental cleaning or to upper-bound some negative effects of pollution.

To get optimality, we follow the Lagrange multipliers approach to transform the original constrained ergodic control problem into an uncostrained problem parametrized by the Lagrange multipliers. Then, we use the regularity properties of the *viscosity solutions* of the degenerate (parametrized) Hamilton-Jacobi-Bellman (HJB) equations associated to the corresponding unconstrained pollution accumulation problem, to find optimal solutions to the latter problem. We shall see that under a suitable choice of a certain multiplier, optimality for the unconstrained problem yields optimality of the original constrained one.

There exist a large number of papers related to the present topic, for instance, to mention just a few, see [7, 8, 10–12] and the references therein. Here we extend the results in [8], since we are now considering constraints as a part of the model, which requires to modify some arguments previously developed in [9]. Indeed, in [9], we have analysed similar problems for nondegenerate controlled diffusions with cost constraints. These results were obtained by using the strong Feller property of the state process under a Markov control, together with the positive recurrence and the uniformly W-exponentially ergodic behavior of the nondegenerate diffusion. Moreover, since the Hamilton-Jacobi-Bellman equation (HJB) is uniformly elliptic, we have used the regularity properties of the HJB equations (see, for instance, [1, Theorem 3.7.11] or [2, Theorem 7.6]). In contrast, in this work, our pollution model satisfies the *degenerate* controlled stochastic differential equation (1) where the uniform ellipticity hypothesis is not satisfied anymore. Therefore, our arguments mainly use the results in [8] to find the existence of *viscosity solutions* to the HJB equations appearing in our context (see Theorem 1 below).

This paper is organized as follows: Section 2 introduces the pollution accumulation model and some of our main assumptions. In Section 3 we define the economic welfare system with constraints, and also present its associated unconstrained problem along with the corresponding HJB equations. Results on the existence and characterizations of solutions to these equations are also provided. In Section 4 we establish and prove our main results regarding the existence of constrained optimal policies. These results are strongly related to the unconstrained problem studied in the previous section. Finally, to illustrate our approach, in Section 5 we give an explicit solution to a control model with a first-order disutility function and a linear cost rate function.

2 Model Description

We represent the stock of pollution $x(t)$ as a controlled diffusion process satisfying

$$dx(t) = [\pi(t) - \eta x(t)]dt + \sigma x(t)dB(t), \quad x(0) = x_0 > 0, \tag{1}$$

defined on a complete pobability space $(\Omega, \mathcal{F}, P)$, where $B(t)$ is a one-dimensional Brownian motion endowed with the natural filtration $\mathcal{F}_t$ generated by $\{B(s) : s \leq t\}$; the constant $\eta > 0$ represents the pollution decay, whereas σ is a given constant associated to the diffusion term. Furthermore, $\pi(t)$ represents the flow of consumption at time t, and it is assumed to be an $\mathcal{F}_t$-progressively measurable process satisfying $0 \leq \pi(t) \leq \gamma$ for all $t \geq 0$, where the constant γ is generally imposed by worldwide protocols. The set of all these consumption processes (control policies) is denoted by $\mathcal{A}$. For $\pi \in \mathcal{A}$, straightforward use of comparison theorems (see, for instance, [4, 5, 10]), we have that (1) admits a unique *positive* solution $\{x(t)\}$.

Let $\mathbb{F}$ be the set of all measurable functions $f : [0, \infty) \to [0, \gamma]$. A control policy of the form $\pi(t) = f(x(t))$, $t \geq 0$ for some $f \in \mathbb{F}$ is called a *stationary policy*. Actually, by an abuse of terminology, f itself will be referred to as a *stationary policy*, and we can assume that $\mathbb{F} \subset \mathcal{A}$.

To complete the description of our constrained control model we introduce the reward rate function

$$r(x, u) := F(u) - D(x) \quad \text{for all } (x, u) \in [0, \infty) \times [0, \infty),$$

where $r(x, u)$ represents the social welfare, with $F \in C^2(0, \infty) \cap C[0, \infty)$ and $D \in C[0, \infty)$ representing the social utility of the consumption u and the social disutility of the pollution x, respectively. These functions satisfy the following standard assumptions (see, for instance, [8, 10–12]):

$$\begin{cases} F' > 0, \quad F'' < 0, \\ F'(\infty) = F(0) = 0, \quad F'(0+) = F(\infty) = \infty, \end{cases} \tag{2}$$

$$\begin{cases} D(x) \text{ is continuous and convex on } [0, \infty) \text{ and, for some integer } m \geq 1, \\ \frac{1}{D_0}x - D_0 \leq D(x) \leq D_0(1 + x^m), \text{ with } D_0 > 0. \\ |D(x) - D(y)| \leq D_0|x - y|(1 + x^{m-1} + y^{m-1}), \quad x, y > 0. \end{cases} \tag{3}$$

Let $c : [0, \infty) \times [0, \gamma] \to \mathbb{R}$ be the cost rate function. We assume that, for some integer $m \geq 1$ and all $0 \leq u \leq \gamma$,

$$\begin{cases} c(x, u) \text{ is continuous and convex on } [0, \infty) \times [0, \gamma], \\ \frac{1}{C_0}x - C_0 \leq c(x, u) \leq C_0(1 + x^m), \text{ with } C_0 > 0. \\ |c(x, u) - c(y, u)| \leq C_0|x - y|(1 + x^{m-1} + y^{m-1}), \quad x, y > 0. \end{cases} \tag{4}$$

Remark 1. We can assume that the integer m in (3) and (4) is the same. Otherwise, we can consider m as the maximun value of both integers. Similarly for D_0, C_0, we can take $C := \max(D_0, C_0)$.

3 The Economic Welfare System with Constraints

The main purpose of the underying economy is to maximize the ergodic expected social welfare of the society with the restriction that the ergodic expected cost does not exceed a given value. In other words, for some fixed constant θ, we are interested in the following constrained problem (CP) associated with the economic welfare system (1)-(4):

$$\text{maximize } R(x_0,\pi) := \varliminf_{T\to\infty} \frac{1}{T}\mathbb{E}\left[\int_0^T \{F(\pi(t)) - D(x(t))\}dt\right] \tag{5}$$

$$\text{subject to: } C(x_0,\pi) := \varlimsup_{T\to\infty} \frac{1}{T}\mathbb{E}\left[\int_0^T c(x(t),\pi(t))dt\right] \le \theta, \tag{6}$$

for all initial state $x_0 > 0$.

Definition 1. (i) *A consumption process* $\pi = \{\pi(t)\}$ *is said to be a* feasible control process *for the CP if it is an* $\mathcal{F}_t$*-progressively measurable process such that* $0 \le \pi(t) \le \gamma$ *for all* $t \ge 0$*, and satisfies the condition (6) for every initial state* $x_0 > 0$*. We denote the set of all feasible controls by* $\mathcal{A}^\theta_{feas}$.

(ii) *A feasible policy* $\pi^* \in \mathcal{A}^\theta_{feas}$ *is called* optimal *for the CP if*

$$R(x_0,\pi^*) = \sup_{\pi\in\mathcal{A}^\theta_{feas}} R(x_0,\pi) =: V^*_\theta(x_0) \quad \textit{for all initial state } x_0 > 0. \tag{7}$$

In this case $V^*_\theta(\cdot)$ *is referred to as the* average optimal social welfare *for the CP, or simply, the* optimal value *for the CP.*

3.1 Lagrange Multipliers.

To find optimal control policies for the CP (5)-(6), we will use the Lagrange multipliers technique. The idea of this method consists in associating an equivalent unconstrained problem to the CP through a suitable multiplier so that by finding optimality for the former, we authomatically get solution for the latter (the CP (5)-(6)).

To introduce this method, consider two constants $\Lambda \le 0$ and θ, with θ as in (6) but satisfiying an additional requirement —see (14) below. Now define the reward rate

$$r_\Lambda(x,u) := r(x,u) + \Lambda\cdot(c(x,u)-\theta) = F(u) - [D(x) - \Lambda\cdot c(x,u) + \Lambda\cdot\theta].$$

In addition, for each fixed $\Lambda \le 0$, every control $\pi = \{\pi(t)\} \in \mathcal{A}$, and every initial state $x_0 > 0$, we consider the Λ*-long-run expected average social welfare* defined by

$$R_\Lambda(x_0,\pi) := \varliminf_{T\to\infty} \frac{1}{T}\mathbb{E}\left[\int_0^T \{F(\pi(t)) - D_\Lambda(x(t),\pi(t))\}dt\right],$$

where $D_\Lambda(x,u) := D(x) - \Lambda \cdot c(x,u) + \Lambda \cdot \theta$. Notice that $D_\Lambda(\cdot,u)$ satisfies conditions similar to those of $D(\cdot)$ in (3), for all $0 \le u \le \gamma$, and $c(\cdot,\cdot)$ as in (4).

Definition 2 (Λ-unconstrained problem (Λ-UP).). *A control policy $\pi^* \in \mathcal{A}$ for which*

$$R_\Lambda(x_0,\pi^*) = \sup_{\pi\in\mathcal{A}} R_\Lambda(x_0,\pi), \quad \textit{for every initial state } x_0 > 0, \tag{8}$$

is called Λ-average optimal *for the Λ-UP. Let $\mathcal{A}_{\Lambda,\theta}$ and $\mathbb{F}_{\Lambda,\theta} := \mathbb{F} \cap \mathcal{A}_{\Lambda,\theta}$ be the set of all Λ-average optimal policies and the set of all Λ-average stationary optimal policies for the Λ-UP, respectively.*

3.2 The HJB Equation.

(a) A pair $(\zeta_\Lambda, h_\Lambda)$ consisting of a constant $\zeta_\Lambda \in \mathbb{R}$ and a fuction $h_\Lambda \in C^2(0,\infty)$ is said to be a *solution to the HJB equation* associated to the Λ-UP, if

$$\zeta_\Lambda = \frac{1}{2}\sigma^2 x^2 h''_\Lambda(x) - \eta x h'_\Lambda(x) + \max_{u\in[0,\gamma]} \{F(u) - D_\Lambda(x,u) + u h'_\Lambda(x)\} \quad \forall x > 0. \tag{9}$$

(b) If $f \in \mathbb{F}$ attains the maximum in (9), that is,

$$\zeta_\Lambda = \frac{1}{2}\sigma^2 x^2 h''_\Lambda(x) - \eta x h'_\Lambda(x) + F(f(x)) - D_\Lambda(x,f(x)) + f(x)h'_\Lambda(x) \quad \forall x > 0,$$

then f is called a *canonical policy* and, furthermore, $(\zeta_\Lambda, h_\Lambda, f)$ is called a *canonical triplet* for the Λ-UP.

We need the following assumption.

Assumption 1. *The integer m in (3)–(4) satisfies that*

$$m \ge 2, \quad \textit{and} \quad 2\eta > m(m-1)\sigma^2,$$

where η and σ are the constants in (1).

The following proposition collects some important facts on the HJB equations (see, for instance, [8, 10])

Proposition 1. *If Assumption 1 holds, then:*

(i) *For every initial condition $x > 0$, there exists a constant $K > 0$ satisfying*

$$\sup_{t\ge 0} \mathbb{E}[x(t)^m] < x^m + K. \tag{10}$$

(ii) *For each $\Lambda \le 0$, the HJB equation (9) admits a unique solution $(\zeta_\Lambda, h_\Lambda) \in \mathbb{R} \times C^2(0,\infty)$. Furthermore, there exists $K_\Lambda > 0$ such that*

$$|h_\Lambda(x)| \le K_\Lambda(1 + x^m) \quad \forall x > 0, \tag{11}$$

and, moreover, the scalar ζ_Λ in (9) coincides with the Λ-average optimal social welfare in (8).

(iii) *There exists a measurable function $f_\Lambda : (0,\infty) \to [0,\gamma]$ such that for every $x > 0$, $f_\Lambda(x)$ attains the maximum in (9). Consequently, $(\rho(\Lambda), h_\Lambda, f_\Lambda)$ becomes a canonical triplet.*

(iv) *The control policy $f_\Lambda \in \mathbb{F}$ is an optimal policy for the Λ-UP; in particular, $\mathbb{F}_{\Lambda,\theta}$ and $\mathcal{A}_{\Lambda,\theta}$ are nonempty sets, and*

$$\rho(\Lambda) = \sup_{\pi\in\mathcal{A}} R_\Lambda(x_0,\pi) = R_\Lambda(x_0, f_\Lambda) \quad \textit{for all initial state } x_0 > 0.$$

(v) *If, in addition, we assume that*

$$\frac{\partial c}{\partial u}(x,u) \le 0, \; \frac{\partial^2 c}{\partial u^2}(x,u) \ge 0,$$

for all $x > 0$ and for all $u > 0$, then f_Λ is unique and is given by

$$f_\Lambda(x) = \begin{cases} {F'_{\Lambda,x}}^{-1}(-h'_\Lambda(x)) & \textit{if } x \in {h'_\Lambda}^{-1}((-\infty, -F'_{\Lambda,x}(\gamma))), \\ \gamma & \textit{if } x \in {h'_\Lambda}^{-1}([-F'_{\Lambda,x}(\gamma), \infty)), \end{cases} \tag{12}$$

for all $x > 0$, where $F_{\Lambda,x}(u) := F(u) - D_\Lambda(x,u)$.

Remark 2. **(a)** Under Assumption 1 and by Proposition 1, for each $\Lambda \le 0$, we obtain that $\sup_{\pi\in\mathcal{A}} R_\Lambda(x_0,\pi)$ is well defined, and that is independent of the initial condition $x_0 > 0$. In other words, R_Λ becomes a constant that will be referred to as the Λ-*average optimal social welfare*, and is denoted by $\rho(\Lambda)$; that is,

$$\rho(\Lambda) := \sup_{\pi\in\mathcal{A}} R_\Lambda(x_0,\pi), \quad \text{for every initial state } x_0 > 0.$$

(b) Assumption 1 is crucial to prove (10) by a simple use of well-known comparison theorems (see [8, Appendix B] or [4]).

(c) Part (ii) of Proposition 1 is fundamental in the proof of our main results in Theorem 1 below. In contrast, in [8, Theorems 4.2, 4.3, and 4.4], the authors prove that the HJB associated with the long-run average problem is obtained as a limit of HJB equations for discounted models when the discount rates converge to zero. In fact, the solutions of these discounted HJB equations are viscosity solutions, but it is argued that under conditions (2)–(4), together with Assumption 1, the limit solution is necessarily a classical solution in $C^2(0,\infty)$.

Consider now

$$\theta_{min}(x_0) := \inf_{\pi\in\mathcal{A}} C(x_0,\pi) \quad \text{for all initial state } x_0 > 0.$$

Similar to Proposition 1, we have the following.

Proposition 2. *Under (4) and Assumption 1, $\theta_{min}(\cdot)$ is constant. Moreover, there exists $h \in C^2(0,\infty)$ such that the following HJB equation holds*

$$\theta_{min} = \frac{1}{2}\sigma^2 x^2 h''(x) - \eta x h'(x) + \min_{u\in[0,\gamma]} \{c(x,u) + uh'(x)\}, \tag{13}$$

for all $x > 0$. Besides, there exists a measurable function $f_\theta : (0,\infty) \to [0,\gamma]$ such that for each $x > 0$, $f_\theta(x)$ attains the minimum in (13). Then $(\theta_{min}, h, f_\theta)$ is a canonical triplet for the HJB (13), and $f_\theta \in \mathbb{F}$ satisfies

$$\theta_{min} = C(x_0, f_\theta) = \min_{\pi\in\mathcal{A}} C(x_0,\pi) \quad \textit{for all initial state } x_0 > 0.$$

To avoid trivial situations, we will consider a constraint θ such that

$$\theta_{min} < \theta < \infty \tag{14}$$

In this case, notice that $f_\theta \in \mathcal{A}^\theta_{feas}$.

4 Main Results

We can now state our main result that ensures the existence of a solution of the CP (5)-(6) associated with the economic welfare system (1)-(4).

Theorem 1. *Suppose that Assumption 1 holds and let θ be as in (14). Then:*

(a) *If the mapping $\Lambda \mapsto \rho(\Lambda)$ is differentiable at a point $\Lambda < 0$, then*

$$\frac{d\rho}{d\Lambda}(\Lambda) = C(x_0, \pi_\Lambda) - \theta \quad \textit{for all initial state } x_0 > 0,$$

where π_Λ is any Λ-average optimal policy in $\mathcal{A}_{\Lambda,\theta}$.

(b) *If $\Lambda^* < 0$ is a critical point of $\rho(\cdot)$, then every Λ^*-average optimal policy $\pi_{\Lambda^*} \in \mathcal{A}_{\Lambda^*,\theta}$ is also optimal for the CP (5)-(6), and $\rho(\Lambda^*)$ coincides with $V^*_\theta(x_0)$ for every initial state $x_0 > 0$, where $V^*_\theta(\cdot)$ is the average optimal social welfare for the CP defined in (7). Moreover, $\rho(\cdot)$ attains a minimum in Λ^*, that is,*

$$\rho(\Lambda^*) = \min_{\Lambda<0} \rho(\Lambda).$$

(c) **Case $\Lambda = 0$:** *If $\pi_0 \in \mathcal{A}_{0,\theta}$ satisfies $C(x_0,\pi_0) \le \theta$ for every initial state $x_0 > 0$, then it is optimal for the CP (5)-(6), and $\rho(0)$ coincides with the optimal value $V^*_\theta(\cdot)$ for the CP. Besides, $\rho(\cdot)$ attains a minimum in 0 satisfying*

$$\rho(0) = \min_{\Lambda\le 0} \rho(\Lambda).$$

To prove Theorem 1, we need some preliminary results. Hereafter, we assume that Assumption 1 holds true.

Notation. We need the following notation. Let $\pi \in \mathcal{A}$ be a control policy, $x_0 > 0$ an arbitrary intial state, and $x(t)$ the corresponding solution of (1). We define:

$$\underline{C}(x_0, \pi) := \varliminf_{T\to\infty} \frac{1}{T}\mathbb{E}\int_0^T c(x(t), \pi(t))dt,$$

$$\overline{R}(x_0, \pi) := \varlimsup_{T\to\infty} \frac{1}{T}\mathbb{E}\int_0^T r(x(t), \pi(t))dt,$$

and

$$\overline{R}_\Lambda(x_0, \pi) := \varlimsup_{T\to\infty} \frac{1}{T}\mathbb{E}\int_0^T r_\Lambda(x(t), \pi(t))dt,$$

where c, r, and r_Λ are the cost and the payoff functions, respectivelly, all of them already defined in previous sections.

Lemma 1. *Let $h_\Lambda \in C^2(0,\infty)$ be as in Proposition 1(ii). Then for any $\pi \in \mathcal{A}$, and any initial condition $x > 0$, the corresponding solution of (1) we have*

$$\lim_{T\to\infty} \frac{1}{T}\mathbb{E}h_\Lambda(x(T)) = 0.$$

Proof. The lemma follows from Proposition 1(i) and (11). ■

Lemma 2. *Fix $\Lambda \leq 0$. Then:*

(a) For every control policy $\pi(\cdot) \in \mathcal{A}$, and for each initial state $x_0 > 0$,

$$\rho(\Lambda) \geq R(x_0, \pi) + \Lambda \cdot [\underline{C}(x_0, \pi) - \theta], \tag{15}$$

and

$$\rho(\Lambda) \geq \overline{R}(x_0, \pi) + \Lambda \cdot [C(x_0, \pi) - \theta]. \tag{16}$$

Hence,

$$\rho(\Lambda) \geq R(x_0, \pi) + \Lambda \cdot [C(x_0, \pi) - \theta]. \tag{17}$$

(b) If $\pi(\cdot) \in \mathcal{A}_{\Lambda,\theta}$, then

$$\rho(\Lambda) = R_\Lambda(x_0, \pi) = \overline{R}_\Lambda(x_0, \pi) = \lim_{T\to\infty} \frac{1}{T}\mathbb{E}\int_0^T r_\Lambda(x(t), \pi(t))dt. \tag{18}$$

Furthermore, we have

$$\rho(\Lambda) = R(x_0, \pi) + \Lambda \cdot [\underline{C}(x_0, \pi) - \theta] = \overline{R}(x_0, \pi) + \Lambda \cdot [C(x_0, \pi) - \theta]. \tag{19}$$

Proof. (a) Fix an arbitrary $T > 0$, and let $\pi(\cdot) \in \mathcal{A}$ be a control policy, and $\{x(t)\}$ the corresponding solution of (1). Furthermore, use $h_\Lambda \in C^2(0, \infty)$ being the solution to the HJB equation (9) established in Proposition 1. Applying Itô's formula to h_Λ, and then using Proposition 1(ii), we obtain

$$\begin{aligned} h_\Lambda(x(T\wedge\tau_n)) &= h_\Lambda(x) + \int_0^{T\wedge\tau_n} \{h'_\Lambda(x(t))(\pi(t)-\eta x(t)) \\ &\quad + \frac{1}{2}\sigma^2 x(t)^2 h''_\Lambda(x(t))\}dt + \int_0^{T\wedge\tau_n} h'_\Lambda(x(t))\sigma x(t)dB(t) \\ &\le h_\Lambda(x) + \int_0^{T\wedge\tau_n} \{\rho(\Lambda) - [F(\pi(t)) - D_\Lambda(x(t),\pi(t))]\}dt \\ &\quad + \int_0^{T\wedge\tau_n} h'_\Lambda(x(t))\sigma x(t)dB(t), \quad \text{a.s.} \end{aligned}$$

where $\{\tau_n\}$ is a sequence of localizing stopping times $\tau_n \uparrow \infty$ of the local martingale. Hence,

$$\begin{aligned} \mathbb{E}[h_\Lambda(x(T\wedge\tau_n))] &\le h_\Lambda(x) + \mathbb{E}\int_0^{T\wedge\tau_n} \{\rho(\Lambda) - [F(\pi(t)) - D_\Lambda(x(t),\pi(t))]\}dt \\ &= h_\Lambda(x) + \mathbb{E}\int_0^{T\wedge\tau_n} \{\rho(\Lambda) - r(x(t),\pi(t)) \\ &\quad - \Lambda\cdot[c(x(t),\pi(t)) - \theta]\}dt. \end{aligned}$$

From (10) and (11) we see that

$$\sup_n \mathbb{E}[|h_\Lambda(x(T\wedge\tau_n))|^{1+1/m}] \le K_\Lambda(1+\sup_{0\le t\le T}\mathbb{E}[x(t)^{m+1}]) < \infty.$$

Thus $\{h_\Lambda(x(T\wedge\tau_n))\}_{n\ge 1}$ is uniformly integrable, which allows us to let n go to ∞ to obtain

$$\begin{aligned} &\mathbb{E}\int_0^T r(x(t),\pi(t))dt + \Lambda\cdot\left[\mathbb{E}\int_0^T c(x(t),\pi(t))dt - \theta\cdot T\right] \\ &\le \rho(\Lambda)\cdot T + h_\Lambda(x) - \mathbb{E}[h_\Lambda(x(T))]. \end{aligned}$$

Dividing both sides of this last expression by T and letting $T\to\infty$, we get after using Lemma 1:

$$R(x_0,\pi) + \Lambda\cdot[\underline{C}(x_0,\pi) - \theta]\cdot\Lambda \le \rho(\Lambda), \tag{20}$$

and

$$\overline{R}(x_0,\pi) + \Lambda\cdot[C(x_0,\pi) - \theta]\cdot\Lambda \le \rho(\Lambda),$$

proving (15), (16), and (17).

(b) Consider $\pi(\cdot)\in\mathcal{A}_{\Lambda,\theta}$. Then $\rho(\Lambda) = R_\Lambda(x_0,\pi)$ for all $x_0>0$. Hence, by the same procedure used in (a), we can easily deduce (18). Moreover,

$$\rho(\Lambda) = R_\Lambda(x_0,\pi) = \varliminf_{T\to\infty}\frac{1}{T}\int_0^T \mathbb{E}\Big\{r(x(t),\pi(t)) + \Lambda\cdot[c(x(t),\pi(t)) - \theta]\Big\}dt$$

$$\leq R(x_0, \pi) + \Lambda \cdot [\underline{C}(x_0, \pi) - \theta].$$

Thus, from (20) we deduce that

$$\rho(\Lambda) = R_\Lambda(x_0, \pi) = R(x_0, \pi) + \Lambda \cdot [\underline{C}(x_0, \pi) - \theta].$$

Similarly, we can show that

$$\rho(\Lambda) = R_\Lambda(x_0, \pi) = \overline{R}(x_0, \pi) + \Lambda \cdot [C(x_0, \pi) - \theta],$$

which proves (19). ■

The following lemma concerns a basic relation between $\rho(\Lambda)$ and the functions $\Lambda \mapsto C(x_0, \pi_\Lambda)$, $\Lambda \mapsto R(x_0, \pi_\Lambda)$ where the policies $\pi_\Lambda(\cdot) \in \mathcal{A}_{\Lambda,\theta}$ are Λ-average optimal policies for the Λ-UP.

Lemma 3. *For each $\Lambda \leq 0$, for every initial state $x_0 > 0$, and every real number δ such that $\Lambda + \delta \leq 0$, we have*

$$\delta \cdot \Big[C(x_0, \pi_\Lambda) - \theta\Big] \leq \rho(\Lambda + \delta) - \rho(\Lambda) \leq \delta \cdot [C(x_0, \pi_{\Lambda+\delta}) - \theta], \tag{21}$$

and

$$\delta \cdot \Big[\underline{C}(x_0, \pi_\Lambda) - \delta\Big] \leq \rho(\Lambda + \delta) - \rho(\Lambda) \leq \delta \cdot [\underline{C}(x_0, \pi_{\Lambda+\delta}) - \theta], \tag{22}$$

where $\pi_\Lambda(\cdot) \in \mathcal{A}_{\Lambda,\theta}$ and $\pi_{\Lambda+\delta}(\cdot) \in \mathcal{A}_{\Lambda+\delta,\theta}$ are Λ-average and $\Lambda + \delta$-average optimal policies, respectively.

Proof. Consider $\Lambda \leq 0$ and δ a real number such that $\Lambda + \delta \leq 0$. Let $\pi_\Lambda(\cdot) \in \mathcal{A}_{\Lambda,\theta}$, and $\pi_{\Lambda+\delta}(\cdot) \in \mathcal{A}_{\Lambda+\delta,\theta}$. From (16) and (19) in Lemma 2, we have

$$\begin{aligned} &\overline{R}(x_0, \pi_\Lambda) + (\Lambda + \delta) \cdot [C(x_0, \pi_\Lambda) - \theta] \leq \rho(\Lambda + \delta) \\ &= \overline{R}(x_0, \pi_{\Lambda+\delta}) + (\Lambda + \delta) \cdot [C(x_0, \pi_{\Lambda+\delta}) - \theta], \end{aligned}$$

and

$$\overline{R}(x_0, \pi_{\Lambda+\delta}) + \Lambda \cdot [C(x_0, \pi_{\Lambda+\delta}) - \theta] \leq \rho(\Lambda) = \overline{R}(x_0, \pi_\Lambda) + \Lambda \cdot [C(x_0, \pi_\Lambda) - \theta].$$

Combining this inequalities, we obtain (21). Similarly, we can prove (22). ■

The following corollary is a direct consequence of Lemma 3.

Corollary 1. *Suppose that Assumption 1 is satisfied, and let x_0 be an initial state. Then:*

(i) *The mappings $\Lambda \mapsto C(x_0, \pi_\Lambda)$ and $\Lambda \mapsto R(x_0, \pi_\Lambda)$ are nondecreasing on $(-\infty, 0]$.*

(ii) *Let $\Lambda \leq 0$. If $C(x_0, \pi_\Lambda) \leq \theta$ then $\rho(\cdot)$ is nonincreasing on $(-\infty, \Lambda]$. If $C(x_0, \pi_\Lambda) \geq \theta$ then $\rho(\cdot)$ is nondecreasing on $[\Lambda, 0]$. In particular, if $C(x_0, \pi_\Lambda) = \theta$ we obtain that $\rho(\cdot)$ is nonincreasing on $(-\infty, \Lambda]$, and nondecreasing on $[\Lambda, 0]$. Hence, $\rho(\Lambda) = \min_{\lambda \leq 0} \rho(\lambda)$.*

(iii) *$\rho(\cdot)$ is continuous on $(-\infty, 0]$.*

(iv) *If $\rho(\cdot)$ is differentiable at $\Lambda < 0$, we obtain*

$$\frac{d\rho}{d\Lambda}(\Lambda) = C(x_0, \pi_\Lambda) - \theta = \underline{C}(x_0, \pi_\Lambda) - \theta, \tag{23}$$

for each Λ-average optimal policy $\pi_\Lambda(\cdot) \in \mathcal{A}_{\Lambda,\theta}$, and any initial state $x_0 > 0$.

(v) *The following relations hold true for any $\pi_\Lambda(\cdot) \in \mathcal{A}_{\Lambda,\theta}$ and $x_0 > 0$.*

$$C(x_0, \pi_\Lambda) = \underline{C}(x_0, \pi_\Lambda) = \frac{d\rho}{d\Lambda}(\Lambda) + \theta, \tag{24}$$

and

$$R(x_0, \pi_\Lambda) = \overline{R}(x_0, \pi_\Lambda) = \rho(\Lambda) - \Lambda \cdot \frac{d\rho}{d\Lambda}(\Lambda). \tag{25}$$

In this case, these values do not depend on the initial state x_0 as well as on the Λ-average optimal policy $\pi_\Lambda(\cdot) \in \mathcal{A}_{\Lambda,\theta}$. Furthermore,

$$\rho(\Lambda) = R(x_0, \pi_\Lambda) + \Lambda \cdot [C(x_0, \pi_\Lambda) - \theta]. \tag{26}$$

Proof. (i) From (21) we see for any fixed initial state $x_0 > 0$, that $\Lambda \mapsto C(x_0, \pi_\Lambda)$ is a nondecreasing function on the parameter $\Lambda \leq 0$. By a similar reasoning, from (22), $\Lambda \mapsto \underline{C}(x_0, \pi_\Lambda)$ is also nondecreasing on $\Lambda \leq 0$.

Next, we prove that $\Lambda \mapsto R(x_0, \pi_\Lambda)$ is also nondecreasing on $(-\infty, 0]$. To this end, we suppose that $R(x_0, \pi_\Lambda)$ is not nondecreasing. Hence, there exist $\Lambda \leq 0$ and $\delta < 0$ such that $R(x_0, \pi_\Lambda) < R(x_0, \pi_{\Lambda+\delta})$. Since $\Lambda \mapsto \underline{C}(x_0, \pi_\Lambda)$ is nondecreasing, then $\underline{C}(x_0, \pi_{\Lambda+\delta}) \leq \underline{C}(x_0, \pi_\Lambda)$. Hence, from (15), (19), and the fact that $\Lambda \leq 0$, we have the contradiction

$$\rho(\Lambda) = R(x_0, \pi_\Lambda) + \Lambda \cdot [\underline{C}(x_0, \pi_\Lambda) - \theta] < R(x_0, \pi_{\Lambda+\delta}) + \Lambda \cdot [\underline{C}(x_0, \pi_{\Lambda+\delta}) - \theta] \leq \rho(\Lambda).$$

(ii) Let $\Lambda \leq 0$ and $\Lambda_1 < \Lambda_2 \leq \Lambda$ be. Since $\Lambda \mapsto C(x_0, \pi_\Lambda)$ is nondecreasing, from the first inequality in (21), we find that if $C(x_0, \pi_\Lambda) \leq \theta$ then $0 \leq (\Lambda_1 - \Lambda_2) \cdot [C(x_0, \pi_{\Lambda_2}) - \theta] \leq \rho(\Lambda_1) - \rho(\Lambda_2)$, which implies that $\rho(\cdot)$ is nonincreasing on $(-\infty, \Lambda]$.

Similarly, if $C(x_0, \pi_\Lambda) \geq \theta$, and $\Lambda \leq \Lambda_1 < \Lambda_2$, then $0 \leq (\Lambda_2 - \Lambda_1) \cdot [C(x_0, \pi_{\Lambda_1}) - \theta] \leq \rho(\Lambda_2) - \rho(\Lambda_1)$, that is, $\rho(\Lambda_1) \leq \rho(\Lambda_2)$, i.e., $\rho(\cdot)$ is nondecreasing on $[\Lambda, 0]$.

In the case of $C(x_0, \pi_\Lambda) = \theta$, then $\rho(\cdot)$ is nonincreasing on $(-\infty, \Lambda]$ and nondecreasing on $[\Lambda, 0]$. Thus, $\rho(\cdot)$ attains a minimum in Λ, that is,

$$\rho(\Lambda) = \min_{\lambda \leq 0} \rho(\lambda).$$

Thus, we have proved (ii).

(iii) Since $\Lambda \mapsto C(x_0, \pi_\Lambda)$ in nondecreasing on the parameter Λ, then it is bounded from above on $(-\infty, 0]$. Furthermore, from (4) and Proposition 1(i), $\Lambda \mapsto C(x_0, \pi_\Lambda)$ is bounded from below. So, $\Lambda \mapsto C(x_0, \pi_\Lambda)$ is bounded on $(-\infty, 0]$. Hence, taking the limit in (21) as $\delta \to 0$, we prove that $\rho(\cdot)$ is continuous in $\Lambda \leq 0$.

(iv). Fix $\Lambda < 0$. From the first inequality of (21), we obtain for each $\delta > 0$ such that $\Lambda + \delta \leq 0$

$$\frac{\rho(\Lambda - \delta) - \rho(\Lambda)}{-\delta} \leq C(x_0, \pi_\Lambda) - \theta \leq \frac{\rho(\Lambda + \delta) - \rho(\Lambda)}{\delta},$$

for each Λ-average optimal policy $\pi_\Lambda(\cdot) \in \mathcal{A}_{\Lambda,\theta}$ and initial state $x_0 > 0$. Taking the limit as $\delta \to 0$, we obtain the first equality in (23). Similarly, from (22), we obtain the second equality in (23). Hence

$$C(x_0, \pi_\Lambda) = \underline{C}(x_0, \pi_\Lambda) = \frac{d\rho}{d\Lambda}(\Lambda) + \theta.$$

(v) From (19) we obtain

$$R(x_0, \pi_\Lambda) = \overline{R}(x_0, \pi_\Lambda) = \rho(\Lambda) - \Lambda \cdot \frac{d\rho}{d\Lambda}(\Lambda),$$

where the last equality was due in virtue of statement (iii). This gives that $\Lambda \mapsto C(x_0, \pi_\Lambda)$ and $\Lambda \mapsto R(x_0, \pi_\Lambda)$ do not depend on the initial state $x_0 > 0$ and on the policies $\pi_\Lambda(\cdot) \in \mathcal{A}_{\Lambda,\theta}$ for each $\Lambda < 0$. Finally, by (25) and (19) in Lemma 2, we obtain (26), including the case when $\Lambda = 0$. ■

Lemma 4. *Suppose that there exist $\Lambda \leq 0$ and a control policy $\hat{\pi} \in \mathcal{A}$ satisfying*

$$R(x_0, \hat{\pi}) = \underline{\lim}_{T\to\infty} \frac{1}{T}\mathbb{E}\left[\int_0^T \{F(\hat{\pi}(t)) - D(x(t), \hat{\pi}(t))\}dt\right] = \rho(\Lambda) \quad \textit{and}$$

$$C(x_0, \hat{\pi}) = \overline{\lim}_{T\to\infty} \frac{1}{T}\mathbb{E}\left[\int_0^T c(x(t), \hat{\pi}(t))dt\right] = \theta, \tag{27}$$

for each initial state $x_0 > 0$. Then $\hat{\pi}$ is in $\mathcal{A}_{\Lambda,\theta}$, and it is an optimal policy for the CP. In this case, $\rho(\Lambda)$ coincides with $V_\theta^*(\cdot)$, the optimal value for the CP defined in (7), and, in addition,

$$\rho(\Lambda) = \inf_{\lambda \leq 0} \rho(\lambda). \tag{28}$$

Proof. Let $\Lambda \leq 0$ and let $\hat{\pi}(\cdot) \in \mathcal{A}$ be a control policy satisfying (27); that is,

$$R(x_0, \hat{\pi}) = \rho(\Lambda) \quad \text{and} \quad C(x_0, \hat{\pi}) = \theta$$

for each initial state $x_0 > 0$. Notice that $\hat{\pi} \in \mathcal{A}^\theta_{feas}$.

Now, from the definition of $\rho(\Lambda)$,

$$\rho(\Lambda) \geq R_\Lambda(x_0, \hat{\pi}) = \underline{\lim}_{T\to\infty} \frac{1}{T}\int_0^T \mathbb{E}\{r(x(t), \hat{\pi}(t)) + \Lambda \cdot [c(x(t), \hat{\pi}(t)) - \theta]\}dt$$

$$\geq R(x_0, \hat{\pi}) + \Lambda \cdot [C(x_0, \hat{\pi}) - \theta] = R(x_0, \hat{\pi}) = \rho(\Lambda),$$

which implies that $\hat{\pi} \in \mathcal{A}_{\Lambda,\theta}$. Furthermore, from (17)

$$R(x_0, \hat{\pi}) = \rho(\Lambda) \geq R(x_0, \pi) + \Lambda \cdot [C(x_0, \pi) - \theta] \geq R(x_0, \pi) \quad \forall \pi \in \mathcal{A}^{\theta}_{feas},$$

which yields

$$\hat{\pi} \in \mathcal{A}^{\theta}_{feas} \quad \text{and} \quad \rho(\Lambda) = R(x_0, \hat{\pi}) = \sup_{\pi \in \mathcal{A}^{\theta}_{feas}} R(x_0, \pi)$$

for each initial state $x_0 > 0$. Then $\hat{\pi}$ is an optimal policy for the CP, and $\rho(\Lambda)$ coincides with the optimal value for the CP. Now, from (17) in Lemma 2 again, for each $\lambda \leq 0$

$$\rho(\lambda) \geq R(x_0, \hat{\pi}) + \lambda \cdot [C(x_0, \hat{\pi}) - \theta] = R(x_0, \hat{\pi}) = \rho(\Lambda),$$

which implies (28). ■

Proof of Theorem 1. (a) This part follows from Corollary 1(iv).

(b) Let $\Lambda^* < 0$ a critical point of the mapping $\Lambda \mapsto \rho(\Lambda)$. From (24) and (25), we have

$$C(x_0, \pi_{\Lambda^*}) = \theta, \quad \text{and} \quad R(x_0, \pi_{\Lambda^*}) = \rho(\Lambda^*),$$

for all initial state $x_0 > 0$ and for every Λ^*-average optimal policy $\pi_{\Lambda^*} \in \mathcal{A}_{\Lambda^*,\theta}$. The rest of the proof follows from Lemma 4.

(c) Assume that there exists $\pi_0 \in \mathcal{A}_{0,\theta}$ such that $C(x_0, \pi_0) \leq \theta$ for every intial state $x_0 > 0$; that is, $\pi_0 \in \mathcal{A}^{\theta}_{feas}$. From Corollary 1(ii), $\rho(\cdot)$ is nonincreasing on $(-\infty, 0]$. Thus

$$\rho(0) = \min_{\Lambda \leq 0} \rho(\Lambda).$$

Now, from (17), we have that

$$\rho(0) = R(x_0, \pi_0) \geq R(x_0, \pi) \quad \forall \pi \in \mathcal{A}^{\theta}_{feas}.$$

Hence,

$$\rho(0) = R(x_0, \pi_0) = \max_{\pi \in \mathcal{A}^{\theta}_{feas}} R(x_0, \pi).$$

Thus, $\pi_0(\cdot)$ is an optimal policy for the CP (5)-(6), and $\rho(0)$ coincides with the optimal value for the CP. Moreover, $\rho(\cdot)$ attains a minimum in 0 satisfying (28). ■

5 Explicit Solutions for First-Order Disutility Function and a Linear Cost Rate Function

In this section, we study the case when both the disutility and the cost rate functions are of the first order.

Assumption 2. **(i)** *The disutility function D is linear; more precisely, $D(x) = ax$ for all $x > 0$, with $a > 0$.*

(ii) *The cost rate function is of the form*

$$c(x,u) = c_1 x + c_2 u \quad \forall (x,u) \in (0,\infty)\times(0,\infty), \tag{29}$$

with $c_1, c_2 \in \mathbb{R}$ *satisfying*

$$c_1 + \eta c_2 > 0. \tag{30}$$

(iii) *The pollution decay* η *is positive and the nonzero diffusion constant* σ *satisfies*

$$\eta > \sigma. \tag{31}$$

Lemma 5. *Suppose that Assumption 2 holds. Then, for each* $\Lambda \leq 0$, *we have:*

(a) *Conditions (2), (3), (4), and Assumption 1 hold with* $m = 2$. *Besides,*

$$\theta_{min} = 0 \quad \text{and} \quad \theta_{max} := \sup_{\pi \in \mathcal{A}} C(x_0, \pi) = \frac{(c_1 + \eta c_2)}{\eta}\gamma.$$

(b) *The function* $h_\Lambda(x) = \frac{(\Lambda \cdot c_1 - a)}{\eta} x$ *and the constant*

$$\rho(\Lambda) = \begin{cases} F(I(a_\Lambda)) - a_\Lambda I(a_\Lambda) - \Lambda \cdot \theta & \text{if } \Lambda < \frac{a - \eta F'(\gamma)}{c_1 + \eta c_2}, \\ F(\gamma) - a_\Lambda \gamma - \Lambda\theta & \text{if } \frac{a - \eta F'(\gamma)}{c_1 + \eta c_2} \leq \Lambda \end{cases} \tag{32}$$

together constitute a solution of the HJB (9) for the Λ*-UP, where* $I(\cdot)$ *is the inverse of the derivative* $F'(\cdot)$, *and* $a_\Lambda := \frac{a - \Lambda(c_1 + \eta c_2)}{\eta}$.

(c) *The map* $\Lambda \mapsto \rho(\Lambda)$ *is differentiable on* $(-\infty, 0)$.

(d) *The policy* $f_\Lambda \in \mathbb{F}_{\Lambda,\theta}$ *that attains the maximum in (9) is unique and it becomes a constant function given by*

$$f_\Lambda = \begin{cases} I(a_\Lambda) & \text{if } \Lambda < \frac{a - \eta F'(\gamma)}{c_1 + \eta c_2}, \\ \gamma & \text{if } \frac{a - \eta F'(\gamma)}{c_1 + \eta c_2} \leq \Lambda. \end{cases} \tag{33}$$

(e) *For each* $\Lambda \leq 0$, *and for every initial state* $x_0 > 0$,

$$C(x_0, f_\Lambda) = \frac{(c_1 + \eta c_2)}{\eta} f_\Lambda =: C(f_\Lambda), \tag{34}$$

and

$$R(x_0, f_\Lambda) = F(f_\Lambda) - \frac{a}{\eta} f_\Lambda =: R(f_\Lambda), \tag{35}$$

with f_Λ *being the function in (33).*

Remark 3. Taking $\Lambda = 0$, expressions (32) and (33) coincide with the explicit solutions for the unconstrained first-order disutility established in [8, Page 706].

Proof of Lemma 5. (a) From Assumption 2(i), we can take D_0 sufficiently large such that $\frac{1}{D_0} < a < D_0$ and, by a direct calculation, we can show that (3) is satisfied with $m = 2$. Similarly, under Assumption 2(ii) we obtain that (4) holds with $m = 2$. Also, Assumption 2(iii) implies that Assumption 1 yields with $m = 2$.

On the other hand, from well-known results (see [1, Theorem 7.3.9] or [2, Theorem 7.6], for instance) and by (29), there exist pairs (θ_{min}, h_*), $(\theta_{max}, h^*) \in \mathbb{R} \times C^2(0, \infty)$, where h_* and h^* are viscosity solutions of the HJB equations

$$\theta_{min} = \min_{u \in [0,\gamma]} [c(x,u) + L^u h_*(x)] \quad \forall x > 0,$$

and

$$\theta_{max} = \max_{u \in [0,\gamma]} [c(x,u) + L^u h^*(x)] \quad \forall x > 0,$$

with $L^u h(x) := (u - \eta x)h'(x) + \frac{\sigma^2 x^2}{2} h''(x)$ for all $h \in C^2(0, \infty)$. Moreover, considering that $h_*(x) = a_* x + b_*$ and $h^*(x) = a^* x + b^*$, we obtain

$$\theta_{min} = (c_1 - \eta a_*)x + \min_{u \in [0,\gamma]} [(c_2 + a_*)u],$$

and

$$\theta_{max} = (c_1 - \eta a^*)x + \max_{u \in [0,\gamma]} [(c_2 + a^*)u].$$

Hence, $a_* = a^* = c_1/\eta$, and by (30), $\theta_{min} = 0$ and $\theta_{max} = (c_1 + \eta c_2)\gamma/\eta$. This proves (a)

(b) Replacing $h_\Lambda(x) = \frac{(\Lambda c_1 - a)}{\eta} x$ into the HJB equation (9), we obtain

$$\rho(\Lambda) = \sup_{u \in [0,\gamma]} \{F(u) - a_\Lambda u\} - \Lambda\theta, \tag{36}$$

with $a_\Lambda := \frac{a - \Lambda(c_1 + \eta c_2)}{\eta}$. Hence, from the properties in (2) of the social utility function F, a direct calculation yields to (32).

(c) Clearly $\rho(\cdot)$ is differentiable in Λ if $\Lambda < \frac{a - \eta F'(\gamma)}{c_1 + \eta c_2}$ or $\Lambda > \frac{a - \eta F'(\gamma)}{c_1 + \eta c_2}$. We only have to proof the differentiability of $\rho(\cdot)$ in $\Lambda_0 := \frac{a - \eta F'(\gamma)}{c_1 + \eta c_2}$. Namely, by a direct calculation, we easily deduce that

$$\lim_{\delta \to 0^+} \frac{\rho(\Lambda_0 + \delta) - \rho(\Lambda_0)}{\delta} = \lim_{\delta \to 0^+} \frac{\rho(\Lambda_0 - \delta) - \rho(\Lambda_0)}{-\delta} = \frac{c_1 + \eta c_2}{\eta} - \theta.$$

(d) Let $f_\Lambda \in \mathbb{F}$ be the function that attains the maximum in (9); in other words, it attains the maximum in (36). Since $F(\cdot)$ is a concave function we obtain that f_Λ is unique, and by using (12) in Proposition 1, we obtain that f_Λ is *constant* and is given by (33).

(e) Plugging $\pi(t) = f_\Lambda(x(t)) = f_\Lambda$ into (1), we obtain

$$dx(t) = [f_\Lambda - \eta x(t)]dt + \sigma x(t) dB(t), \quad x(0) = x > 0, \tag{37}$$

which is a *scalar linear stochastic differential equation.* Thus, the solution $x(t) = x^{f_\Lambda}(t)$ of (37) with initial state $x > 0$, is given by (see, for instance, [3])

$$x(t) = \Phi_t \left(x + f_\Lambda \int_0^t \Phi_s^{-1} ds \right),$$

where

$$\Phi_t = \exp\left[-(\eta + \sigma^2/2)t + \sigma(B(t) - B(0))\right] \quad \text{for all } t \geq 0.$$

Hence, by the properties of the stochastic integrals, we have

$$\mathbb{E}x(t) = \frac{f_\Lambda}{\eta} + \frac{(\eta x - f_\Lambda)}{\eta} \exp(-\eta t). \tag{38}$$

Therefore, (34) and (35) follow from (38), combined with the explicit form of $F(\cdot)$ in (2), and the definitions of $D(\cdot)$, and $c(\cdot, \cdot)$ in Assumption 2. ∎

We have arrived at the main result of the section regarding to show explicit solutions for optimal consumption strategies as well as an analytical expression of the optimal social welfare.

Theorem 2. *Suppose that Assumption 2 holds. Let θ be such that $\theta_{min} < \theta < \theta_{max}$, with θ_{min} and θ_{max} as in Lemma 5(a), and consider the CP (5)-(6).*

(a) *If $F'(\frac{\eta\theta}{c_1+\eta c_2}) > \frac{a}{\eta}$, then there exists $\Lambda^* < 0$ a critical point of $\rho(\cdot)$. Furthermore, $f_{\Lambda^*} = \frac{\eta\theta}{c_1+\eta c_2} \in \mathbb{F}_{\Lambda^*,\theta}$ is an optimal policy for the CP (5)-(6), with optimal value given by*

$$\rho(\Lambda^*) = R(f_{\Lambda^*}) = F\left(\frac{\eta\theta}{c_1 + \eta c_2}\right) - \frac{a\theta}{c_1 + \eta c_2}.$$

In addition, $C(f_{\Lambda^*}) = \theta$.

(b) *If $F'(\frac{\eta\theta}{c_1+\eta c_2}) \leq \frac{a}{\eta}$, then $f_0 = I(\frac{a}{\eta}) \in \mathbb{F}_{0,\theta}$ and $C(f_0) = \frac{(c_1+\eta c_2)}{\eta} I(\frac{a}{\eta}) \leq \theta$. Moreover, f_0 is an optimal policy for the CP (5)-(6), with optimal value:*

$$\rho(0) = R(f_0) = F\left(I\left(\frac{a}{\eta}\right)\right) - \frac{a}{\eta} I\left(\frac{a}{\eta}\right).$$

Proof. (a) Let $\Lambda^* \in \mathbb{R}$ be such that

$$a_{\Lambda^*} = F'\left(\frac{\eta\theta}{c_1 + \eta c_2}\right) > \frac{a}{\eta}, \tag{39}$$

with $a_{\Lambda^*} = \frac{a - \Lambda^*(c_1+\eta c_2)}{\eta}$. Thus,

$$\Lambda^* = \frac{a - \eta F'(\frac{\eta\theta}{c_1+\eta c_2})}{c_1 + \eta c_2} < 0. \tag{40}$$

Since $\theta_{min} = 0 < \theta < \theta_{max} = \frac{(c_1+\eta c_2)}{\eta}\gamma$, then $0 < \frac{\eta\theta}{c_1+\eta c_2} < \gamma$. Because $F'(\cdot)$ is a strictly decreasing function, we have $F'(\gamma) < F'(\frac{\eta\theta}{c_1+\eta c_2}) = a_{\Lambda^*}$. Then, there exists V an open neighborhood of Λ^*, such that $F'(\gamma) < a_\Lambda$ for all $\Lambda \in V$, or equivalently, $\Lambda < \frac{a-\eta F'(\gamma)}{c_1+\eta c_2}$ for all $\Lambda \in V$. From (32),

$$\rho(\Lambda) = F(I(a_\Lambda)) - a_\Lambda I(a_\Lambda) - \Lambda\theta \quad \forall \Lambda \in V. \tag{41}$$

Hence, $\rho(\cdot)$ is differentiable on V, with continuous derivative

$$\frac{d\rho}{d\Lambda}(\Lambda) = \frac{(c_1+\eta c_2)}{\eta} I(a_\Lambda) - \theta \quad \forall \Lambda \in V.$$

Replacing $\Lambda = \Lambda^*$, with Λ^* given by (40), in this latter equation, from (39) and from the fact that $I(\cdot)$ is the inverse of the derivative $F'(\cdot)$, we deduce that Λ^* is a critical point of $\rho(\cdot)$. In this case, based on (33), $f_{\Lambda^*} \in \mathbb{F}_{\Lambda^*,\theta}$ takes the form:

$$f_{\Lambda^*} = I(a_{\Lambda^*}) = \frac{\eta\theta}{c_1+\eta c_2}.$$

Thus, from Theorem 1(b), f_{Λ^*} is an optimal policy for the CP (5)-(6), with average optimal social welfare for the CP $\rho(\Lambda^*) = R(f_{\Lambda^*})$ given by

$$\rho(\Lambda^*) = F\Big(\frac{\eta\theta}{c_1+\eta c_2}\Big) - \frac{a\theta}{c_1+\eta c_2}.$$

On the other hand, if we replace again $\Lambda = \Lambda^*$ into (34) and (35), we obtain that $R(f_{\Lambda^*})$ coincides with $\rho(\Lambda^*)$, and $C(f_{\Lambda^*}) = \theta$.

(b) Notice that

$$F'(\gamma) < F'(\frac{\eta\theta}{c_1+\eta c_2}) \leq \frac{a}{\eta} = a_0;$$

that is,

$$0 < \frac{a-\eta F'(\gamma)}{c_1+\eta c_2}.$$

From (32) and (33), and the fact that $a_0 = a/\eta$,

$$\rho(0) = F\Big(I\Big(\frac{a}{\eta}\Big)\Big) - \frac{a}{\eta} I\Big(\frac{a}{\eta}\Big) \quad \text{and} \quad f_0 = I\Big(\frac{a}{\eta}\Big). \tag{42}$$

Replacing $f_0 = I(\frac{a}{\eta})$ into (34), and using the fact of that $I(\cdot) = (F')^{-1}(\cdot)$ is decreasing, we get

$$C(f_0) = \frac{(c_1+\eta c_2)}{\eta} I\Big(\frac{a}{\eta}\Big) \leq \frac{(c_1+\eta c_2)}{\eta} I\Big(F'\Big(\frac{\eta\theta}{c_1+\eta c_2}\Big)\Big) = \theta.$$

Thus, from Theorem 1(c), f_0 is an optimal policy for the CP (5)–(6), with optimal value $\rho(0) = R(f_0)$. By a direct calculation, we can replace f_0 given by (42) into (35), to comfirm that $\rho(0) = R(f_0)$. ∎

6 Acknowledgement

This research was partially supported by CONACyT grants 221291 and 238045.

References

1. Arapostathis, A., Borkar, V.S., and Ghosh, M.K. *Ergodic Control of Diffusion Processes. Encyclopedia of Mathematics and its Applications*, Cambridge University Press, Cambridge, 2012.
2. Arapostathis, A. and Borkar, V.S.: Uniform recurrence properties of controlled diffusions and applications to optimal control. *SIAM J. Control Optim.* **48** (2010) 4181-4223.
3. Arnold L. *Stochastic Differential Equations: Theory and Applications.* John Wiley & Sons, Inc., NY-London-Sydney, 1974.
4. Ferreyra, G. and Sundar, P.: Comparison of solutions of stochastic differential equations and applications. *Stoch. Anal. Appl.* **18** (2000) pp. 211-229.
5. Ikeda, N. and Watanabe, S.: A comparison theorem for solutions of stochastic differential equations and its applications, *Osaka J. Math.* **14** (1977) 619-633.
6. Jasso-Fuentes, H. and Hernández-Lerma, O.: Characterizations of overtaking optimality for controlled diffusion processes. *Appl. Math. Optim.* **57** (2008) 349-369.
7. Jorgensen, S., Martin-Herran, G. and Zaccour, G.: Dynamic games in the economics and management of pollution. *Environ. Model. Assess.* **15** (2010) 433-467.
8. Kawaguchi, K., Morimoto, H. Long-run average welfare in a pollution accumulation model. *J. Econom. Dyn. Control 31* (2007) 703-720.
9. Mendoza-Pérez, A.F., Jasso-Fuentes, H. and Hernández-Lerma, O.: The Lagrange approach to ergodic control of diffusions with cost constraints. *Optimization* **64** (2015) 179-196.
10. Morimoto, H. *Stochastic Control and Mathematical Modeling.* Cambridge University Press, Cambridge, 2010.
11. Ploeg, F. and van der Withagen, C.: Pollution control and the Ramsey problem. *Environmental and Resourse Economics* **1** (1991) 215-236.
12. Toman, M. and Withagen, C.: Accumulative pollution, “clean technology”, and policy design. *Resource and Energy Economics* **22** (2000) 367-384.

On the Optimal Stopping of a Skew Geometric Brownian Motion

Pui Chan Lon[1], Neofytos Rodosthenous[2] and Mihail Zervos[1]

[1] London School of Economics, Department of Mathematics
Houghton Street, London WC2A 2AE, UK
lonpuichan@gmail.com
mihalis.zervos@gmail.com

[2] Queen Mary University of London, School of Mathematical Sciences,
London E1 4NS, UK
n.rodosthenous@qmul.ac.uk

Abstract. The purpose of this paper is to present the explicit solution to an optimal stopping problem that involves a skew geometric Brownian motion and arises in the context of pricing perpetual American options. The problem that we study is the first non-trivial example of optimally stopping a diffusion with generalised drift that has been fully solved. It turns out that the optimal strategy takes several qualitatively different forms, depending on parameter values.
Keywords: optimal stopping, skew Brownian motion, perpetual American options.
AMS 2000 subject classification: Primary 60G40, Secondary 60J55, 60J60

1 Introduction

We consider the problem of optimally stopping the skew geometric Brownian motion, which is the solution to the stochastic differential equation

$$dX_t = bX_t\,dt + \beta\,dL_t^z + \sigma X_t\,dW_t, \quad X_0 = x > 0, \tag{1}$$

where L^z is the symmetric local time of X at a given level $z > 0$ and W is a standard one-dimensional Brownian motion. In particular, we solve the optimal stopping problem whose value function is defined by

$$v(x) = \sup_\tau \mathbb{E}\left[e^{-r\tau}(X_\tau - K)^+\right], \tag{2}$$

which corresponds to the value of a perpetual American option written on an underlying asset with dynamics modelled by X.

The process X introduced by (1) is an example of a diffusion with generalised drift. It is closely related to the skew Brownian motion that was introduced by [6, Section 4.2, Problem 1] who constructed it by means of a reflected standard

Brownian motion by changing the sign of each excursion independently with probability $1-p$, so that a given excursion is positive with probability p and negative with probability $1-p$. The construction of a skew Brownian motion from its scale function and speed measure was further discussed by [13] and was studied by means of stochastic differential equations involving local times, such as the one given by (1), by [5]. A wide class of diffusions with generalised drift that include the skew geometric Brownian motion that we consider here as well as the skew Brownian motion were exhaustively studied by [3].

The theory of optimal stopping has a well-developed body of theory that has been documented in several references, including the monographs [2, 4, 8, 9] and [12]. Apart from results of a general nature, there are several problems involving the optimal stopping of diffusions that have been explicitly solved. To the best of our knowledge, the optimal stopping of diffusions with generalised drift has been considered in the literature but no example has been solved in its full generality. In particular, [9, Section IV.9.3] provided three examples of optimally stopping such processes, which show that the so-called "principle of smooth fit", namely, the C^1 regularity of the value function along the free-boundary, may fail. In a similar spirit, [1] considered two examples of optimally stopping a skew Brownian motion that are associated with one-sided optimal stopping rules. Furthermore, [10] discussed a general method for deriving the solution to optimal stopping problems that can be applied to ones involving diffusions with generalised drift. The results that we present in this paper are exhaustive. In particular, our analysis is based on the use of variational inequalities (see [7] for some recent relevant developments in this direction).

The paper is organised as follows. In Section 2, we state the precise definition of the problem we study. In Section 3, we prove a verification theorem that provides sufficient conditions for an appropriate solution to a variational inequality to identify with the optimal stopping problem's value function. In Section 4, we present a detailed study of the increasing minimal r-excessive function of the process X, which plays a fundamental role in the optimal stopping problem's solution. Finally, we discuss the solution to the optimal stopping problem that we consider in Section 5.

2 The Optimal Stopping Problem

We fix a filtered probability space $(\Omega, \mathcal{F}, (\mathcal{F}_t), \mathbb{P})$ satisfying the usual conditions and supporting a standard one-dimensional $(\mathcal{F}_t)$-Brownian motion W. Also, we denote by $\mathcal{T}$ the set of all $(\mathcal{F}_t)$-stopping times.

We consider the skew geometric Brownian motion that satisfies the stochastic differential equation

$$dX_t = bX_t\,dt + \beta\,dL_t^z + \sigma X_t\,dW_t, \quad X_0 = x > 0, \tag{3}$$

for some constants $b \in \mathbb{R}$, $\beta \in]-1,1[\setminus\{0\}$, $z > 0$ and $\sigma \neq 0$. The process L^z appearing here is the symmetric local time of X at level z (see [11, Exercise VI.1.25] for the precise definition). The stochastic differential equation (3) has a unique strong solution that is a strictly positive process (see [3] and [5]). The value function of the optimal stopping problem that we study is defined by

$$v(x) = \sup_{\tau \in \mathcal{T}} \mathbb{E}\left[e^{-r\tau}(X_\tau - K)^+\right], \tag{4}$$

for some constants $r, K > 0$. We make the following assumption.

Assumption 1. *$b \in \mathbb{R}$, $\beta \in]-1,1[\setminus \{0\}$, $z > 0$, $\sigma \neq 0$, $r, K > 0$ and $r > b$.*

3 A Verification Theorem

We solve the optimal stopping problem defined by (3)–(4) by constructing an appropriate solution to the variational inequality given by

$$\max\left\{\frac{1}{2}\sigma^2 x^2 w''(x) + bxw'(x) - rw(x),\ (x-K)^+ - w(x)\right\} = 0 \tag{5}$$

inside $]0, z[\, \cup\,]z, \infty[$, and

$$\max\left\{(1+\beta)w'_+(z) - (1-\beta)w'_-(z),\ (z-K)^+ - w(z)\right\} = 0. \tag{6}$$

The following result, which we prove in the Appendix, provides sufficient conditions under which the optimal stopping problem's value function v identifies with a solution $w : \mathbb{R}_+ \to \mathbb{R}_+$ to (5)–(6).

Theorem 1. *Consider the optimal stopping problem formulated in Section 2 and suppose that Assumption 1 holds true. Let $w : \mathbb{R}_+ \to \mathbb{R}_+$ be a function satisfying the variational inequality (5)–(6) in the sense that*

(I) *w is C^1 inside $]0, z[\, \cup\,]z, \infty[$ and C^2 inside $]0, z[\, \cup\,]z, \infty[\setminus S$, where S is a finite set,*

(II) *w satisfies (5) inside $]0, z[\, \cup\,]z, \infty[\setminus S$, and*

(III) *w satisfies (6),*

(IV) *w satisfies*

$$\sup_{y \in]0,n[} |w'_-(y)| < \infty \text{ for all } n \geq 1 \quad \text{and} \quad \lim_{y \to \infty} \frac{w(y)}{\psi(y)} = 0, \tag{7}$$

where w_- is the left-hand derivative of w and ψ is the increasing minimal r-excessive function of X that we study in the next section.

Then $w(x) = v(x)$ for all $x > 0$ and

$$\tau_\star = \inf\left\{t \geq 0 \mid w(X_t) = (X_t - K)^+\right\} \tag{8}$$

defines an optimal stopping time.

4 The Increasing Minimal r-Excessive Function ψ

The increasing minimal r-excessive function ψ plays a central role in the solution to the optimal stopping problem that we derive. This function is defined by

$$\psi(x) = \psi(y)\mathbb{E}\left[e^{-rT_y}\right] \quad \text{for all } y > x \tag{9}$$

and all initial values $x > 0$ of (3), where T_y is the first hitting time of $\{y\}$, namely,

$$T_y = \inf\{t \geq 0 \mid X_t = y\}.$$

In particular, ψ is unique, up to multiplicative strictly positive constants, and strictly increasing. In view of standard theory of one-dimensional diffusions (e.g., see [6]), ψ should satisfy the Euler ODE

$$\frac{1}{2}\sigma^2 x^2 w''(x) + bxw'(x) - rw(x) = 0 \tag{10}$$

as well as the identity

$$(1+\beta)w'_+(z) = (1-\beta)w'_-(z). \tag{11}$$

It is well-known that every solution to the ODE (10) is given by

$$w(x) = Ax^n + Bx^m,$$

for some constants $A, B \in \mathbb{R}$, where $m < 0 < n$ are the solutions to the quadratic equation

$$\frac{1}{2}\sigma^2 k^2 + \left(b - \frac{1}{2}\sigma^2\right) k - r = 0,$$

which are given by

$$m, n = \frac{-\left(b - \frac{1}{2}\sigma^2\right) \mp \sqrt{\left(b - \frac{1}{2}\sigma^2\right)^2 + 2\sigma^2 r}}{\sigma^2}.$$

Furthermore, we can check that a solution to the ODE (10) that satisfies (11) is given by

$$\psi(x) = \begin{cases} x^n, & \text{if } x < z, \\ Ax^n + B(z)x^m, & \text{if } x \geq z, \end{cases} \tag{12}$$

for

$$A = \frac{n(1-\beta) - m(1+\beta)}{(n-m)(1+\beta)} \begin{cases} > 1, & \text{if } \beta < 0, \\ \in \,]0,1[, & \text{if } \beta > 0, \end{cases} \tag{13}$$

and

$$B(z) = \frac{2n\beta}{(n-m)(1+\beta)} z^{n-m} \begin{cases} < 0, & \text{if } \beta < 0, \\ > 0, & \text{if } \beta > 0. \end{cases} \tag{14}$$

We prove the following result in the Appendix.

Lemma 1. *Suppose that the problem data is as in Assumption 1. The function defined by (12)–(14) identifies with the minimal r-excessive function ψ defined by (9).*

It turns out that the function ψ can take qualitetively different forms, depending on parameter values (see Figures 1.a-c).

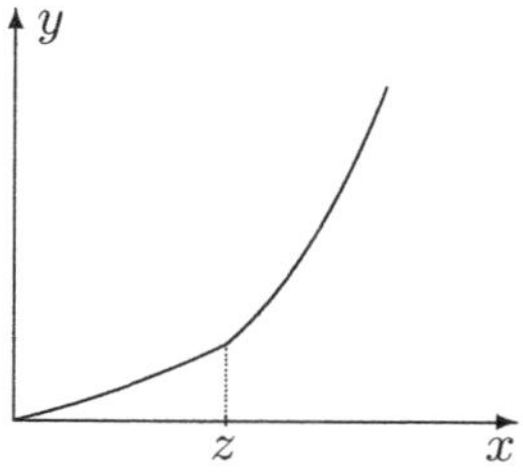

Figure 1.a

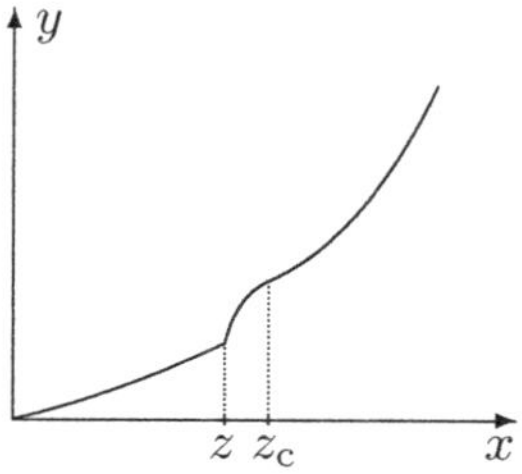

Figure 1.b

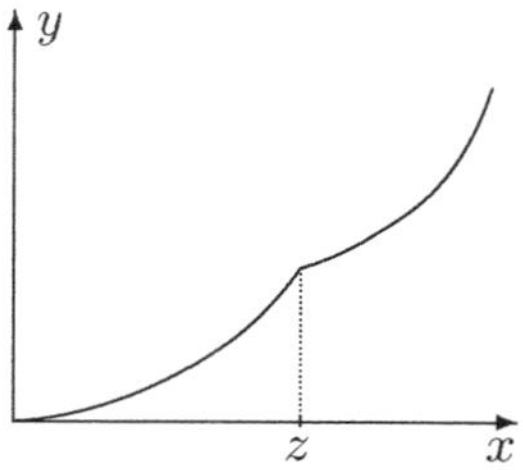

Figure 1.c

The following result, which we prove in the Appendix, makes use of the facts that

$$r > b \quad \Leftrightarrow \quad n > 1, \tag{15}$$

$$n + m - 1 = -\frac{2b}{\sigma^2} \quad \text{and} \quad nm = -\frac{2r}{\sigma^2}. \tag{16}$$

Lemma 2. *Suppose that the problem data is as in Assumption 1 and consider the function ψ defined by (12)–(14). The equivalence*

$$\psi'_-(z) = nz^{n-1} < nAz^{n-1} + mB(z)z^{m-1} = \psi'_+(z) \quad \Leftrightarrow \quad \beta < 0 \tag{17}$$

holds true. Furthermore,

$$\textit{if } b > 0, \textit{ then } \beta_c := \frac{n-1}{n+2m-1} \in \,]-1, 0[, \tag{18}$$

while

(I) *if ($b \leq 0$ and $\beta \in \,]-1, 0[$) or ($b > 0$ and $\beta \in [\beta_c, 0[$), then*

$$(n-1)(1-\beta) - 2m\beta \geq 0 \tag{19}$$

and ψ is convex (see Figure 1.a);

(II) *if $b > 0$ and $\beta \in \,]-1, \beta_c[$, then*

$$(n-1)(1-\beta) - 2m\beta < 0, \tag{20}$$

the restrictions of ψ in $[0, z]$ as well as in $[z_c, \infty[$ are convex, while the restriction of ψ in $[z, z_c]$ is concave, where

$$z_c = \left(\frac{2m(m-1)\beta}{(n-1)\big[m(1-\beta) - n(1+\beta)\big]}\right)^{\frac{1}{n-m}} z > z$$

(see Figure 1.b);

(III) *if $\beta \in \,]0, 1[$, then the restrictions of ψ in $[0, z]$ as well as in $[z, \infty[$ are convex but ψ is not convex in its entire domain (see Figure 1.c).*

5 The Solution to the Optimal Stopping Problem

In this section, we discuss the solution to the optimal stopping problem formulated in Section 2. The rich structure of the minimal r-excessive function ψ that we studied in the previous section results in a rather involved and lengthy analysis and in several possible optimal stopping rules. In what follows, we have organised the presentation of our results based on the qualitative nature of ψ, as this is depicted by Figures 1.a-c. Due to space limitations, we cannot include a detailed analysis or any proofs here: these will appear in a future research paper.

5.1 The Case when $-1 < \beta < 0$ and ψ is Convex (see Figure 1.a)

In this case, the convexity of ψ suggests that the solution to the optimal stopping problem we consider should have the qualitative form that arises in the pricing of a perpetual American option written on a standard geometric Brownian motion. Accordingly, we expect that there exists a point $a > 0$ such that the interval $]0, a[$ is the waiting region, while the interval $[a, \infty[$ is the stopping region (see Figure 2).

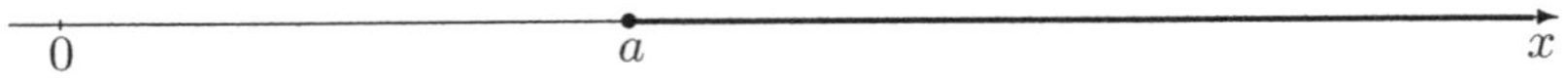

Figure 2

Indeed, it turns out that the function defined by

$$w(x) = \begin{cases} \Gamma\psi(x), & \text{if } x \leq a, \\ x - K, & \text{if } x > a, \end{cases} \tag{21}$$

satisfies the requirements of Theorem 1 and identifies with the value function v for appropriate choices for the constant $\Gamma > 0$ and the free-boundary point $a > 0$. For sufficiently small values of z, we expect that $a > z$, while, for sufficiently large values of z, we expect that $a < z$. In both of these cases, we use the so-called "principle of smooth fit", namely, the requirement that the value function should be C^1 along the free-boundary point a, to determine Γ, a. Such a requirement yields the system of equations

$$\begin{cases} \Gamma\psi(a) = a - K, \\ \Gamma\psi'(a) = 1, \end{cases} \quad \Leftrightarrow \quad \begin{cases} \Gamma = (a - K)/\psi(a) = 1/\psi'(a), \\ a\psi'(a) - \psi(a) - K\psi'(a) = 0. \end{cases} \tag{22}$$

In view of the definition (12) of ψ, we can see that the possibility that $a < z$ implies that

$$a = \frac{nK}{n-1} > 0 \tag{23}$$

solves the equation for a in (22). On the other hand, we can check that, if $z < a$, then the equation for a in (22) is equivalent to

$$F(a; z) = 0, \tag{24}$$

where

$$F(a;z) = a^{-m+1}\big[(a-K)\psi'(a) - \psi(a)\big]$$
$$= \big[(n-1)a - nK\big]Aa^{n-m} + \big[(m-1)a - mK\big]B(z). \qquad (25)$$

Finally, we note that for intermediate values of z such that $a = z$, the function w given by (21) is only C^0 at a. We can show that equation (24) has a unique solution $a > \frac{rK}{r-b} \vee z$ if $z < nK/(n - \frac{1+\beta}{1-\beta})$. Furthermore, we can establish the following result.

Theorem 2. *Consider the optimal stopping problem formulated in Section 2 and suppose that the problem data is as in Assumption 1. Also, suppose that $\beta \in \,]-1, 0[$ and that (19) in Lemma 2.(I) holds true. Consider the following possibilities:*

(I) *if $z < \frac{nK}{n - \frac{1+\beta}{1-\beta}}$, then let $\Gamma > 0$ be given by (22) and let $a > z$ be the unique solution to (24);*

(II) *if $z \in \left[\frac{nK}{n - \frac{1+\beta}{1-\beta}}, \frac{nK}{n-1}\right]$, then define $\Gamma = (z-K)z^{-n} > 0$ and $a = z$;*

(III) *if $z > \frac{nK}{n-1}$, then let $\Gamma > 0$ be given by (22) and $a < z$ be given by (23).*

The function w defined by (21) for these choices of parameters identifies with the value function v of the discretionary stopping problem, while the $(\mathcal{F}_t)$-stopping time defined by

$$\tau_\star = \inf\big\{t \geq 0 \mid X_t \geq a\big\} \qquad (26)$$

is optimal.

5.2 The Case when $-1 < \beta < 0$ and ψ is Not Convex (see Figure 1.b)

In this case, we expect that the solution to the optimal stopping problem we study has the same qualitative form as in the previous section (see Figure 2) for at least a range of possible values of z. In particular, we can prove the following result, where $\overline{z} < \frac{rK}{r-b}$ is the unique solution to the equation

$$\frac{\overline{z} - K}{\overline{z}^n}\psi\big(a(\overline{z})\big) - a(\overline{z}) + K = 0. \qquad (27)$$

Theorem 3. *Consider the optimal stopping problem formulated in Section 2 and suppose that the problem data is as in Assumption 1. Also, suppose that $\beta \in \,]-1, 0[$ and that (20) in Lemma 2.(II) holds true. Consider the following possibilities:*

(I) *if $z < \overline{z}$, then let $\Gamma > 0$ be given by (22) and $a > z$ be the unique solution to (24);*

(I) *if $z \in \left[\frac{rK}{r-b}, \frac{nK}{n-1}\right]$, then define $\Gamma = (z-K)z^{-n} > 0$ and $a = z$;*

(I) *if $z > \frac{nK}{n-1}$, then let $\Gamma > 0$ be given by (22) and $a < z$ be given by (23).*

The function w defined by (21) for these choices of parameters identifies with the value function v of the discretionary stopping problem, while the $(\mathcal{F}_t)$-stopping time defined by (26) is optimal.

To proceed further, we consider the problem that arises when $z \in \left[\overline{z}, \frac{rK}{r-b}\right[$. In this case, a careful inspection of (21)–(22), (27) and the nature of the solution studied in Theorem 3 suggests that z becomes an isolated point of the stopping region. Accordingly, we look for a point $\xi > z$ such that the stopping region is the set $\{z\} \cup [\xi, \infty[$, while the waiting region is the set $]0, z[\cup]z, \xi[$ (see Figure 3).

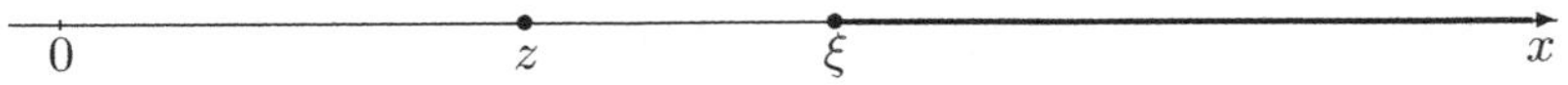

Figure 3

Indeed, it turns out that, if $z \in \left[\overline{z}, \frac{rK}{r-b}\right[$, then the function defined by

$$\overline{w}(x) = \begin{cases} z^{-n}(z-K)x^n, & \text{if } x \le z, \\ Cx^n + Dx^m, & \text{if } x \in \,]z, \xi[, \\ x - K, & \text{if } x \ge \xi, \end{cases} \tag{28}$$

satisfies the requirements of Theorem 1 and identifies with the value function v for appropriate choices for the constants C, D and the free-boundary point $\xi > 0$. To determine these constants, we require that $\overline{w}$ should be continuous at z and should satisfy the so-called "principle of smooth fit", namely, be C^1, at ξ, which yields

$$C = -\frac{1}{n-m}\big[(m-1)\xi - mK\big]\xi^{-n}, \quad D = \frac{1}{n-m}\big[(n-1)\xi - nK\big]\xi^{-m}, \tag{29}$$

and

$$J(\xi; z) = 0, \tag{30}$$

where

$$\begin{aligned} J(\xi; z) = {} & \big[(n-1)\xi - nK\big]\xi^{-m} \\ & - \big[(m-1)\xi - mK\big]z^{n-m}\xi^{-n} - (n-m)(z-K)z^{-m}. \end{aligned} \tag{31}$$

We can show that, if $z \in \left[\overline{z}, \frac{rK}{r-b}\right[$, then equation (30) has a unique solution $\xi > \frac{rK}{r-b}$. Furthermore, we can prove the following result.

Theorem 4. *Consider the optimal stopping problem formulated in Section 2 and suppose that the problem data is as in Assumption 1. Also, suppose that $\beta \in \,]-1, 0[$ and that (20) in Lemma 2.(II) holds true. Given any $z \in \left[\overline{z}, \frac{rK}{r-b}\right[$, where $\overline{z}$ is defined by (27), the function $\overline{w}$ defined by (28) for $C, D > 0$ being given by (29) and $\xi > z$ being the unique solution to equation (30) identifies with the value function v of the discretionary stopping problem. In particular, the $(\mathcal{F}_t)$-stopping time defined by*

$$\tau_\star = \inf\big\{t \ge 0 \mid \; X_t \in \{z\} \cup [\xi, \infty[\big\} \tag{32}$$

is optimal.

5.3 The Case when $\mathbf{0 < \beta < 1}$ (see Figure 1.c)

In the case that arises when $\beta \in \,]0,1[$, the form of the function ψ suggests that the function w defined by (21) may identify with the value function v for appropriate values of the constant Γ and the free-boundary point a, provided that $a > z$. Once again, we appeal to the "principle of smooth fit", which yields the system of equations (22) and equation (24) for a. In particular, we can establish the following result.

Theorem 5. *Consider the optimal stopping problem formulated in Section 2, suppose that the problem data is as in Assumption 1 and assume that $\beta \in \,]0,1[$. There exists a unique z^* such that the function w that is defined by (21), for $\Gamma > 0$ being given by (22) and $a > \frac{rK}{r-b} \vee z$ being the unique solution to (24) identifies with the value function v of the discretionary stopping problem if and only if $z < z^*$. In particular, the $(\mathcal{F}_t)$-stopping time defined by (26) is optimal.*

A careful inspection of (21)–(22) and the nature of the solution studied in Theorem 5 suggests that the stopping region involves two disjoint intervals if $z \geq z^*$. In particular, we expect that there exist two free-boundary points $\frac{nK}{n-1} < \gamma < \zeta$ such that the set $\left[\frac{nK}{n-1}, \gamma\right] \cup [\zeta, \infty[$ is the stopping region, while the set $]0, \frac{nK}{n-1}[\,\cup\,]\gamma, \zeta[$ is the waiting region (see Figure 4).

Figure 4

Indeed, it turns out that, if $z \geq z^*$, then the function defined by

$$\widetilde{w}(x) = \begin{cases} \frac{1}{n}\left(\frac{nK}{n-1}\right)^{-n+1} x^n, & \text{if } x \leq \frac{nK}{n-1}, \\ C_\ell x^n + D_\ell x^m, & \text{if } x \in [\gamma, z], \\ C_r x^n + D_r x^m, & \text{if } x \in \,]z, \zeta], \\ x - K, & \text{if } x \in \,]\frac{nK}{n-1}, \gamma[\, \cup \,]\zeta, \infty[, \end{cases} \tag{33}$$

satisfies the requirements of Theorem 1 and identifies with the value function v for appropriate choices for the constants C_ℓ, D_ℓ, C_r, D_r and the free-boundary points $\gamma \in \,]\frac{nK}{n-1}, z[$ and $\zeta > z$.

The requirements that $\widetilde{w}$ should be continuous at z and the fact that it should satisfy the identity

$$(1+\beta)\widetilde{w}'_+(z) = (1-\beta)\widetilde{w}'_-(z) \tag{34}$$

yield the identities

$$C_r = \frac{n(1-\beta) - m(1+\beta)}{(n-m)(1+\beta)} C_\ell - \frac{2m\beta}{(n-m)(1+\beta)} D_\ell z^{-(n-m)} \tag{35}$$

and

$$D_r = \frac{2n\beta}{(n-m)(1+\beta)} C_\ell z^{n-m} + \frac{n(1+\beta) - m(1-\beta)}{(n-m)(1+\beta)} D_\ell. \tag{36}$$

On the other hand, C^1-regularity of $\widetilde{w}$ at the free-boundaries γ, ζ implies that

$$C_\ell = -\frac{1}{n-m}\big[(m-1)\gamma - mK\big]\gamma^{-n}, \tag{37}$$
$$C_r = -\frac{1}{n-m}\big[(m-1)\zeta - mK\big]\zeta^{-n}, \tag{38}$$
$$D_\ell = \frac{1}{n-m}\big[(n-1)\gamma - nK\big]\gamma^{-m} \tag{39}$$
$$\text{and} \quad D_r = \frac{1}{n-m}\big[(n-1)\zeta - nK\big]\zeta^{-m}. \tag{40}$$

Substituting these expressions for the constants C_ℓ, D_ℓ, C_r, D_r in (35)–(36), we obtain the system of equations

$$\begin{aligned}&\big[(n-1)\zeta - nK\big]z^m\zeta^{-m}\\ &+\frac{2n\beta}{(1+\beta)(n-m)}\big[(m-1)\gamma - mK\big]z^n\gamma^{-n}\\ &-\frac{n(1+\beta)-m(1-\beta)}{(1+\beta)(n-m)}\big[(n-1)\gamma - nK\big]z^m\gamma^{-m} = 0,\end{aligned} \tag{41}$$

$$\begin{aligned}&\big[(m-1)\zeta - mK\big]z^n\zeta^{-n}\\ &-\frac{n(1-\beta)-m(1+\beta)}{(1+\beta)(n-m)}\big[(m-1)\gamma - mK\big]z^n\gamma^{-n}\\ &-\frac{2m\beta}{(1+\beta)(n-m)}\big[(n-1)\gamma - nK\big]z^m\gamma^{-m} = 0.\end{aligned} \tag{42}$$

We conclude with the following result.

Theorem 6. *Consider the optimal stopping problem formulated in Section 2, suppose that the problem data is as in Assumption 1, and assume that $\beta \in \,]0,1[$ and $z \geq z^*$. The system of equations (41)–(42) has a unique solution (γ, ζ) such that $\gamma < z < \zeta$. Furthermore, the function $\widetilde{w}$ defined by (33) for these values of γ, ζ and for $C_\ell, D_\ell, C_r, D_r > 0$ being given by (37)–(40) identifies with the value function v of the discretionary stopping problem. In particular, the $(\mathcal{F}_t)$-stopping time defined by*

$$\tau_\star = \inf\left\{t \geq 0 \mid X_t \in \left[\frac{nK}{n-1}, \gamma\right] \cup [\zeta, \infty[\right\} \tag{43}$$

is optimal.

6 Appendix

Proof of Theorem 1. We use the Itô-Tanaka-Meyer and the occupation times formula (see [11, Exercise VI.1.25]) to calculate

$$\begin{aligned}
& w(X_t) - r\int_0^t w(X_s)\,ds \\
&= w(x) + \frac{1}{2}\int_0^t (w'_+ + w'_-)(X_s)\,dX_s + \frac{1}{2}\int_0^\infty L_t^y\, w''(dy) - r\int_0^t w(X_s)\,ds \\
&= w(x) + \int_0^t \left[\frac{1}{2}\sigma^2 X_s^2 w''(X_s) + bX_s w'(X_s) - rw(X_s)\right] \mathbf{1}_{\{X_s \neq z\}}\,ds \\
&\quad + \frac{1}{2}\left[(1+\beta)w'_+(z) - (1-\beta)w'_-(z)\right] L_t^z + \sigma\int_0^t X_s w'_-(X_s)\,dW_s.
\end{aligned}$$

Using the integration by parts formula, we obtain

$$\begin{aligned}
& e^{-rt}w(X_t) \\
&= w(x) + \int_0^t e^{-rs}\left[\frac{1}{2}\sigma^2 X_s^2 w''(X_s) + bX_s w'(X_s) - rw(X_s)\right] \mathbf{1}_{\{X_s \neq z\}}\,ds \\
&\quad + \frac{1}{2}\left[(1+\beta)w'_+(z) - (1-\beta)w'_-(z)\right] \int_0^t e^{-rs}\,dL_s^z + M_t, \qquad (44)
\end{aligned}$$

where

$$M_t = \sigma\int_0^t e^{-rs} X_s w'_-(X_s)\,dW_s.$$

If we define

$$T_n = \inf\{t \geq 0 \mid X_t \geq n\},$$

then we can see that Itô's isometry implies that

$$\begin{aligned}
\mathbb{E}\left[M_{t\wedge T_n}^2\right] &= \sigma^2\mathbb{E}\left[\int_0^t \left[\mathbf{1}_{\{s \leq T_n\}} e^{-rs} X_s w'_-(X_s)\right]^2 ds\right] \\
&\leq \sigma^2 n^2 \sup_{y\leq n}\left[w'_-(y)\right]^2 t \\
&< \infty,
\end{aligned}$$

which proves that the stopped process M^{T_n} is a square integrable martingale. This observation and (44) imply that, given any $(\mathcal{F}_t)$-stopping time τ,

$$\begin{aligned}
& \mathbb{E}\left[e^{-r\tau}(X_\tau - K)^+ \mathbf{1}_{\{\tau \leq T_n\}}\right] + \mathbb{E}\left[e^{-rT_n} w(X_{T_n})\mathbf{1}_{\{\tau > T_n\}}\right] \\
&= w(x) + \mathbb{E}\left[e^{-r\tau}\{(X_\tau - K)^+ - w(X_\tau)\}\mathbf{1}_{\{\tau \leq T_n\}}\right] \\
&\quad + \mathbb{E}\left[\int_0^{\tau\wedge T_n} e^{-rs}\left[\frac{1}{2}\sigma^2 X_s^2 w''(X_s) + bX_s w'(X_s) - rw(X_s)\right] \mathbf{1}_{\{X_s \neq z\}}\,ds\right] \\
&\quad + \frac{1}{2}\left[(1+\beta)w'_+(z) - (1-\beta)w'_-(z)\right] \mathbb{E}\left[\int_0^{\tau\wedge T_n} e^{-rs}\,dL_s^z\right]. \qquad (45)
\end{aligned}$$

If τ is any $(\mathcal{F}_t)$-stopping time, then (45) and the fact that w satisfies the variational inequality (5)–(6) imply that

$$\mathbb{E}\left[e^{-r\tau}(X_\tau - K)^+ \mathbf{1}_{\{\tau \leq T_n\}}\right] + w(n)\mathbb{E}\left[e^{-rT_n}\mathbf{1}_{\{\tau > T_n\}}\right] \leq w(x).$$

Similarly, we can see that, if $\tau_\star$ is defined by (8), then

$$\mathbb{E}\left[e^{-r\tau_\star}(X_{\tau_\star} - K)^+ \mathbf{1}_{\{\tau_\star \leq T_n\}}\right] + w(n)\mathbb{E}\left[e^{-rT_n}\mathbf{1}_{\{\tau_\star > T_n\}}\right] = w(x).$$

Combining these observations with the calculations

$$\begin{aligned} 0 &\leq \lim_{n\to\infty} w(n)\mathbb{E}\left[e^{-rT_n}\mathbf{1}_{\{\tau > T_n\}}\right] \\ &\overset{(9)}{=} \lim_{n\to\infty} \frac{w(n)\psi(x)}{\psi(n)} \frac{\mathbb{E}\left[e^{-rT_n}\mathbf{1}_{\{\tau > T_n\}}\right]}{\mathbb{E}\left[e^{-rT_n}\right]} \\ &\leq \lim_{n\to\infty} \frac{w(n)\psi(x)}{\psi(n)} \\ &\overset{(7)}{=} 0, \end{aligned}$$

which hold true for any $(\mathcal{F}_t)$-stopping time τ, we can see through the monotone convergence theorem that $v(x) = w(x)$ and that $\tau_\star$ is optimal. ■

Proof of Lemma 1. Consider the function ψ defined by (12)–(14). Combining the Itô-Tanaka-Meyer and the occupation times formulae (see [11, Exercise VI.1.25]) with the fact that ψ satisfies (10)–(11), we calculate

$$\begin{aligned} &\psi(X_t) - r\int_0^t \psi(X_s)\,ds \\ &= \psi(x) + \frac{1}{2}\int_0^t (\psi'_+ + \psi'_-)(X_s)\,dX_s + \frac{1}{2}\int_0^\infty L_t^y\,\psi''(dy) \\ &\quad - r\int_0^t \psi(X_s)\,ds \\ &= \psi(x) + \int_0^t \left[\frac{1}{2}\sigma^2 X_s^2 \psi''(X_s) + bX_s\psi'(X_s) - r\psi(X_s)\right]\mathbf{1}_{\{X_s \neq z\}}\,ds \\ &\quad + \frac{1}{2}\left[(1+\beta)\psi'_+(z) - (1-\beta)\psi'_-(z)\right] L_t^z + \sigma\int_0^t X_s\psi'_-(X_s)\,dW_s \\ &= \psi(x) + \sigma\int_0^t X_s\psi'_-(X_s)\,dW_s. \end{aligned}$$

Using the integration by parts formula, we obtain

$$e^{-rt}\psi(X_t) = \psi(x) + \sigma\int_0^t e^{-rs}X_s\psi'_-(X_s)\,dW_s. \tag{46}$$

In view of the fact that

$$\mathbb{E}\left[\left(\int_0^{t\wedge T_y} e^{-rs} X_s \psi'_-(X_s)\, dW_s\right)^2\right] = \mathbb{E}\left[\int_0^t \left[\mathbf{1}_{\{s\leq T_y\}} e^{-rs} X_s \psi'_-(X_s)\right]^2 ds\right]$$
$$\leq y^2 \sup_{u\leq y}\left[\psi'_-(u)\right]^2 t$$
$$< \infty \quad \text{for all } t\geq 0,$$

which follows from Itô's isometry, we can see that the stochastic integral in (46) is a square integrable martingale if stopped at T_y. It follows that

$$\psi(x) = \psi(y)\mathbb{E}\left[e^{-rT_y}\mathbf{1}_{\{T_y\leq t\}}\right] + \mathbb{E}\left[e^{-rt}\psi(X_t)\mathbf{1}_{\{t<T_y\}}\right].$$

Using the monotone and the dominated convergence theorems, we can pass to the limit as $t\to\infty$ to conclude that ψ satisfies (9), as claimed. ∎

Proof of Lemma 2. The equivalence stated in (17) is straightforward to establish. To show the rest of the claims, we first note that ψ is always convex in $[0, z]$. On the other hand, Assumption 1, (15)–(16) and the calculation

$$\begin{aligned}\psi''(x) &= x^{m-2}\left[n(n-1)Ax^{n-m} + m(m-1)B(z)\right]\\ &= \frac{nx^{m-2}}{(n-m)(1+\beta)}\Big\{(n-1)\left[n(1-\beta) - m(1+\beta)\right]x^{n-m}\\ &\qquad + 2m(m-1)\beta z^{n-m}\Big\}, \quad \text{for } x>z, \end{aligned} \tag{47}$$

imply that

$$\begin{aligned}&\psi''(x)>0 \quad \text{for all } x>z\\ &\quad\Leftrightarrow\quad \psi''(z+) \equiv \frac{n\left[(n-1)(1-\beta) - 2m\beta\right]z^{n-2}}{1+\beta} \geq 0\\ &\quad\Leftrightarrow\quad n-1-(n+2m-1)\beta\geq 0\\ &\quad\Leftrightarrow\quad n-1+\left(\frac{2b}{\sigma^2}-m\right)\beta\geq 0. \end{aligned}\tag{48}$$

We now distinguish between two possible cases:
(a) If $\frac{2b}{\sigma^2} - m = -(n+2m-1) < 0$, then

$$n-1+\left(\frac{2b}{\sigma^2}-m\right)\beta > n-1+\frac{2b}{\sigma^2}-m \overset{(16)}{=} -2m>0 \quad \text{for all } \beta\in\,]-1,1[.$$

(b) If $\frac{2b}{\sigma^2} - m = -(n+2m-1) \geq 0$, then we note that

$$n-1+\left(\frac{2b}{\sigma^2}-m\right)\beta\geq 0 \quad\Leftrightarrow\quad \beta\geq\frac{n-1}{n+2m-1}.$$

In this case, we can see that

$$\frac{n-1}{n+2m-1} < 0$$

and

$$\frac{n-1}{n+2m-1} \leq -1 \quad \Leftrightarrow \quad b \leq 0.$$

A careful consideration of the results above reveals that

$$\text{if } b \leq 0, \text{ then (48) holds true for all } \beta \in]-1,1[,$$

while

$$\text{if } b > 0, \text{ then } \frac{n-1}{n+2m-1} \in]-1,0[$$

$$\text{and (48) holds true if and olny if } \beta \in \left] \frac{n-1}{n+2m-1}, 1 \right[.$$

Combining these observations with (17) and (47), we derive (18) as well as the results stated in parts (I)–(III) of the lemma. ■

References

1. Crocce, F. and Mordecki, E.: Explicit solutions in one-sided optimal stopping problems for one-dimensional diffusions. *Stochastics An International Journal of Probability and Stochastic Processes*, **86** (2014) 491-509.
2. El Karoui, N.: Les aspects probabilistes du contrôle stochastique. In *Ninth Saint Flour Probability Summer School - 1979, Lecture Notes in Mathematics*, **876**, Springer, 1981, 73-238.
3. Engelbert, H.J. and Schmidt, W.: On one-dimensional stochastic differential equations with generalized drift. In *Lecture Notes in Control and Information Sciences* **69**, Springer, 1985, 143-155.
4. Friedman, A. *Stochastic Differential Equations and Applications.* Reprint of the 1975 and 1976 original, Dover Publications, 2004.
5. Harrison, J.M. and Shepp, L.A.: On skew Brownian motion. *The Annals of Probability*, **9** (1981) 309-313.
6. Itô, K. and McKean Jr, H.P. *Diffusion Processes and their Sample Paths.* Springer, 1965.
7. Lamberton, D. and Zervos, M.: On the optimal stopping of a one-dimensional diffusion. *Electronic Journal of Probability*, **18** (2013) 1-49.
8. Krylov, N.V. *Controlled Diffusion Processes.* Springer, 1980.
9. Peskir, G. and Shiryaev, A. *Optimal Stopping and Free-Boundary Problems.* Lectures in Mathematics, ETH Zürich, Birkhäuser, 2006.
10. Presman, E.: Solution of the optimal stopping problem for one-dimensional diffusion based on a modification of the payoff function. In *Prokhorov and contemporary probability theory*, Springer Proceedings in Mathematics & Statistics **33**, Springer, 2013, 371-403.

11. Revuz, D. and M. Yor, M. *Continuous Martingales and Brownian Motion.* 2nd edition, Springer, 1994.
12. Shiryayev, A.N. *Optimal Stopping Rules.* Springer, 1978.
13. Walsh, J. B.: A diffusion with a discontinuous local time. *Astérisque, Société Mathématique de France*, **52** (1978) 37-45.

Coordination in Networked Control Processes through Gossip-Like Local Interactions

Vivek S. Borkar[1] and K. Suresh Kumar[2]

[1] Department of Electrical Engineering
Indian Institute of Technology, Powai, Mumbai 400076, India. borkar.vs@gmail.com

[2] Department of Mathematics
Indian Institute of Technology Powai, Mumbai 400076, India.
suresh@math.iitb.ac.in

Abstract. A population of interacting controlled diffusions is considered, with each diffusion being associated with the node of an irreducible directed graph. The interaction is on a faster time scale and is limited to a gossip-like averaging of the neighbors' values. In the limit as the time scale separation goes to infinity, followed by the limit as the graph grows to an infinite graph, we derive the limiting dynamics which has the averaging effect of the time scale separation as well as the decoupling aspect of mean field limits. This leads to a system of identical deterministic controlled differential equations, leading in turn to a simple control problem that can be handled by classical techniques.
Keywords: interacting diffusions, controlled diffusions, graph limits, two time scales, averaging.
AMS 2000 subject classification: Primary 60K35, Secondary 93E20, 93C15, 60H30.

1 Introduction

We consider a population of agents sitting on the nodes of an irreducible directed graph, each controlling a stochastic dynamical system associated with that node. We assume that the agents belong to finitely many 'types' which are independent and identically distributed across agents. The agents interact through a gossip-like [10] component whereby they modify their own state process depending on where their own state stands vis-a-vis the average state of their neighbors, trying to bring it closer to the said average. This, however, forms but one component of the dynamics. Another component is contributed by their individual dynamics, which corresponds to a controlled diffusion described by a controlled stochastic differential equation driven by independent Brownian motions. The drift of this controlled diffusion depends on the type. The averaging is done on a fast time scale of order $\frac{1}{\epsilon}$ for a small $\epsilon > 0$, leading to a two time scale phenomenon [7]. We analyze the system first in the $\epsilon \downarrow 0$ limit, leading to an 'averaged system' wherein the dynamics become *identical* individual dynamics coupled through a measure-valued process that captures the control profile across the agent population. This is followed by the limit as the graph itself grows to an infinite graph subject to certain conditions. Then the dynamics decouple into separate identical *deterministic* dynamics. We formulate a population level control (or 'team') problem in this framework and analyze it.

The paper is organized as follows. In the next section we describe the model in detail. Section 3 analyzes the $\epsilon \downarrow 0$ limit for a fixed finite graph. Section 4 in turn looks at the further limit as the graph grows to an infinite graph. Section 5 describes the control problem.

2 The Model

We consider an increasing family of directed graphs G^N with vertex set V^N and edge set $\mathcal{E}^N$ resp., for $N \geq 1$. We assume that these are irreducible, i.e., there is a directed path from any node to any other node, and $M(N) := |V^N| \uparrow \infty$. Fix N for the time being. Let $\mathcal{N}(i) := \{j \in V^N : (i,j) \in \mathcal{E}^N\}$, i.e., the set of successors of i in G^N. Associated with G^N is an irreducible stochastic matrix $P^N = [[p^N(i,j)]]_{i,j \in V^N}$ with unique stationary distribution $\pi^N = [\pi^N(1), \cdots, \pi^N(M(N))]$. P^N is assumed to be compatible with G^N, i.e., $p^N(i,j) = 0$ for $j \notin \mathcal{N}(i)$. Let $\epsilon > 0$. Each node is classified into one of the L types from $\{1, 2, \cdots, L\}$. The types of the nodes are given by an i.i.d. sequence $\{\xi^i\}_{i \in V^N}$ of random variables, each ξ^i being associated with node i with $P\{\xi^1 = l\} = p_l, \ l = 1, \cdots, L$.

With each node (say, i) we associate a d-dimensional controlled stochastic differential equation

$$dX^{i,N,\varepsilon}(t) = \frac{1}{\varepsilon}\Big(\sum_{j \in \mathcal{N}(i)} p^N(i,j) X^{j,N,\varepsilon}(t) - X^{i,N,\varepsilon}(t) \Big) dt \qquad (1)$$
$$+ m_{\xi^i}(X^{i,N,\varepsilon}(t), u_{i,\varepsilon}(t))dt + dW^i(t), \ t \geq 0.$$

Here:

- $X^{i,N,\epsilon}(0)$ have prescribed laws which have uniformly bounded fourth moments as i, N, ϵ vary (in particular rendering them tight),
- for a compact metric action space A,

$$m_l = [m_{l1}, \cdots, m_{ld}]^T : \mathcal{R}^d \times A \mapsto \mathcal{R}^d, l = 1, \cdots, L,$$

  is continuous and $m_l(\cdot, u)$ are Lipschitz uniformly in u,
- $W^i, i \geq 1$, are independent standard Brownian motions in $\mathcal{R}^d$,
- $u_{i,\epsilon}$ are A-valued control processes with measurable paths, satisfying the non-anticipativity condition: for $t > s$, $W^k(t) - W^k(s), k \geq 1$, are independent of $\sigma(X^{i,N,\epsilon}(y), u_{i,\epsilon}(y), W^i(y), \xi^i, i \geq 1, y \leq s)$. We do not impose any other conditions on $u_{i,\epsilon}$ for the time being.

Let $\mathcal{P}(\mathcal{X})$:= the Polish space of probability measures on a Polish space $\mathcal{X}$ with Prohorov topology. We shall use the relaxed control framework ([1], section 2.3), i.e., we take A to be the form $\mathcal{P}(A')$ for a compact metric space A'. Further, we take m_l to be of the form $m_l(x,u) = \int m_l'(x,y)u(dy)$ for some $m_l' = [m_{l1}', \cdots, m_{ld}']^T : \mathcal{R}^d \times A' \to \mathcal{R}^d$ where the integration is componentwise and m_l' satisfies identical regularity conditions as those for m_l. Define the space $\mathcal{A}$ of measurable A-valued functions with the coarsest topology that renders continuous the maps

$$u(\cdot) \in \mathcal{A} \to \int_s^t h(y) \int_{A'} f(z)u(y)(dz)dy$$

for all $t > s, h \in L_2[s,t]$ and $f \in C(A')$. This space is compact and metrizable ([1], p. 50).

Our interest is in the $\varepsilon \downarrow 0, N \uparrow \infty$ asymptotics of this system, performed in that order. We take this up next.

3 The $\varepsilon \downarrow 0$ Limit

We keep N fixed throughout this section and write $X^{i,\varepsilon}$ for $X^{i,N,\varepsilon}$. Let

$$q^N(i,j) := p^N(i,j) - \delta_{ij} \text{ and } Q^N := \text{ the matrix } [[q^N(i,j)]]_{i,j \in V^N},$$

where δ_{ij} is the Kronecker delta. Then by the variation of constants formula, we have for $i \in V^N$,

$$X^{i,\varepsilon}(t) = \sum_{j=1}^{M(N)} e^{Q^N \frac{t}{\varepsilon}}(i,j)X^{j,\epsilon}(0) + \sum_{j=1}^{M(N)} \int_0^t e^{Q^N \frac{(t-s)}{\varepsilon}}(i,j)m_{\xi^j}(X^{j,\varepsilon}(s), u_{j,\varepsilon}(s))ds$$
$$+ \sum_{j=1}^{M(N)} \int_0^t e^{Q^N \frac{(t-s)}{\varepsilon}}(i,j)dW^j(s), \ t \geq 0. \tag{2}$$

Let $P^{N,*}$:= the rank one matrix with identical rows $\equiv \pi^N$. Then under our hypotheses,

$$\lim_{\varepsilon \downarrow 0} e^{Q^N \frac{t}{\varepsilon}} = \lim_{t \uparrow \infty} e^{Q^N t} = P^{N,*}.$$

Using standard estimates (see, e.g., [1], p. 55) one can get a bound

$$E\left[\|X^{i,\epsilon}(t) - X^{i,\epsilon}(s)\|^4\right] \leq C|t-s|^2 \tag{3}$$

for suitable $C > 0$ and all $t > s, |t-s| \leq 1$. By a well known test for tightness ([4], p. 95), it then follows that the laws of individual $X^{i,\epsilon}$ are tight as probability measures on $C([0,\infty); \mathcal{R}^d)$. Since $\mathcal{A}$ is compact, so is $\mathcal{P}(\mathcal{A})$ by Prohorov's theorem. Thus the laws of $u_{i,\epsilon}$ are tight for free. Hence the entire collection $\{X^{i,\epsilon}, u_{i,\epsilon}, i \geq 1\}$, has tight laws as $\epsilon \downarrow 0$. Passing to any subsequential limit in law in (2) and using classical martingale arguments (see, e.g., Corollary 2.3.9, p. 55, [1]), we get the limiting dynamics

$$dX^{i,N}(t) = \sum_{k=1}^{M(N)} \pi^N(k) m_{\xi^k}(X^{k,N}(t), u_k(t))dt + \sum_{k=1}^{M(N)} \pi^N(k) dW^k(t), \quad (4)$$
$$X^{i,N}(0) = \overline{X}^N,$$

where $\overline{X}^N$ = a limit point (in law) of $\sum_{k=1}^{M(N)} \pi^N(k) X^{k,\epsilon}(0)$ as $\epsilon \downarrow 0$. Since the initial condition and dynamics are the same for all i, it follows that $X^{i,N} = X^{j,N}$ $\forall\, i, j$, a.s. Hence we drop the superscript i and write the above s.d.e. as

$$dX^N(t) = \sum_{k=1}^{M(N)} \pi^N(k) m_{\xi^k}(X^N(t), u_k(t))dt + \left(\sum_{k=1}^{M(N)} \pi^N(k)^2\right)^{\frac{1}{2}} d\widetilde{W}(t), \quad (5)$$
$$X^N(0) = \overline{X}^N,$$

for a d-dimensional Brownian motion

$$\widetilde{W} := \left(\sum_{j=1}^{M(N)} \pi^N(j)^2\right)^{-\frac{1}{2}} \sum_{k=1}^{M(N)} \pi^N(k) W^k.$$

For $f \in C(\{1, \cdots, L\} \times A')$, define $t \mapsto \mu_t^N(\{l\} \times du)$ as a measurable process taking values in $\mathcal{P}(\{1, \cdots, L\} \times A')$, given by:

$$\sum_{l=1}^{L} \int f(l, y) \mu_t^N(\{l\} \times dy) := \sum_{k=1}^{M(N)} \pi^N(k) \int_{A'} f(\xi^k, y) u_k(t)(dy). \quad (6)$$

One can rewrite (5) as follows:

$$X^N(t) = \overline{X}^N + \sum_{l=1}^{L} \int_0^t \int m_l'(X^N(s), y) \mu_s^N(\{l\} \times dy) ds \quad (7)$$
$$+ \left(\sum_{k=1}^{M(N)} \pi^N(k)^2\right)^{\frac{1}{2}} \widetilde{W}(t).$$

4 The $N \uparrow \infty$ Limit

We now consider the $N \uparrow \infty$ limit. For this purpose, we suppose that G^N grows to an infinite irreducible directed graph in such a manner that

$$\lim_{N \uparrow \infty} \sum_{i=1}^{M(N)} \pi^N(i)^2 = 0. \quad (8)$$

As an example, let G^N be undirected, identified with a directed graph with bidirectional edges. Let P^N be the transition matrix of the simple random walk on G^N, i.e., $p^N(i,j) = \frac{1}{d(i)}$, where $d(i) :=$ the degree of node i. Then $\pi^N(k) = \frac{d(k)}{2|\mathcal{E}^N|}$, as can be easily verified from the local balance conditions. Suppose that as $N \uparrow \infty$, the degrees $\{d(i)\}$ remain bounded by some $K < \infty$. Then

$$\sum_k \pi^N(k)^2 \leq \frac{K^2}{2(M(N)-1)} \overset{N\uparrow\infty}{\rightarrow} 0,$$

implying (8).

Define $\mathcal{U}$ as the space of measurable functions $t \mapsto \mu_t \in \mathcal{P}(\{1,\cdots,L\} \times A')$ topologized as follows. Give $\mathcal{U}$ the coarsest topology that renders continuous the maps

$$\mu \in \mathcal{U} \mapsto \int_s^t h(y) \sum_{l=1}^{L} \int f(l,u)\mu(y)(\{l\} \times du)dy$$

for any $t > s, f \in C(\{1,\cdots,L\} \times A'), h \in L_2[s,t]$. Then by the arguments of [1], p. 55, $\mathcal{U}$ is a compact metric space. By Prohorov's theorem, $\mathcal{P}(\mathcal{U})$ is also compact. We view $t \mapsto \mu_t^N$ as $\mathcal{U}$-valued processes.

As before, standard estimates (see, e.g., [1], p. 55) enable us to get a bound

$$E[\|X^N(t) - X^N(s)\|^4] \leq C|t-s|^2, \quad |t-s| \leq 1, \tag{9}$$

for some constant $C > 0$. Again, by the test for tightness in ([4], p. 95), it follows that the laws of X^N are tight as probability measures on $\mathcal{P}(C([0,\infty);\mathcal{R}^d))$. By dropping to a subsequence if necessary, let $(X^N, \mu_\cdot^N)$ converge to $(X, \mu_\cdot)$. By abuse of terminology, we index this subsequence by $\{N\}$ again. Using Skorohod theorem ([5], p. 23), the convergence may be taken to be a.s. on a common probability space.

Now taking $N \to \infty$ in (7) it follows from (8) that X is given by

$$X(t) = X(0) + \int_0^t \sum_{\ell=1}^{L} \int_{A'} m_l'(X(s),u)\mu_s(\{l\} \times du)ds. \tag{10}$$

Disintegrate $\mu_t(\{l\} \times du)$ as $\varphi_t(du|l)\nu_t(l)$ with $\varphi : \{1,\cdots,L\} \to A$ being the regular conditional law. Then (10) takes the form

$$X(t) = X(0) + \int_0^t \sum_{l=1}^{L} \int_{A'} m_l'(X(s),u)\varphi_s(du|l)\nu_s(l)ds. \tag{11}$$

Note that $\nu_t \in \mathcal{P}(\{1, \cdots, L\})$ is a limit point of the (random) probability measures $\{\nu_t^N\}$ given by

$$\sum_{\ell=1}^{L} f(\ell)\nu_t^N(\ell) := \sum_{k=1}^{M(N)} \pi^N(k) f(\xi^k), \ f : \{1, \cdots, L\} \to \mathcal{R}, \ t \geq 0.$$

We claim that $\nu_\cdot^N \to \nu_\cdot^*$ where $\nu_t^*(\ell) = p_\ell \ \forall t, \ell$. To see this, note that for any $t > s \geq 0, \ 1 \leq \ell \leq L$, and $g \in L_2[s, t]$,

$$E\left[\left|\int_s^t g(y)\nu_y^N(l)dy - p_l \int_s^t g(y)dy\right|^2\right]$$

$$= E\left[\left|\sum_{k=1}^{M(N)} \pi^N(k) I\{\xi^k = l\} \int_s^t g(y)dy - p_l \int_s^t g(y)dy\right|^2\right]$$

$$= \left(\int_s^t g(y)dy\right)^2 E\left[\left|\sum_{k=1}^{M(N)} \pi^N(k)(I\{\xi^k = l\} - p_l)\right|^2\right]$$

$$= \left(\int_s^t g(y)dy\right)^2 E\left[\sum_{k=1}^{M(N)} \pi^N(k)^2 (I\{\xi^k = l\} - p_l)^2\right]$$

$$= \left(\int_s^t g(y)dy\right)^2 p_l(1 - p_l) \sum_{k=1}^{M(N)} \pi^N(k)^2$$

$$\to 0$$

by (8), proving the claim. Hence (10) takes the form

$$X(t) = X(0) + \int_0^t \sum_{l=1}^{L} p_l \int_{A'} m_l'(X(s), u)\varphi_s(du|l)ds. \tag{12}$$

We summarize our results as follows.

Theorem 1. *Under (8), every subsequential limit in law of $X^{i,N}$ as $N \uparrow \infty$ is of the form X as in (12). That is, every subsequential limit in law of $X^{i,N,\epsilon}$ as $\epsilon \downarrow 0$ followed by $N \uparrow \infty$ is of the form X as in (12).*

5 The Control Problem

Since the entire agent population obeys an identical deterministic dynamics in the limit, we can consider a single control problem of controlling (12) to minimize a cost

$$\int_0^T \left(\sum_l p_l c(X(t), u)\varphi_t(du|l)\right) dt. \tag{13}$$

Here $c : \mathcal{R}^d \times A' \mapsto \mathcal{R}$ is a bounded continuous 'running cost'. We assume that the partial derivatives $\frac{\partial c}{\partial x_k}, k = 1, \cdots, d$ exist and are continuous and bounded uniformly with respect to the A'-valued control.

Before we analyze this control problem, we have some important observations concerning this limit as follows.

1. The relaxed open loop control $\varphi.(du|l))$ is to be chosen in an identical manner by each agent of class l. Thus the controls get identified with a class, not with individual agents, so that we need specify controls only on a per class basis and not individually. Since the dynamics of each agent is identical, this reduces to a single control problem with A^L-valued control process $\hat{\varphi}.(du) := [\varphi.(du|1), \cdots, \varphi.(du|L)]$.

2. The limiting dynamics (12) was in principle a subsequential limit. But it is identical regardless of the subsequence once the class-wise control processes are fixed, as long as the initial conditions $\{X^{i,N,\epsilon}(0)\}$ have a unique (necessarily common) limit in law as $\epsilon \downarrow 0$, followed by $N \uparrow \infty$. That is, under the latter condition, it is a legitimate limit in law, not just a limit point in law. For this to work, it suffices, e.g., if the initial conditions are i.i.d., but it can also happen under significantly weaker hypotheses.

3. This limit does not depend on the choice of weights $p(i, j)$ and the corresponding stationary distributions $\pi^N, N \geq 1$, as long as (8) holds. Therefore it has *universality* properties independent of modelling specifics. In particular, optimizing over choice of weights $p(i, j)$ won't affect the optimum, though it can affect the rate of convergence in the $\epsilon \downarrow 0$ limit and hence the quality of approximation by the limiting dynamics.

4. It is instructive to compare Theorem 1 with the kind of mean field limits one gets in the McKean-Vlasov equation [11]. There the interaction is averaged over *all* agents equally and is explicitly incorporated in the dynamics. But the initial conditions do not get averaged in the limit, nor does the local state which also enters the local dynamics separately as a non-interacting term. As a result, the limiting equations, while decoupled and identical, do not represent identical trajectories and the interaction is through a measure-valued process that represents the overall population profile. The gossip dynamics we consider averages everything including the local state and the initial condition, leading to exactly identical trajectories. This is not surprising when one recalls that gossip dynamics has been traditionally used to ensure consensus across agents.

5. We have taken a common cost function independent of the class because if the different classes have different cost functions (which would make it a non-cooperative game), then there is no reason why they would opt for

gossip-based averaging.

6. We can also replace the class by (or combine it with) one or more local characteristics such as degree, as long as it has appropriate averaging properties in the $N \uparrow \infty$ limit.

We now replace the Lipschitz continuity of $m'_l, l = 1, \cdots, L$, with the stronger condition that $m'_l(\cdot, u)$ are continuously differentiable with the partial derivatives bounded uniformly in $u \in A'$. The above control problem is then very much amenable to classical treatment. Define the value function $V : \mathcal{R}^d \times [0, T] \mapsto \mathcal{R}$ by

$$V(x,t) := \inf_{X(0)=x, \hat{\varphi}.(du)} \int_t^T \left(\sum_l p_l \int c(X(s), y) \varphi_s(dy|l) \right) ds.$$

Then we have:

Theorem 2. *(i) V is the unique continuous viscosity solution of the Hamilton-Jacobi equation*

$$0 = \frac{\partial V}{\partial t} + \min_{u_1, \cdots, u_L \in A'} \left[\sum_{k=1}^{L} p_k \left(c(x, u_k) + \left\langle \nabla_x V(x,t), m'_k(x, u_k) \right\rangle \right) \right], \ t \in [0, T], \tag{14}$$

with terminal condition $V(\cdot, T) \equiv 0$. Here ∇_x denotes the gradient in x, i.e., the space variables.

(ii) An optimal control $\hat{\varphi}^*_\cdot(du)$ exists and is characterized by: $\forall\ k, t$,

$$support\Big(\varphi^*_\cdot(du|k)\Big) \subset Argmin\left(c(x, \cdot) + \left\langle \nabla_x V(x, \cdot), m_k(x, \cdot) \right\rangle \right), \quad a.e. \tag{15}$$

Proof. Consider the equation

$$0 = \frac{\partial V}{\partial t} + \min_{\mu_1, \cdots, \mu_L \in A} \left[\sum_{k=1}^{L} p_k \left(\int c(x,y) \mu_k(dy) \right.\right.$$
$$\left.\left. + \left\langle \nabla_x V(x,t), \int m'_k(x,y) \mu_k(dy) \right\rangle \right) \right], \ t \in [0, T], \tag{16}$$

with terminal condition $V(\cdot, T) \equiv 0$.

Using the standard arguments based on 'vanishing viscosity' method as in [2], [8], it follows that the equation (16) has a unique continuous viscosity solution. Existence of an optimal relaxed control also follows by a standard 'compactness-continuity' argument based on the Arzela-Ascoli theorem, see, e.g., [9]. Let X, X' denote solutions to (12) with identical control processes, but different initial conditions x, y resp. Let $K > 0$ denote the common bound of $\frac{\partial c}{\partial x_k}(\cdot, u), u \in$

$A', k = 1, \cdots, d$. Then

$$\begin{aligned}
&|V(x,t) - V(y,t)| \\
&\leq \sup_{X(0)=x, X'(0)=y, \hat{\varphi}.(du)} \Big| \int_t^T \left(\sum_l p_l \int c(X(s),u)\varphi_s(du|l) \right) ds \\
&\quad - \int_t^T \left(\sum_l p_l \int c(X'(s),u)\varphi_s(du|l) \right) ds \Big| \\
&\leq K \int_0^T \|X(t) - X'(t)\| dt \\
&\leq KCTe^{C'T}\|x - y\|,
\end{aligned}$$

where the last inequality follows from Gronwall inequality and the Lipschitz property of the m_i's. Thus V is Lipschitz and by Rademacher's theorem, a.e. differentiable. The necessity of (15) now follows from the results of [3], Theorem 4. Note that in [3], the control set is assumed to be a subset of some Euclidean space, but we have the control set as $\mathcal{P}(A')$, a compact Polish space which is also convex and hence connected and locally connected. Therefore by the Hahn-Mazurkiewicz theorem ([6], Theorem 3-30, p. 129), $\mathcal{P}(A')$ is a continuous image of $[0, 1]$. This enables us to use [3], Theorem 4, in the present set-up. The sufficiency follows from the results of [12], Theorem 3.1. Since the minimum in (16) is attained at the extreme points of the convex set $\mathcal{P}(A')$ and these correspond to Dirac measures, the theorem follows. ■

Acknowledgements The idea of this work arose from a conversation of VSB with Prof. Roland Malhame of Ecole Polytechnique de Montreal. The research of VSB is supported in part by a J. C. Bose Fellowship and the project SB/S3/EECE/0182/2014, 'Approximation of High Dimensional Optimization and Control Problems' from the Department of Science and Technology, Government of India. The work of KSK was supported in part by the project SR/S4/MS:751/12, 'Risk-sensitive stochastic control' from the Department of Science and Technology, Government of India.

References

1. Arapostathis, A., Borkar, V. S. and Ghosh, M. K. *Ergodic Control of Diffusion Processes.* Cambridge University Press, Cambridge, UK, 2012.
2. Bardi, M. and Capuzzo-Dolcetta, I. *Optimal Control and Viscosity Solutions of Hamilton-Jacobi-Bellman Equations.* Birkhäuser, Boston, 2000.
3. Barron, E. N. and Jensen, R.: The Pontryagin maximum principle from dynamic programming and viscosity solutions to first-order partial differential equations. *Trans. American Math. Society,* **258** (1987) 635-641.
4. Billingsley, P. *Convergence of Probability Measures.* John Wiley and Sons, New York, 1968.

5. Borkar, V. S. *Probability Theory: An Advanced Course.* Springer Verlag, New York, 1995.
6. Hocking, J. G. and Young, G. S. *Topology.* Dover, New York, 1988.
7. Kabanov, Y. and Pergamenschikov, S. *Two-scale Stochastic Systems.* Springer Verlag, Berlin-Heidelberg, 2002.
8. Lions, P. L. *Generalized solutions of Hamilton-Jacobi equations.* Pitman, London, 1982.
9. Roxin, E.: The existence of optimal controls. *Michigan Math. Journal,* **9** (1962) 109-119.
10. Shah, D.: Gossip algorithms. *Foundations and Trends in Networking,* **3** (2009) 1-125.
11. Sznitman, A.-S.: Topics in propagation of chaos. In *Ecole d'Ete de Probabilites de Saint-Flour XIX - 1989,* (P.-L. Hennequin, ed.), Lecture Notes in Math. No. 1464, Springer Verlag, Berlin-Heidelberg, 1991, 164-251.
12. Zhou, X. Y.: Verification theorems within the framework of viscosity solutions. *J. Optim. Theory and Appl.,* **177** (1993) 208-225.

A Multidimensional Comparison Theorem for SDE with Monotone Drift. Applications to Control of a Group of Independent Identical Agents

Svetlana V. Anulova

V.A. Trapeznikov Institute of Control Sciences RAS
Laboratory of Statistical Information Treatment, Russian Federation 117997, Moscow, Profsoyuznaya, 65 IPU RAN, RF
anulovas@ipu.ru

Abstract. A problem of controlling a group of independent identical agents is considered. The dynamics of an agent is described by a stochastic differential equation. The general method of the Bellman equation [6] cannot be applied to this problem. The optimal control is found by the comparison method for solutions of SDE. The results of the preceding author's papers [1, 2] are generalized: the dynamics of an agent is extended to state dependent by means of drift. Comparison with partial ordering (in $\mathbb{R}^d$) is a pioneering approach.
Keywords: comparison theorems for SDE, multi-agents control.
AMS 2000 subject classification: Primary 60H20, Secondary 93E20

1 Introduction

1.1 Problem Origin

In [7] the following problem is considered. There are two identical independent agents in the non-negative semiaxis, the control action drives the agents in the positive direction, the goal is to maximize the moment of attaining zero by any of the agents. The free dynamics of the agents system is described by two independent Wiener processes, the control action — by a drift term with the sum of coordinates bounded by one. Starting from the Bellman equation, the authors have found analytically the value function, which has determined the optimal policy: regardless of the agents configuration, the whole control resource should be directed at the agent nearest to zero.

1.2 My Previous Results

In my paper [2] the described result was extended to the multidimensional case and to the agents with a more general free dynamics, and even to a more general problem formulation. The system level goal remains to maximize the moment of exiting the positive semiaxis for the agent configuration. The total sum of the control actions on the agents is bounded. The policy of [7] proves optimal in this case also, and what is more, not only for the criterion of [7], but for a whole family of criteria. In paper [1] I investigated a more complicated dynamics of the agents. This paper was a continuation of [2]. As planned in [2], I made a step in the direction of state dependent free dynamics of an agent. I have not managed to add a general drift term, but I have succeeded with a special singular one. Description: the state space turned to a half-line $(-\infty, 1]$, at the base of it the particle was instantaneously reflected into the interior. Control action consisted as before in adding a drift amount, and the aim was to hold the group as close to the base as possible — criterion was the distance of the very remote particle to the base. I have given a detailed description of the applied partial ordering for the general dimension through specifying the corresponding cone.

1.3 My New Result

I continue the progress in the direction of state dependent dynamics of the system. I try to add a general drift term in the free dynamics of an agent, again moving in $\mathbb{R}^1$. But the drift term can't be arbitrary, or else the strategy described above becomes non-optimal. Example: suppose that having reached a certain high level the agent cannot (or almost cannot) return back down. Then the highest one of all agents should be driven quite away, so that afterwards the whole control resource might be directed at the remaining agents. The optimality of such a policy seems plausible. Therefore I have proved the optimality of the policy acting on the lowest agent only with a constraint on the free dynamics drift: a special monotonicity.

2 Problem Formulation and Main Results

Pathwise comparison theorems for solutions of stochastic differential equations remain an active subject, see [3, 9, 14–16], where also 1-dimensional reflected processes are studied. I consider comparison of multidimensional processes with respect to a partial ordering of the Euclidean space.

Now I shall describe (in terms of [4, Section 2.4.1]) two our basic objects — cones K and C in $\mathbb{R}^d$. K is a solid (that is, with nonempty interior) polyhedral cone. C is a proper cone. Denote $\pi : \mathbb{R}^d \to K$ the orthogonal projection on K and Φ the Skorokhod operator for K with normal reflection on the boundary (cf. [11]). The operator Φ maps a continuous function x to $\Phi(x) : [0, \infty) \to K$, and there exists a continuous $\phi : [0, \infty) \to \mathbb{R}^d$ with bounded variation $|\phi|$ such that

$$\Phi(x_t) = x_t + \phi_t,$$
$$d\phi_t = I_{\partial K}(\Phi(x_t))d\phi_t \in \mathcal{N}_{\Phi(x_t)}(K)d|\phi|_t,\ t \in [0, \infty),$$

$\mathcal{N}_x$ being the cone of unit inward normal vectors at a point x of the boundary ∂K of K, cf. [12].

Define an order $\preceq$. For points $x^1, x^2 \in \mathbb{R}^d$ $x^1 \preceq x^2$ if $x^2 - x^1 \in C$. For functions $y^1, y^2 : [0,\infty) \to \mathbb{R}^d$ $y^1 \preceq y^2$ means: $y^1(0) \preceq y^2(0)$ and $y^2 - y^1$ has a locally bounded variation $\mathrm{var}(y^2 - y^1)$, satisfying the following condition:

$$\frac{d(y^2 - y^1)}{d\,\mathrm{var}(y^2 - y^1)}(t) \in C \text{ a.e. for } t \in [0,\infty).$$

Theorem 1. *If the operator π is monotonic, then the operator Φ is monotonic (both with respect to the partial order $\preceq$).*

Example 1. Let $d = 1$ and $K = C = [0,\infty)$. For any two functions y^1, y^2 satisfying the condition "$y^2 - y^1 \geq 0$ and nondecreasing" holds $\Phi(y^2) \geq \Phi(y^1)$.

This theorem helps us to establish the optimality of the described strategy. Note that Ikeda and Watanabe were the first to use comparison theorems in the optimal control theory, see §2 ch.VI [5].

Now we shall give a rigorous description of particles with forming a controlled multi-agents group.

Let $(\Omega, \mathcal{F}, F, \mathbf{P})$ be a standard stochastic basis, $W : [0,\infty) \to \mathbb{R}^d$ an F-Wiener process, $u = \{u(t) \in [0,1]^d,\ \sum_1^d u_i(t) \leq 1,\ t \in [0,\infty)\ \text{a.s.}\}$ a d-dimensional F-well measurable process — a control policy (cf. [5, Chapter VI, Section 2]), $U = \{u\}$. The free dynamics of an agent $x(t) \in \mathbb{R}^1$, $t \in [0,\infty)$, is described by a stochastic differential equation

$$dx(t) = b_{\text{agent}}(x(t))dt + dW_1(t),\ t \in [0,\infty),$$

with $b_{\text{agent}} : \mathbb{R}^1 \to \mathbb{R}^1$ and W_1 the one-dimensional Wiener process. The dynamics of the system corresponding to the strategy u is described by the equation with drift function $b : \mathbb{R}^d \to \mathbb{R}^d$ (cf. [5, Chapter IV, Section 7]):

$$y^u(t) = y + \int_0^t (b(y^u(s)) + u(s))ds + W(t),\ \ t \in [0,\infty).$$

This drift function b is generated by the drift term b_{agent} in the free dynamics of an agent:

$$b_i(x) \equiv b_{\text{agent}}(x_i), i = 1, \dots, d.$$

Define $u^* : \mathbb{R}^d \to \mathbb{R}^d$:

$$u_i^*(x) = \begin{cases} 1, & \text{if } i = \min\{\arg\min\{x_j, j = 1, \dots, d\}\}; \\ 0 & \text{otherwise.} \end{cases}$$

We shall prove that the function u^* generates the optimal policy. This policy is defined correctly, because the equation

$$y^*(t) = y + \int_0^t (b(y^*(s)) + u^*(y^*(s)))ds + W(t),\ \ t \in [0,\infty), \tag{1}$$

has a strong solution, cf. [5] and [13, Theorem 1]).

The following theorem is my conjecture proved below only in a particular case: two agents, a markov strategy u, and Lipschitz property of b and u.

Denote by K the cone $\{x_1 \leq x_2 \ldots \leq x_d\}$ and by pr_K the projection operation onto the cone K.

Assumption There exists a proper cone $C \subseteq \mathbb{R}^d$ with the following properties:

- for every $x, y \in K$ satisfying $y - x \in \partial C$, and $\mathbf{n} \in \mathcal{N}_{(y-x)}(C)$ holds $\langle b(y) - b(x), \mathbf{n} \rangle \geq 0$;
- for each $a \in \mathbb{R}^d$ the projection of the shifted cone $\{C + a\}$ on the cone K is included in the shifted cone $\mathrm{pr}_K\, a + C$.

Theorem 2. *Let $y \in \mathbb{R}^d$, $u \in U$ and equation*

$$y^u(t) = y + \int_0^t (b(y^u(s)) + u(s))ds + W(t),\ t \in [0, \infty),$$

have a solution. Then there exists an extension of the original stochastic basis $(\Omega', \mathcal{F}', F', \mathbf{P}')$ and an (F')-Wiener process W' on it such that the solution of the equation

$$y^*(t) = y + \int_0^t (b(y^*(s)) + u^*(y^*(s)))ds + W'(t),\ t \in [0, \infty),$$

satisfies the inequality

$$\min_i y_i^*(t) \geq \min_i y_i^u(t),\ t \in [0, \infty). \tag{2}$$

Remark 1. This theorem implies: for $T \in [0, \infty)$ and real bounded measurable functions $G_1, G_2 : \mathbb{R}^d \to \mathbb{R}^1$ non-increasing in the second argument with respect to $\preceq$

$$\int_0^T G_1(t, y^u(t))dt + G_2(T, y^u(T))$$

is maximal for the policy u^* in probability (see [10] for the detailed description of this concept).

Example 2. Suppose $d = 2$ and the drift term b_{agent} in the free dynamics of an agent is non-decreasing with additional property $b_{\text{agent}}(x_1 + h) - b_{\text{agent}}(x_1) \geq b_{\text{agent}}(x_2 + h) - b_{\text{agent}}(x_2)$ for every $x_1 \leq x_2$ and $h \geq 0$. Then Assumption holds for the cone C with edge vectors $(-1, -1)$ and $(-1, 1)$.

The objects we have introduced describe a group of d independent identical agents, each with a controlled drift. The theorem produces the optimal policy to hold the group as high as possible.

3 Proofs of Theorems

Proof of Theorem 1 is given in [1]. ■

Proof of Theorem 2 in a particular case. I shall prove the inequality (2) for $d = 2$, the cone C with orthogonal edge vectors $(-1,-1)$ and $(-1,1)$, b being a bounded Lipschitz function, and a markov Lipschitz strategy u. The proof is based on the multidimensional comparison theorem from [8]. Change the coordinates in $\mathbb{R}^2$: the new coordinate vectors are $e_1 = (-1,1)$ and $e_2 = (-1,-1)$. Denote the vector functions in the original coordinates $((b+u^*)(x_1),(b+u^*)(x_2))$ and $((b+u)(x_1),(b+u)(x_2))$ by f and $\bar{f}$ in the new coordinates. I insert the following lemma so as to use [8]:

Lemma 1. *For any couple* (x_1, x_2), $(\bar{x}_1, \bar{x}_2) \in K$ *holds:*
1) if $x_1 = \bar{x}_1$ *and* $x_2 \le \bar{x}_2$ *then* $f_1(x_1,x_2) \le \bar{f}_1(\bar{x}_1,\bar{x}_2)$;
2) if $x_2 = \bar{x}_2$ *and* $x_1 \le \bar{x}_1$ *then* $f_2(x_1,x_2) \le \bar{f}_2(\bar{x}_1,\bar{x}_2)$.

This lemma follows directly from the Assumption (see the proof below) and allows to apply Theorem 2 of [8]. According to this theorem for any initial condition $y \in K$ solutions y^* and y^u satisfy $y^u - y^* \in C$ until the moment of hitting the boundary ∂K. Using Theorem 1 we obtain the following: these solutions with added reflection on the boundary ∂K satisfy the same relation $y^u - y^* \in C$ (and consequently $y_1^u \le y_1^*$) on the whole time interval $[0,\infty)$, cf. [1, III. PROOFS OF THEOREMS] (in the present model approximations of reflection are to be made with the help of projection from ∂K onto $\{x \in K : \text{dist}(x,\partial K) \ge \varepsilon > 0\}$ with $\varepsilon \to 0$). Finallly we proceed from K to $\mathbb{R}^2$ according to the scheme [1, III. PROOFS OF THEOREMS], see "Proof of Theorem 2 for d = 2. ...In order to return from K in the plane...". ■

Proof of Lemma 1. Consider statement 1). In the original coordinates it means: for points $x, \bar{x} \in K$, $\bar{x} = (x_1 - h, x_2 - h), h \ge 0$, and the unit inward normal vector $\mathbf{n} = (-1,1)$ of the cone C face holds: $\langle b(x) + (1,0), \mathbf{n}\rangle \le \langle b(\bar{x}) + u(x), \mathbf{n}\rangle$. This follows from the next two inequalities. As defined, $u_1(x), u_2(x) \ge 0, u_1(x) + u_2(x) = 1$, thus $\langle u(x) - (1,0), \mathbf{n}\rangle = \langle u(x) - (1,0), (-1,1)\rangle \ge 0$. And according to the Assumption item one $\langle b(\bar{x}) - b(x), \mathbf{n}\rangle \ge 0$. ■

Statement 2) is based on the same proof.

4 Future Works

I hope to generalize paper [8] for random coefficients of SDE. Then my conjecture Theorem 2 will be at once proved for general, not markov strategies for $d = 2$. To proceed to many dimensions I have already (cf. [1, 4.1 Additional details to Proof of Theorem 2]) a cone C satisfying Theorem 1, for an arbitrary d. It is d-dimensional and its edge vectors are all orthogonal: they are represented by columns of the following matrix

$$\begin{pmatrix} -1 & -1 & -1 & -1 & \dots & -1 & -1 & -1 \\ 1 & -1 & -1 & -1 & \dots & -1 & -1 & -1 \\ 0 & 2 & -1 & -1 & \dots & -1 & -1 & -1 \\ 0 & 0 & 3 & -1 & \dots & -1 & -1 & -1 \\ \vdots & \vdots & \vdots & \vdots & \dots & \vdots & \vdots & \vdots \\ 0 & 0 & 0 & 0 & \dots & d-2 & -1 & -1 \\ 0 & 0 & 0 & 0 & \dots & 0 & d-1 & -1 \end{pmatrix}.$$

So I shall be able to prove fully Theorem 2, in the same way as it was done now for $d = 2$, but already with general strategies.

5 Acknowledgments

I am very grateful to Professors A. Shiryayev who attracted my attention to the paper of McKean and Shepp [7]. I also gratefully acknowledge the outstanding organization of the workshop "Modern trends in controlled stochastic processes: theory and applications 2015" by Prof. Piunovskiy and the University of Liverpool.

This work was supported by Russian Foundation for Basic Research under grant 14-01-00739.

References

1. Anulova, S.: A Multidimensional Comparison Theorem for SDE with Reflection. Applications to Control of a Group of Independent Identical Agents. In *2014 European Control Conference (ECC), June 24-27, Strasbourg, France*, 2014, 564-568.
2. Anulova, S.: A multidimensional comparison theorem for solutions of the Skorokhod problem in a wedge with applications to control of a group of independent identical agents. In *Decision and Control (CDC), 2010 49th IEEE Conference on*, 2010, 4177-4179.
3. Bo, L. and Yao, R.: Strong comparison result for a class of reflected stochastic differential equations with non-Lipschitzian coefficients. *Frontiers of Mathematics in China*, **2** (2007) 73-85.
4. Boyd, S. and Vandenberghe, L. *Convex Optimization.* Cambridge University Press, Cambridge, 2004.
5. Ikeda, N. and Watanabe, S. *Stochastic Differential Equations and Diffusion Processes.* North Holland Mathematical Library **24**, 1989.
6. Krylov, N.V. *Controlled Diffusion Processes.* Springer Verlag, 2008.
7. McKean, H.P. and Shepp, L.A.: The advantage of capitalism vs. socialism depends on the criterion. *Journal of Mathematical Sciences*, **139** (2006) 6589-6594.
8. Milian, A.: Stochastic viability and a comparison theorem. *Colloq. Math.*, Polish Academy of Sciences (Polska Akademia Nauk - PAN), Institute of Mathematics (Instytut Matematyczny), Warsaw, **68** (1995) 297-316.

9. Peng, Sh. and Zhu, X.: Necessary and sufficient condition for comparison theorem of 1-dimensional stochastic differential equations. *Stochastic Processes Appl.*, **116** (2006) 370-380.
10. Shaked, M. and Shantikumar, J.G. *Stochastic Orders.* Springer Series in Statistics, 2007.
11. Situ, R. *Theory of Stochastic Differential Equations with Jumps and Applications: Mathematical and Analytical Techniques with Applications to Engineering.* Springer Verlag, 2005.
12. Tanaka, H.: Stochastic differential equations with reflecting boundary condition in convex regions. *Hiroshima Math. J.*, **9** (1979) 163-177.
13. Veretennikov, A.Yu.: On strong solutions and explicit formulas for solutions of stochastic integral equations. *Math. USSR, Sb.*, **39** (1981) 387-403.
14. Yang, Zh., Mao, X. and Yuan, Ch.: Comparison theorem of one-dimensional stochastic hybrid delay systems. *Systems & Control Letters*, **57** (2008) 56 - 63.
15. Yang, Zh., Wei, L. and Elliott, R.J.: Multiple solutions to stochastic differential delay equations and a related comparison theorem. *Stochastic Analysis and Applications*, **31** (2013) 539-551.
16. Zhao, Sh. and Gao, F.: A necessary condition on comparison theorem for a one-dimensional stochastic differential equation. *Wuhan Univ. J. Nat. Sci.*, **15** (2010) 13-15.

Steps Towards a Management Toolkit for Central Branch Risk Networks, Using Rational Approximations and Matrix Scale Functions

Florin Avram[1] and Andreea Minca[2]

[1] Laboratoire de Mathématiques Appliquées, Université de Pau, France
`florin.avram@orange.fr`,

[2] Cornell University, Ithaca, NY, US
`acm299@cornell.edu`

Abstract. This paper attempts to exploit the extensive one-dimensional machinery available nowadays for one-dimensional risk models towards managing simple central branch risk networks. More specifically, we
a) introduce a concept of efficient subsidiary,
b) find explicitly value functions resulting from the allocation of a fixed sum by a deterministic central branch to one subsidiary, and
c) compute approximate value functions for non-deterministic central branches with one subsidiary by applying rational approximation, and by using recently developed matrix scale methodology.
Keywords: multi-dimensional risk process, Sparre Andersen process, Markov additive process, matrix exponential approximation, optimal allocation, ruin probability.
AMS 2000 Math. Subject Classification: Primary 60G51, Secondary 60K30, 60J75

1 Introduction

1.1 Multi-Dimensional Risk Networks

Multi-dimensional risk networks (**MRN**) is an emerging discipline, which awakes considerable interest in mathematical finance and risk theory.

A risk network is defined by:

$$X(t) = u + ct - S(t) = (X_i(t),\ t \geq 0, i \in \mathcal{I}),$$

where $\mathcal{I}$ is a finite set, the vector u represents the capital of the **MRN** at time 0, the vector c represents a constant cash inflow rate, and $S(t)$ is a process representing cash outflows at time t, which may include both Levy and Sparre-Andersen renewal components.

If no boundary condition is specified, we will call this a *free spectrally negative **MRN***.

Remark 1. The minus sign comes from the one-dimensional case most studied historically, the spectrally negative Cramér-Lundberg process, but the case when $X(t)$ is spectrally positive is also interesting. The case of spectrally two-sided $X(t)$ is of course interesting, but harder.

Remark 2. The simplest case is when the component are i.i.d. compound renewal Sparre-Andersen processes, generated by i.i.d. pairs of inter-arrival times and claims $(A_j^{(i)}, C_j^{(i)}), j = 1, 2, ...$

$$X_i(t) = u_i + c_i\, t - S_i(t),\ S_i(t) = \sum_{j=1}^{N_i(t)} C_j^{(i)},\quad i = 1, ..., I,$$

where $N_i(t) = \max\{k : T_k^{(i)} := \sum_{j=1}^{k} A_j^{(i)} \leq t\}$ are renewal counting processes associated to the independent inter-arrivals [9], with intensity $\lambda_i = E(A_1^{(i)})^{-1}$, and the claims sizes $C_j^{(i)}, j \geq 1$ are nonnegative i.i.d. random variables with arbitrary marginal distribution functions denoted by $F_i(x), i \in \mathcal{I}$, with finite expectation, denoted m_i. However, the restriction to jump processes is not essential.

Remark 3. After reaching special subsets exterior to the state space, several continuation/regulation mechanisms are possible, like absorption, reflection, or jumping to the interior. These correspond to various possible interactions between the components at times of distress.

Example 1. A toy example with one absorbing boundary and several reflecting boundaries. Consider a central branch which must simultaneously lay the basis for several subsidiaries, in various environments.

The central branch will keep the subsidiaries solvent by bail-outs until the moment of its bankruptcy, or until the moment when a subsidiary is deemed non-profitable. The subsidiaries will make payments to the common fund. Finally, the expected present value benefit to the central branch consists in the difference between the expected discounted payments and bail-outs.

This example suggests the following model:

Definition 1. *A central branch (CB) network is formed from:*

1. *A unit, called central branch, with reserves denoted by $X_0(t)$, whose ruin time*
$$\tau_0 = \inf\{t \geq 0 : X_0(t) < 0\}$$
causes the ruin of the whole network.
2. *Several subsidiaries $X_i(t), i = 1, \ldots, I$ that must be kept nonnegative or above certain prescribed levels, by transfers from the CB.*

For this network, the boundaries $u_i = 0, i = 1, ..., I$ are reflecting and $u_0 = 0$ is absorbing.

Remark 4. There are many applications of the central branch concept: a government/central bank, a reinsurance company, an insurance group, a central clearinghouse, etc.

An important economic issue is modeling the *patience* of the central branch (which can be in terms of the cumulated bail-out cost, the number of bailouts, the cumulated net income from the subsidiary, etc.).

Another interesting application is that of a coalition or default fund created by several institutions, for bailing them out when they go bankrupt. Interesting issues here are determining fair conditions for merging into (profit participation schemes) and splitting out of the coalition.

Notation. Denote the ruin times and ruin probabilities (finite time and eventual) of the components when isolated from the network by

$$\tau_i(u_i) = \inf\{t \geq 0 : X_i(t) < 0\}, i = 1, 2, ...$$
$$\Psi_i(t, u_i) = P(\tau_i(u_i) < t), \quad \Psi_i(u_i) := P(\tau_i(u_i) < \infty).$$

The ruin probability of the CB and its Laplace transform will be denoted respectively by

$$\Psi(t, u) = \Psi(t, u, c) = P_u[\tau_0 < t], \ \widehat{\psi}_q(u) := \widehat{\psi}_q(u, c) = E_u\left[e^{-q\tau_0}\right].$$

1.2 Multi-Dimensional First Passage Problems

One-dimensional first passage problems have been very extensively studied; typically, Laplace transforms are available, especially when either A_i or C_i have a matrix exponential distribution, and explicit inversion of the Laplace transforms is also possible sometimes, especially when at least one of A_i or C_i have an exponential distribution.

Building upon the case of exponential arrivals A_i, it was found that the solution of a large gamut of one dimensional first passage problems for *spectrally one-sided Lévy* processes (dividends, drawdowns, exotic options, Parisian options, etc...) reduces to the study of a couple of scalar *scale functions* – see for example [12, 47]. This idea was extended to Lévy processes in a modulated Markovian environment, which include the case of phase-type A_i, and finally to spectrally negative Markov additive processes (SNMAP). The end result is a *matrix scale methodology*, based on computing a couple of *matrix scale functions* [40, 44, 45, 48].

Multi-dimensional first passage problems are considerably harder than one dimensional ones, and one cannot expect general formulas[4].

We can attempt however to exploit the extensive one-dimensional machinery for providing approximations in the multidimensional case. We take some steps in this direction by:

---

[4] One exception is a Pollaczek-Khinchine type formula for the transform of ruin probabilities $\Psi(u)$ of spectrally negative networks provided in the foundational paper [22]. However, this formula involves several unknown functions (the Laplace transforms over each boundary facet of the state space), and it isn't at all obvious how to exploit this formula numerically.

1. Defining a concept of efficient subsidiaries.
2. Computing the value function resulting from the allocation of a fixed sum by a *deterministic central branch* to one subsidiary, a case which can solved exactly[4].
3. Computing the value function for *non-deterministic central branches* with one subsidiary, a case which does not admit an exact solution due to its complex dependent Sparre-Andersen structure. This is achieved by applying the classic idea of rational approximation to replace the Sparre-Andersen structure by a Markov modulated Lévy structure, and by using subsequently the matrix scale methodology.

1.3 Efficient Subsidiaries

A crucial issue for a coalition is how to accept *efficient members* and eventually reject them if they are not fully efficient. For that it is natural to evaluate each member separately, by classic one dimensional risk measures like ruin probabilities, or the value of future dividend payments made to the coalition.

Efficiency as readiness. The choice of an economic principle for evaluating efficiency is not at all evident. We make the ad-hoc proposal to consider optimizing discounted dividends of rate $d = c\gamma$, $\gamma \in (0, 1]$ taken above a constant threshold b (γ represents the proportion of income taken above the threshold) – see for example [4, 10], and to define efficiency as *local optimality of* $b = 0$ over some interval $[0, \epsilon), \epsilon > 0$.

The motivation is that subsidiaries are functional from the start and can contribute cash-flows to the central branch without having to wait first until its reserves build out; effectiveness is thus translated in this paper as *readiness.*

In conclusion, we propose to postulate that a subsidiary is:

1. *Non-efficient* and rejected immediately if its loading factor $\frac{c}{\lambda E[C_1]} - 1$ is not nonnegative, since this implies an infinite number of bail-outs.
2. *Totally efficient* and accepted for ever in the coalition iff

$$k \leq f(q)$$

   where q is the discount rate, and $f(q)$ is an increasing function of q, obtained as optimality of $b = 0$ for some specific dividends distribution scheme, and $k \geq 1$ captures the cost associated with capital infusions towards a subsidiary.
3. *Partially efficient* if the loading condition $\rho = \lambda E[C_1]/c < 1$ is satisfied, and $k > f(q)$. These subsidiaries will also be accepted, but only with an *impatience rate* θ, resulting in killing the subsidiary after its time or lowest value in an orange zone exceeds an exponential r.v. of rate θ. The impatience rate θ is chosen so that

$$\widetilde{f}(q, \theta) = k,$$

[4] For other cases which may be solved explicitly see for example [15, 18].

where $\widetilde{f}(q, \theta)$ is computed from the optimality of $b = 0$ for the *impatience modified* value function. This is illustrated in example 3 below, where $\widetilde{f}(q, \theta) = f(q + \theta)$.

1.4 Judging Efficiency by Optimizing Bail-outs and Dividends

For bail-out intervention times, one may consider the classic ruin time τ, and also several interesting alternatives generalizing it:

1. The classic De Finetti objective is maximizing expected discounted dividends until the ruin time. In our setup, it is natural to take into account also the final bail-out of the subsidiary, resulting in the optimization objective:
$$V^{(w)}(x) = \sup_{\pi} E_x \left[\int_0^{\tau} e^{-qt} dD^{\pi}(t) + e^{-q\tau} w(U(\tau)) \right], \tag{1}$$
where $w(u)$ is the so called Gerber-Shiu penalty function.
The optimal dividend distribution is of *multi-barrier* type [35], and the end result may be expressed in terms of scale functions [14, 16]. Further conditions are necessary to ensure that *single constant barrier* strategies suffice [17, 14, 56, 57].
2. One may replace τ in (1) by the *bankruptcy time* τ_B – see for example [64, Sec. 4].
3. One may replace τ in (1) by the Parisian ruin time τ_P – see for example [51].
4. Poissonian observed ruin times – see for example [6, 5].
5. For each of these and other possibilities, one may consider the objective after one intervention, or over an infinite horizon, over which the subsidiary will possibly need to be bailed out a number $N_I \geq 0$ of times. In the case of linear transaction costs $ku - K$, the optimization objective (of particular interest in a bail-out setting) becomes the expectation over an infinite horizon of a linear combination of discounted dividends $D(t)$, cumulative bailouts $Z(t)$, and number of interventions $N_I^{\pi}(t)$ up to time t:
$$\begin{aligned} &V^{(k)}(x) \\ &= \sup_{\pi} E_x \left[\int_0^{\infty} e^{-qt} dD^{\pi}(t) - k \int_0^{\infty} e^{-qt} dZ^{\pi}(t) - K \int_0^{\infty} e^{-qt} dN_I^{\pi}(t) \right]. \end{aligned} \tag{2}$$
Since in a diffusion setting this objective has first been considered by Shreve, Lehoczky, and Gaver (SLG) [66] – see also Lokka and Zervos [58] – we will call it the SLG objective.
For spectrally negative Levy processes, the optimal dividend distribution for the SLG objective is always of *constant barrier* type, and the end result may be expressed in terms of scale functions [14].

Example 2. A simple, but unsatisfactory definition of efficiency. Instead of *readiness*, consider defining efficiency as nonnegativity of the SLG objective for the 0 barrier, when dividends are $D(t) = d\,t$. By [14, Thm. 1, (4.4)]

$$V^{(k)}(0) = \frac{c}{q} - k\frac{\lambda m_1}{q} \geq 0 \tag{3}$$

yields $k \leq \frac{c}{\lambda m_1} = \rho^{-1}$. This generalizes easily for the 0 threshold, yielding $k \leq \frac{d}{\lambda m_1} = \rho^{-1}\gamma$. Unfortunately, in both cases the discount factor cancels.

Example 3. SLG readiness. Consider now efficiency defined as the optimality of $b = 0$ for the SLG objective *constant reflecting barrier.* This problem is fully analyzed in [14, Thm. 3], and in particular [14, Lem. 2] shows that the optimal SLG constant barrier is $b^* = 0$ iff $k \leq 1 + \frac{q}{\lambda}$.

By this criterion, a subsidiary i with

$$k_i \leq 1 + \frac{q}{\lambda_i} \tag{4}$$

will be deemed totally efficient and accepted for ever in the coalition. Subsidiaries with loading condition $\rho_i < 1$ and

$$k_i > 1 + \frac{q}{\lambda_i}$$

will be deemed *partially efficient* and accepted only for a random time with law $\mathcal{E}(\theta_i)$, where

$$\theta_i + q = \lambda_i(k_i - 1) \tag{5}$$

(rendering thus $b_i^* = 0$ optimal with respect to the total discount rate $q_i = q+\theta_i$).

Unfortunately, the criterion (4) does not take into account the claim size law. Further alternatives of dividend payment strategies, (two-step premia, tax, linear reflecting barriers,...) are being investigated in the parallel paper [13].

1.5 Contents and Contributions

The central branch model is described in more detail in Section 2.

1. Our first contribution is to propose a hierarchical approximate optimization approach, based on setting first impatience parameters on each subsidiary viewed in isolation from the network, followed by setting the dividend barrier parameters and the optimal allocation using a *decoupled* objective of the form $V(u) = \sum_i V_i(u_i)$. When transaction costs are present, they may finally be incorporated via the reduction result (6).
2. Our second contribution, Theorem 1 in Section 3, applies to the case of one subsidiary and a *purely deterministic CB* $\widetilde{X}_0(t) = u_0 + c_0 t$. In this case, the computation of the finite time ruin probabilities and other performance measures (including total subsidiary dividends until ruin) reduces to the corresponding computation for a subsidiary with modified initial capital $u_0 + u_1/k_1$ and initial income rate $c_0 + c_1/k_1$. More precisely,

$$\begin{aligned} \Psi(t,u,c) &= \Psi_1(t, u_0/k + u_1, c_0/k + c_1), \\ V(u,c) &= V_1(u_0/k + u_1, c_0/k + c_1), \end{aligned} \tag{6}$$

without any distributional assumptions! For example, the subsidiaries may be dual risk processes, or spectrally two-sided Levy processes,... The proof of this result, via a pathwise argument, yields also an upper bound when $I > 1$, and an extension to hierarchical networks is given in Corollary 1.

3. Our third contribution, in section 4, deals with *non-deterministic CB*'s with one subsidiary. We propose an approximation approach which uses *bivariate phase-type* approximations for the *joint law of the downward ladder time and height* to obtain a *SNMAP* (spectrally negative Markov additive process) approximation for our non-MAP central branch. The advantage of this approach is that once a SNMAP approximation is obtained, many similar problems may be solved just by applying the scale matrix methodology developed by [40, 45, 48], and using the SNMAP Mathematica package of J. Ivanovs[42]. Different problems are thus solved simultaneously!
A numeric illustration is performed in section 5, where we consider the problem of choosing a barrier B maximizing CB dividends until ruin, in the case of one subsidiary with exponential claims (and without dividends).
In this case, *univariate phase-type* approximations of the downward ladder density, obtained via a continued fraction expansion –see Section 4, provide a SNMAP approximation of the CB process. Subsequently, using the SNMAP package of Ivanovs provides the optimal barrier.
Finally, some useful background on Sparre-Andersen risk processes with exponential claims is included for completeness in Section 6. This material is used in the parallel paper [11], which among others generalizes the dependence structure of central branches with one Sparre-Andersen subsidiary having phase-type claims.

2 Minimizing the Ruin Probability under Transaction Costs for Central Branch Networks

The bail-out policy of a central branch with SA subsidiaries will consist in infusing capital into subsidiary i every time its surplus level drops below 0 for the j-th time, resetting it to predefined levels $\chi_j^{(i)} \geq 0$, by transferring an amount $\zeta_j^{(i)} = \chi_j^{(i)} + y_j^{(i)}$, where $y_j^{(i)}$ is the severity of ruin at the j'th default.

We assume that only the central branch may intervene in bailouts, and that transaction costs of $k_i\zeta_j^{(i)} + K_i$ will be incurred at each capital infusion of $\zeta_j^{(i)}$, *except at the start time* $t = 0$, *when the subsidiaries are set up.*

Assumption A: Below, the reset levels are $\chi_j^{(i)} = 0, \forall i, j$, *and there are no fixed costs* K_i.

Remark 5. Taking $S_i(t)$ to be spectrally positive Levy perturbed compound processes *SPLPCP* [32, 54, 73] (for example with a Brownian motion perturbation) poses often no problem.

In this case, the subsidiaries may be kept nonnegative using *minimal Skorohod regulation.* Then:

$$X_0(t) = u_0 + c_0 t - S_0, S_0 = \sum_{i=1}^{I} k_i I_i(t), \quad I_i(t) = -\inf_{s \leq t}\{X_i(s), 0\},$$

where the *regulator process* I_i is the minimal process whose addition to X_i ensures that the sum is non-negative.

Studying multidimensional CB networks is quite challenging. Hence, we will restrict from now on mainly to exponential or phase-type subsidiary claims.

With $(\boldsymbol{\beta}, B)$ subsidiary claims, the bailout (time, size) pairs $(\widetilde{A}_k, \widetilde{C}_k)_k$ are IID random variables with joint distributions of the special form

$$P(A_k \in dt, C_k \in dx) = \boldsymbol{\alpha}(t)\, e^{B_k x}\, b_k\; dx\, dt, \quad t, x \in \mathbb{R}_+,$$

where $B_k := k^{-1}B$, and $\boldsymbol{\alpha}(t) = (\alpha_1(t), ..., \alpha_i(t), ...)$ contains the densities of the ladder time joint with ruin in phase i.

The CB is thus itself a Sparre-Andersen process with phase-type claims, exhibiting however a *non-standard* dependence (7).

Remark 6. The classic Sparre-Andersen model with jumps of phase-type $(\boldsymbol{\beta}, B)$ and independent inter-arrivals with density $a(t)$ is obtained by taking densities of the form

$$\boldsymbol{\alpha}(t) = a(t)\boldsymbol{\beta}$$

Remark 7. The CB SPMAP process. It will be convenient to represent our Sparre-Andersen process as the workload of a Ph/G/1 queue, as in Lemma 3. The essential difference from the classic independent Sparre-Andersen process is that here the initial phase of a service(claim) period is decided at the bottom of the up-jump representing its inter-arrival time, and decides therefore also the size of the jump.

Remark 8. Consider a CB network with several independent subsidiaries starting all at $u_i = 0, , i = 1, ..., I$, having claims of phase-type $\boldsymbol{\beta}^{(i)}, B^{(i)}, i = 1, ..., I$, and let $\rho^{(i)}(t), \bar{R}^{(i)}(t)$ denote the respective down ladder densities and survival functions.

By conditioning, we find that the density of the the first bailout of the CB is

$$P(\widetilde{A}_k \in dt, C_k \in dx) = \sum_{i=1}^{I} \left(\prod_{j \neq i} \bar{R}^{(j)}(t) \right) \boldsymbol{\alpha}^{(i)}(t)\, e^{B_{k_i}^{(i)} x}\, b_{k_i}\; dx\, dt, \quad t, x \in \mathbb{R}_+$$

where $\boldsymbol{\alpha}^{(i)}(t)$ have Laplace transforms satisfying Kendall equations.

However, the fact that $u_i = 0, i = 1, ..., I$ stops being true after the first bailout. To keep track of what happens after that time, we are forced to introduce supplementary variables which render the problem harder.

3 Linear Networks: Reduction to One Dimension

It turns out that as long as the CB in isolation is a *deterministic drift* with parameters u_0, c_0, the *ruin probability* of the CB equals that of a subsidiary with modified parameters $u_1' = u_1 + u_0/k, \quad c_1' = c_1 + c_0/k$, where $k = k_1$, independent of the reset policy!

The result is the same as if the CB transfers everything at the time 0_+. This is also the case with several other problems involving a drift CB with no extra liabilities, which ends up liquidated totally.

To see this, introduce a pooled assets combining X_0 and the reflected processes $\widetilde{X}_i(t) = X_i(t)+I_i(t)$ in such a way that the transfers and regulation cancel out. Putting $u = u_0 + \sum_{i=1}^{I} k_i u_i, \quad c = c_0 + \sum_{i=1}^{I} k_i c_i$, we find:

$$\begin{aligned} X(t) &= X_0(t) + \sum_{i=1}^{I} k_i \left(X_i(t) + I_i(t)\right) \\ &= u_0 + c_0 t + \sum_{i=1}^{I} k_i X_i(t) = u + ct - \sum_{i=1}^{I} k_i S^{(i)}(t). \end{aligned}$$

Remark 9. Note that with spectrally negative Levy subsidiaries $X(t)$ is also a *spectrally negative Levy process*, while $X_0(t)$ is a complicated superposition of SA processes. However, when $I = 1$, the ruin time of X_0 and U coincide. Furthermore, the pooled reserves from the point of view of the subsidiary

$$\frac{X(t)}{k_1} = u_1 + \frac{u_0}{k} + t(c_1 + \frac{c_0}{k}) - S^{(1)}(t)$$

has the same law as the subsidiary with *combined initial value and income rate*!

Remark 10. With several subsidiaries,

$$\tau = \inf\{t : X(t) < 0\}. \tag{7}$$

represents the ruin time if the subsidiaries may start helping each other at no cost, once the CB is ruined.

These remarks yield the following:

Theorem 1. *Let X_0 be a CB with deterministic drift and arbitrary structure subsidiaries.*

A) Assume $I = 1$ and put $k = k_1$. Then, the ruin time τ_0 of the MRN equals a.s. and in distribution the time τ of the pooled process defined in (7), and equals furthermore the ruin time of the subsidiary with modified initial reserve $u_0/k+u_1$ and premium rate $c_0/k + c_1$, $F_i(x)$, namely

$$\Psi(u, c, t) = \Psi_1(u_0/k + u_1, c_0/k + c_1, t), \tag{8}$$

independently of the reset policy!

B) For $I > 1$, the time τ is an upper bound for the ruin time τ_0 of the CB:

$$\tau_0 \leq \tau.$$

C) Statements similar to A) hold for any Gerber-Shiu objective, with or without dividends to the subsidiary, as long as the CB is a deterministic drift.

Remark 11. Optimal allocation of total reserves $u_+ = u_0 + u_1$ and premium rate $c_+ = c_0 + c_1$. Note that:

1. When $k \geq 1$, $u_0 = c_0 = 0, u_1 = x, c_1 = c$ achieve the *minimal ruin probability.*
2. For $k = 1$, the ruin probability is independent of the amount $u_1 \in [0, u_+]$, as well as of the amount of premium rate $c_1 \in [0, c_+]$.

This optimization result fits the intuitively clear fact that with one subsidiary and no expenses, it is optimal to take advantage of the first transfer without cost to transfer everything to the subsidiary.

Corollary 1. *Let $X_0, ..., X_{I-1}$ denote a linear chain of CB's with deterministic drift. Assume $X_i, i = 0, ..., I-1$ must pay proportional costs k_i for bailing out X_{i+1}.*

Then, the probability of ruin of the MRN satisfies

$$\Psi(u, c, t) = \Psi_I(\frac{u_0}{k_0...k_{I-1}} + \frac{u_1}{k_1...k_{I-1}} + ...u_I, \frac{c_0}{k_0...k_{I-1}} + \frac{c_1}{k_1...k_{I-1}} + ...c_I, t),$$

where Ψ_I is the ruin probability of the last subsidiary in the chain.

Remark 12. One interesting feature of this result is that it does not require any assumption on the probabilistic structure of the subsidiary risk process.

Another interesting feature is that similar reductions hold for other problems, as long as $I = 1$ and the main branch is a deterministic drift (in the absence of subsidiaries). For example, one may add subsidiary dividends, ruin observed only at Poissonian times, Parisian ruin, etc.

4 Two Point Padé Approximations for the Downward Ladder Time of the Cramér-Lundberg Process with Exponential Claims

Our approach is based on the idea that approximating the excursions of a process ensures approximating the process, and in particular various functionals of the process [50, 72].

With phase-type jumps, one would need to provide *bivariate matrix-exponential approximations* for the *joint law of the downward ladder time and height* of a SA process.

We start with the simplest case of *exponential claims*, when the density of the downward ladder time may be expressed as a hypergeometric function:

$$\rho(t) = \tilde{\rho}(c\mu t)c\mu, \quad \tilde{\rho}(t) = \rho e^{-(1+\rho)t} \; {}_0F_1(2, \rho t^2). \tag{9}$$

However, what we need is a *phase-type* approximation of this. This topic has already been considered in [2] (at order two), as one of many possible methods for approximating the M/M/1 busy period density.

We recall now some basic facts on this case, summarizing from Section 6:

Lemma 1. *a) With Poisson arrivals of rate l and exponential claims of rate μ, the Laplace transform of the downward ladder time density satisfies a quadratic equation*

$$\widehat{\rho}_q = \widehat{a}(q + c\mu - c\mu\widehat{\rho}_q) = \frac{\lambda}{\lambda + q + c\mu - c\mu\widehat{\rho}_q} = \frac{\rho}{1 + \delta + \rho - \widehat{\rho}_q}, \ \rho := \frac{\lambda}{c\mu}, \ \delta = \frac{q}{c\mu}$$

with solution

$$\widehat{\rho}_q = \widehat{\widehat{\rho}}_\delta = \frac{1}{2}\left(\delta + \rho + 1 - \sqrt{(\delta + \rho + 1)^2 - 4\rho}\right). \tag{10}$$

b) The Laplace transform (10) *may be computed iteratively by the continued fraction expansion*

$$\widehat{\widehat{\rho}}_\delta = \frac{\rho}{1 + \delta + \rho - \frac{\rho}{1+\delta+\rho-\frac{\rho}{1+\delta+\rho-s\dots}}}.$$

The convergents $\widehat{\widehat{\rho}}_\delta^{(n)}, n = 1, 2, 3...$ with s constant satisfy $\widehat{\widehat{\rho}}_\delta^{(n)} = \frac{\rho P_{n-1}(\delta+\rho+1)}{P_n(\delta+\rho+1)}$ [61, (75)] where $P_n(x)$ are Chebyshev polynomials [61, (77)]. When $s = 0$, the first three are

$$\left\{\frac{\rho}{\delta + \rho + 1}, \frac{\rho(\delta + \rho + 1)}{\delta^2 + 2\rho\delta + 2\delta + \rho^2 + \rho + 1}, \frac{\rho\left(\delta^2 + 2\rho\delta + 2\delta + \rho^2 + \rho + 1\right)}{(\delta + \rho + 1)\left(\delta^2 + 2\rho\delta + 2\delta + \rho^2 + 1\right)}, \dots\right\}$$

Decomposing in partial fraction the third convergent yields an order three rational approximation of the Laplace transform of the ladder time:

$$\widehat{\rho}_q \approx \frac{\rho}{\delta + \rho + 1}\frac{(\delta + \rho + 1)^2 - \rho}{(\delta + \rho + 1)^2 - 2\rho} =$$
$$\frac{\rho}{2}\left(\frac{1}{\delta + \rho + 1} + \frac{1/2}{\delta + \rho + 1 + \sqrt{2\rho}} + \frac{1/2}{\delta + \rho + 1 - \sqrt{2\rho}}\right).$$

Inverting the Laplace transform yields a hyperexponential density approximation:

$$\widetilde{\rho}(t) \approx \rho e^{-(1+\rho)t}\left(\frac{1}{2} + \frac{1}{4}(e^{-\sqrt{2\rho}t} + e^{+\sqrt{2\rho}t})\right) := \sum_{i=0}^{2} \alpha_i \lambda_i e^{-\lambda_i t}$$
$$= \frac{\rho}{2}e^{-(1+\rho)t} + \frac{\rho}{4}e^{-(1+\rho+\sqrt{2\rho})t} + \frac{\rho}{4}e^{-(1+\rho-\sqrt{2\rho})t}, \tag{11}$$

where $\alpha_0 = \frac{\rho}{2(1+\rho)}$ and $\alpha_{1,2} = \frac{\rho}{4(1+\rho\pm\sqrt{2\rho})}$ are nonnegative for any $\rho \in [0, \infty)$. Furthermore, $\sum_i \alpha_i < \frac{3}{4}\rho$ iff $\rho < 1$, providing us thus with a valid approximations for any ρ in this range.

Remark 13. A further simplification of the Laplace transform (10) may be obtained factoring $\delta + \rho + 1$ and changing variables $a = \rho(1 + \rho + \delta)^{-2}$:

$$\widehat{\rho}_q = \widehat{\widehat{\rho}}_\delta = \frac{\delta + \rho + 1}{2}\left(1 - \sqrt{1 - 4\frac{\rho}{(\delta + \rho + 1)^2}}\right)$$
$$= \frac{\rho}{\delta + \rho + 1}\frac{1 - \sqrt{1 - 4a}}{2a}. \tag{12}$$

The second factor put thus in evidence is the generating function of the famous Catalan numbers

$$\frac{1-\sqrt{1-4a}}{2a} = \sum_{k=0}^{\infty} \frac{\binom{2k}{k}}{k+1} a^k = 1 + a + 2a^2 + 5a^3 + 14a^4 + ...$$

and a continued fraction (cf) representation

$$\frac{1-\sqrt{1-4a}}{2a} = \frac{1}{1-\frac{a}{1-\frac{a}{1+\dots}}} \tag{13}$$

may be found for example in [26, (7.7.5)].[4] The lowest order approximations (13) are

$$\frac{1}{1-a}, \frac{1-a}{1-2a}, \frac{1-2a}{1-3a+a^2}, \frac{1-3a+a^2}{1-4a+3a^2}, \frac{1-4a+3a^2}{1-5a+6a^2-a^3}, \dots$$

Lemma 2. *The rational convergents R_n of the continued fraction* (13) *increase towards* $\frac{1-\sqrt{1-4a}}{2a}, \forall a \in (0, 1/4)$.

Proof. This is immediate by the positivity of a. ■

Remark 14. Alternatively, we may use two point Padé approximations which ensure also the equality of the derivatives around 0. [2, Sec 3] provide an in-depth numerical comparison of several hyper-exponential approximations of order two, and find that fitting the derivatives yields excellent results around 0, while fitting the moments is less satisfactory, since better results may be obtained with asymptotic approximations.

Let us invert now the lowest order two-point Padé approximation of $\widehat{\rho}(q)$ which ensures also the condition $\widetilde{\widehat{\rho}}_{q=0} = \rho$, $\rho \in (0, 1]$:

$$\frac{\rho\left(\delta^2+2\rho\delta+2\delta+\rho+1\right)}{(\delta+\rho+1)\left(\delta^2+2\rho\delta+2\delta+1\right)} = \frac{\rho}{\rho+2}\left(\frac{\rho+1}{\delta+\rho+1}+\frac{\delta+\rho+1}{\delta^2+2\delta(\rho+1)+1}\right) \tag{14}$$

$$= \frac{\rho}{\rho+2}\left(\frac{\rho+1}{\delta+\rho+1}+\frac{1}{\lambda_1-\lambda_2}\left(\frac{\lambda_1-(\rho+1)}{\delta+\lambda_1}+\frac{\rho+1-\lambda_2}{\delta+\lambda_2}\right)\right),$$

$$\lambda_{1,2} = \rho + 1 \pm \sqrt{\rho(\rho+2)}$$

This yields a density approximation:

$$\widetilde{\rho}(t) \approx \sum_{i=0}^{2} \alpha_i \lambda_i e^{-\lambda_i t}, \tag{15}$$

where $\alpha_0 = \frac{\rho}{\rho+2}, \alpha_1 = \frac{\rho}{\rho+2}\frac{1-(\rho+1)/\lambda_1}{\lambda_1-\lambda_2} = \frac{\rho^{1/2}\left(\sqrt{\rho(\rho+2)}+1\right)}{2(\rho+2)^{3/2}\left(\rho+1+\sqrt{\rho(\rho+2)}\right)}$,

[4] As well known, the Padé approximations obtained by truncating continued fractions have good properties, like larger domains of convergence than the corresponding power series.

$\alpha_2 = \frac{\rho}{\rho+2}\frac{(\rho+1)/\lambda_2-1}{\lambda_1-\lambda_2} = \frac{\rho^{1/2}\left(\sqrt{\rho(\rho+2)}-1\right)}{2(\rho+2)^{3/2}\left(\rho+1-\sqrt{\rho(\rho+2)}\right)}$. Note that α_i are nonnegative for any $\rho \in [0,\infty)$, and $\sum_i \alpha_i = \rho$.

5 Padé Based SNMAP Approximations for $X(t)$ when C_i are Exponential Random Variables, and the Optimal Dividend Barrier.

In this section, even though finite time ruin probabilities have an explicit Bessel density with exponential claims, we will replace them by matrix exponential approximations, since this allows solving network problems by the SNMAP methodology.

After applying the order three approximation (15) to the subsidiary's ladder time, the central branch becomes a MAP with three states, with transition rates $Q_{ij} = \lambda_i \alpha_j$ accompanied by exponential jumps of rate μ/k translated by K (we could include here *phase-type jumps to the CB*, and *fixed costs*, since these pose no problem to the MAP methodology).

When $K = 0$, $\widehat{f}(s) = \frac{\mu/k}{s+\mu/k}$, and we find from the general formula that the symbol of the approximated CB is

$$\kappa(s) = diag(c_0 s - \lambda_i) + \lambda\boldsymbol{\alpha}\widehat{f}(s) \tag{16}$$
$$= \begin{pmatrix} c_0 s - \lambda_0 + \lambda_0\alpha_0\frac{\mu/k}{s+\mu/k} & \lambda_0\alpha_1\frac{\mu/k}{s+\mu/k} & \lambda_0\alpha_2\frac{\mu/k}{s+\mu/k} \\ \lambda_1\alpha_0\frac{\mu/k}{s+\mu/k} & c_0 s - \lambda_1 + \lambda_1\alpha_1\frac{\mu/k}{s+\mu/k} & \lambda_1\alpha_2\frac{\mu/k}{s+\mu/k} \\ \lambda_2\alpha_0\frac{\mu/k}{s+\mu/k} & \lambda_2\alpha_1\frac{\mu/k}{s+\mu/k} & c_0 s - \lambda_2 + \lambda_2\alpha_2\frac{\mu/k}{s+\mu/k} \end{pmatrix}.$$

To apply the scale based MAP methodology, it is convenient to transform this MAP with exponential jumps into a continuous MMBM:

$$\widetilde{\kappa}(s) = \begin{pmatrix} c_0 s - \lambda_0 & 0 & 0 & \lambda_0\alpha_0 & \lambda_0\alpha_1 & \lambda_0\alpha_2 \\ 0 & c_0 s - \lambda_1 & 0 & \lambda_1\alpha_0 & \lambda_1\alpha_1 & \lambda_1\alpha_2 \\ 0 & 0 & c_0 s - \lambda_2 & \lambda_2\alpha_0 & \lambda_2\alpha_1 & \lambda_2\alpha_2 \\ \mu/k & 0 & 0 & -s-\mu/k & 0 & 0 \\ 0 & \mu/k & 0 & 0 & -s-\mu/k & 0 \\ 0 & 0 & \mu/k & 0 & 0 & -s-\mu/k \end{pmatrix} \tag{17}$$

After obtaining an SNMAP approximation for $X(t)$, we may solve approximatively various problems related to this process, using the package [42].

6 Review of the SA Process with Exponential Claims: Kendall's Functional Equation and the Takacs Series

We recall first a *risk-queueing duality* result relating busy periods in queueing and ruin times of SA processes, obtained by reversing the meaning of claims and interarrivals, which we lifted from [31] and [41, 4.2] – see also [9, VI.4, VII.7] for related results and references.

Lemma 3. *The downward ladder time τ of a zero-delayed SA process $X(t)$ with independent pairs (A_i, C_i), and starting at $u = 0$ has the same distribution as the busy period of a dual queueing process starting at $u = 0$ with a jump (service demand) A_1*

$$\widetilde{X}(t) = A_1 - \frac{t}{c} + \sum_{i=2}^{M_t} A_i,$$

with drift $\frac{1}{c}$, arrival-jump pairs (C_i, A_i), and renewal process M_t associated to C_i, divided by c.

Proof. Construct an auxiliary Markovian process in which the meaning of A_i and C_i are interchanged, so that the inter-arrival times A_i become claims (or simply jumps) going up, and C_i become *linear depletion/service* of the jumps A_i, at a rate $\frac{1}{c}$. The resulting spectrally positive process is a spectrally positive Levy process (called in queueing theory workload, or *net input* process), and its ruin time (*busy period*) τ' is achieved by smooth linear crossing of 0. Furthermore, we may check that $\tau' = \sum_{i=1}^{N_\tau} C_i + U_\tau = c\sum_{i=1}^{N_\tau} A_i = c\tau$ ∎

Remark 15. This elementary observation renders available in ruin theory the classic results for the G/G/1 busy period obtained by Pollaczek [63], Wishart [71], Conolly [24], Takacs [67] and [28], as well as related results on branching processes (in which case "ruin" is replaced by "extinction").

Remark 16. Imagine now another world ("Alpha-Centauri") where insurance claims are paid at linear rates, over time, according to some priority rules, and where servers give instantaneous service to customers. In that world, risk and queueing theories would be interchanged!

Note that the number of claims still unpaid under the "Alpha-Centauri linear rate reimbursement regime" (or of jumps still unserviced) provides us with an integer valued birth and death queueing process associated to a SA process. Essentially, the *M/G/1 busy period and SA ruin theories coincide, and may be embedded in the theory of Crump-Mode-Jagers branching processes!*[4].

The proof of the next result is more natural in queueing theory (or in the the "Alpha-Centauri" world).

Proposition 1. *For $X(t)$ a zero-delayed Sparre-Andersen process with*

a) i.i.d. inter-arrivals-claims pairs $\{(A_i, C_i),\ i \geq 0\}$

b) A_i independent of C_i, and

c) exponential marginal $C_i \sim \mathcal{E}(\mu)$, it holds that:

---

[4] An especially beautiful interpretation of (18), due to [53, 55], emerges when one considers the busy period of the M/G/1 LIFO (or LFCS-PR) service discipline. Then, the first root jump arrived is the last to be serviced, and all the other interrupting jumps arrive according to a Poisson process of rate μ, during the service time A_1 of the root, and each interruption is an independent downward ladder time for the auxiliary process, yielding (22).

1. *The Laplace transform* $\widehat{\rho} = \widehat{\rho}_{q,z} = \int_0^\infty e^{-qt}\rho_{q,z}(t)dt$ *of the density of the doubly killed downward ladder time satisfies the functional equation*

$$\widehat{\rho} = z\,\widehat{a}\,(q + c\mu(1-\widehat{\rho})) \in (0,1),\ \forall q > 0, z \in (0,1]^4. \tag{18}$$

2. *The Laplace transform may be inverted explicitly yielding the Takacs series [68, 69]*

$$\rho(t) = \sum_{n=1}^{\infty} a^{n,*)}(t)p_{n-1}(t)\frac{1}{n},\quad p_{n-1}(t) = e^{-c\mu t}\frac{(c\mu t)^{n-1}}{n!} \tag{19}$$

see Remark 18.

Remark 17. The functional equation (18) is known to describe the total size in the theory of branching processes [46, (59)], and when $z = 1 = c$, it describes the Laplace transform of the density of the M/G/1 busy period in queueing theory, where it is called *Kendall's equation.* If the busy period is started with an initial workload x, the Laplace transform of the busy period $T(x)$ becomes

$$Ee^{-qT(x)} = e^{-x(q+\mu(1-\widehat{\rho}))},$$

which implies (18) (since the first jump has density $a(x)$).

Kendall's functional equation (18), has been intensively studied numerically by iterative methods, contour integrals, convergence acceleration, saddle point expansions, etc. – see for example [1, 20, 23, 29, 30, 59, 60]. These methods admit natural generalizations to non-exponential claims; in particular, with phase-type claims one enters the realm of Markovian trees, recently explored by [19, 38, 39].

Note that in the case $q = 0$, $\widehat{\rho}(z) := \widehat{\rho}_{0,z} = Ez^{N_\tau} = \sum_{n=1} z^n P[N_\tau = n]$ is the pgf of the number of claims N_τ before ruin. One may also write the RHS of Kendall's equation as

$$\widehat{\rho}(z) = z\; g(\widehat{\rho}(z)),\quad g(z) := \widehat{a}\,(c\mu(1-z))\,, \tag{20}$$

and recognize that $g(z)$ is the probability generating function of the *number of workload jumps arriving during a service time*, and (20) is the formula of the pgf of the total number of progeny in a Galton-Watson process with progeny pgf given by $g(z)$.

When $z = 1$, (20) reduces to the famous equation yielding the extinction probability of a branching process [62], [70, (6)].

These branching connections are not surprising, in view of the well known connection of risk and queueing (see lemma 3), and of the connection of branching to queueing [34, 46, 49]. The net result is that risk quantities can be read off from the richer branching process model.

In particular, the number of claims causing ruin N_τ equals the total number of progeny in the branching model, and has therefore a Lagrange type distribution computable by Lagrange inversion [25]. For example, with Erlang (and rational) arrivals, the result is hypergeometric – see (24).

Remark 18. A probabilistic proof of (19) may be obtained conditioning on the independent events that a) the downward ladder has been achieved by n arrivals and has length t, and that b) the renewal process associated to claims has exactly $n-1$ claims with sum in $[0, ct)$, yielding

$$\rho_n(t) = a^{n,*)}(t) e^{-c\mu t} \frac{(c\mu t)^{n-1}}{(n-1)!} \frac{1}{n}. \tag{21}$$

The correction term $\frac{1}{n}$ is a consequence of the Sparre-Andersen identity [8, 43], applied to the sums $\sum_{i=1}^{n} cA_i - C_i$.

Proof of Proposition 1.
1. The equation (18) may be written as a distributional equality

$$\tau_0 \equiv A_1 + \sum_{i=1}^{N_{c\mu}(A_1)} \tau_i \tag{22}$$

where A_1 is the first inter-arrival time killed with probability $1-z$, $N_{c\mu}(t)$ is an independent Poisson arrival process of rate $c\mu$, and τ_i are i.i.d copies of τ_0.

This probabilistic decomposition is obvious for the M/G/1 busy period under work preserving service disciplines (FIFO, LIFO, etc), since the busy period equals the service time of the first jump A_1 + other busy periods initiated by the jumps arriving (as Poisson process) during A_1. Note also the interpretation as the total lifetime of a continuous time branching process with descendants arriving at exponential rates, well-known since [37, 36, 46].

2. Put $g(\rho) = q + c\mu - c\mu\rho$ and note [37, 68, 69], [3, (1.4)], [52] that the Kendall equation (18)

$$\widehat{\rho}_{q,z} = z\widehat{a}(q + c\mu - c\mu\widehat{\rho}_{q,z})) = z\widehat{a}(g(\widehat{\rho}_{q,z})))$$

may be solved for $\widehat{\rho}_{q,z}$ in a Lagrange power series in z, each term of which is a Laplace transform in q:

$$\begin{aligned}
\widehat{\rho}_q &= \sum_{n=1}^{\infty} \frac{z^n}{n!} \frac{D^{n-1}}{d\rho^{n-1}} [\widehat{a}^n (q + c\mu - c\mu\rho))]_{\rho=0} \\
&= \sum_{n=1}^{\infty} \frac{(-c\mu)^{n-1} z^n}{n!} \frac{D^{n-1}}{dg^{n-1}} [\widehat{a}^n(g)]_{g=q+c\mu} \\
&= \sum_{n=1}^{\infty} \frac{(c\mu)^{n-1} z^n}{n!} \frac{(-D)^{n-1}}{dg^{n-1}} [\widehat{a^{n,*)}(t)}(g)]_{g=q+c\mu} \\
&= \sum_{n=1}^{\infty} \frac{(c\mu)^{n-1} z^n}{n!} [\widehat{t^{n-1} a^{n,*)}}(t)(g)]_{g=q+c\mu} \\
&= \int_0^{\infty} e^{-qt} dt \sum_{n=1}^{\infty} z^n a^{n,*)}(t) \left(e^{-c\mu t} \frac{(c\mu t)^{n-1}}{n!} \right).
\end{aligned} \tag{23}$$

Inverting yields the Takacs series:

$$\rho_z(t) := \sum_{n=1}^{\infty} z^n a^{n,*)}(t) e^{-c\mu t} \frac{(c\mu t)^{n-1}}{n!}$$

(and finally the downward ladder density (19) by setting $z = 1)^4$. ■

Example 4. With exponential claims of rate μ and Poisson arrivals of rate l, the Laplace transform of the downward ladder time density satisfies a quadratic equation $\widehat{\rho} = z\widehat{a}(q + c\mu - c\mu\widehat{\rho}) = z\frac{\lambda}{l+q+c\mu-c\mu\widehat{\rho}} = \frac{\rho z}{1+\rho+\delta-\widehat{\rho}}$, $\delta := \frac{q}{c\mu}$ with solution

$$\widehat{\rho}_{q,z} = \widehat{\widehat{\rho}}_{\delta,z} = \frac{1}{2}\left(1 + \rho + \delta - \sqrt{(1+\rho+\delta)^2 - 4\rho z}\right), \ \delta = \frac{q}{c\mu}.$$

Iterating yields the converging continued fraction

$$\widehat{\rho} = \frac{\rho z}{1 + \rho + \delta - \frac{\rho z}{1+\rho+\delta-\frac{\rho z}{1+\rho+\delta-\widehat{\rho}\ldots}}}$$

To compute the approximation to a given order, $\widehat{\rho}$... above may be replaced for example by $\widehat{\rho}_{0,1} = \rho$. The resulting approximation is rational (and will continue to be so for any arrivals with rational transform $\widehat{a}(s)$). Inverting $\widehat{\rho}(q) := \widehat{\rho}_{q,1}$ will yield matrix exponential approximations for the density $\rho(t)$, and inverting yields

$$\widehat{\rho}(z) := \widehat{\rho}_{0,z} = \frac{\rho z}{1 + \rho - \frac{\rho z}{1+\rho-\frac{\rho z}{1+\rho-\widehat{\rho}\ldots}}} = z\frac{\rho}{1+\rho} + z^2\frac{\rho^2}{(1+\rho)^3} + 2z^3\frac{\rho^3}{(1+\rho)^5} + \ldots$$

In conclusion [33, (3.5)], [52, (24)],

$$P[N = k] = \frac{(2k-2)!}{k!(k-1)!}(\frac{\rho}{1+\rho})^k(\frac{1}{1+\rho})^{k-1}$$

is a defective extended negative binomial (ENB) law. Note the probabilistic interpretation of

More generally, with Erlang arrivals of parameters (n, l), $a(t) = \gamma_{n,\lambda}(t) = \frac{\lambda^n t^{n-1}}{(n-1)!}e^{-\lambda t}$, $a^{k,*}(t) = \gamma_{nk,\lambda}(t)$, and [33, (3.15)] show that

$$P[N = k] = \frac{(nk+k-2)!}{(nk-1)!k!}\,(\frac{\rho}{1+\rho})^{nk}\,(\frac{1}{1+\rho})^{k-1}, \rho = \frac{\lambda}{c\mu}. \tag{24}$$

Note that the GLE with $\delta = 0$ is now

$$\widehat{\rho}(z) = \frac{z\rho^n}{(1 + \rho - \widehat{\rho}(z))^n}$$

4 Multiplying by $c\mu t$ and taking Laplace transform yields an integral representation [1, 3] [20, (39)]

and (24) may be obtained either using the continued fraction, or the inverse Lagrange series.

Convolutions are now explicit, and the Takacs formula yields the particular case $u = 0$ of the formula [21, (8)]

$$\psi(t;u) = \frac{\lambda^n t^{n-1} e^{-\lambda t}}{(n-1)!} e^{-\mu(u+ct)}$$
$$(\frac{ct}{u+ct} HypergeometricPFQ[, \frac{n+1}{n}, \frac{n+2}{n}, ..., \frac{2n-1}{n}, 2, \frac{(\lambda t)^n \ \mu(u+ct)}{n^n}]$$
$$+ \frac{u}{u+ct} HypergeometricPFQ[, \frac{n}{n}, \frac{n+1}{n}, ..., \frac{2n-1}{n}, \frac{(\lambda t)^n \ \mu(u+ct)}{n^n}]),$$

where HypergeometricPFQ is the Fox-Wright generalized hypergeometric function.

We review now in Proposition 2 the fundamental case of SA processes with exponential claims and possible dependence between (A_i, C_i) (which by Lemma 3 may be viewed as M/G/1 workload processes, a particular case of Levy process conditioned to start by a jump), when the killed ruin probability/Laplace transform of the ruin time is exponential in the initial reserves[4].

Proposition 2. *1. The doubly killed ruin probability of a zero-delayed Sparre-Andersen process with i.i.d. inter-arrivals-claims pairs* $\{(A_i, C_i),\ i\ \geq 0\}$, *with possibly dependent components and exponential marginal* $C_i \sim \mathcal{E}(\mu)$ *is:*

$$\widehat{\Psi}(q,z;u) := E_u\left[z^{N_\tau} e^{-q\tau}, \tau < \infty\right] = \widehat{\rho} e^{-\gamma u}, \tag{25}$$

where $\gamma = \gamma_{q,z}$ *is the unique positive solution of the generalized Lundberg equation*

$$Ee^{-(q+\gamma c)A_i + \gamma C_i} = 1, \tag{26}$$

[9], and the prefactor $\widehat{\rho} = \widehat{\rho}_{q,z}$ *is related to* γ *via*

$$\widehat{\rho} = 1 - \frac{\gamma}{\mu}.$$

2. Delayed SA process. When moreover the initial arrival A_0 *has a different law, with Laplace transform* $\widehat{a}_0(\delta)$, *exponential behavior with the same exponent* γ *as in* (25) *holds, but with modified prefactor:*

$$\widehat{\Psi}(q,z;u) = z\widehat{a}_0\left(q + c\gamma\right) e^{-\gamma u}. \tag{27}$$

Remark 19. Replacing equation (18), by

$$\widehat{\rho}_n(q) = z\widehat{a}(q + c\mu - c\mu\widehat{\rho}_{n+1}(q)))$$

suggests considering recursive approximations [61]:

$$\widehat{\rho}_{q,z} \approx z\widehat{a}(q + c\mu - c\mu z\widehat{a}(q + c\mu - c\mu...)). \tag{28}$$

---

[4] see for example [65, Thm 6.4.5, Cor 6.5.2], who use the Wiener-Hopf decomposition, and the memoryless property of the exponential implying that $m_+(s) = \rho\frac{\mu}{\mu+s}$.

Example 5. [7, Example 1] finds that the ultimate ruin probability for Moran and Downton's bivariate exponential with joint Laplace transform

$$E[e^{-s_1A_1-s_2C_1}] = g(s_1,s_2)^{-1}, g(s_1,s_2) = (1+s_1/\lambda)(1+s_2/\mu - \alpha\frac{s_1}{\lambda}\frac{s_2}{\mu}))$$

is given by (25) with $\gamma = \frac{\mu-\lambda/c}{1-\alpha}$. The independent case is recovered by setting the correlation $\alpha = 0$.

Finally, $\widehat{\rho}(q)$ may be computed by iterations (28) with $\widehat{a}$ being replaced by the function

$$h(s) = \left(1 + \frac{cs+q}{\lambda} - \alpha(cs+q)\frac{s}{\lambda(\mu+s)}\right)^{-1}$$

Proof of Proposition 2. Both results are easy consequences of the optional stopping theorem applied to the discrete time *De-Finetti exponential martingale*

$$M_n = e^{-\sum_{i=1}^n((q+c_1\gamma)A_i-\gamma C_i)}, n = 0,1,2,...$$

Indeed, consider the more general killed ruin probability with claims killed with probability $(1-z)$ [41]

$$\widehat{\Psi}(q,z;u) = E[z^{N_\tau}e^{-q\tau}\mathbf{1}_{\{\}}\tau < \infty],$$

and the more general martingale

$$M_n = z^n e^{-\sum_{i=1}^n((q+c\gamma)A_i-\gamma C_i)}, n = 0,1,2,... \tag{29}$$

where $\gamma = \gamma_{q,z}$ is a positive root of the *killed generalized Lundberg equation*

$$\begin{aligned} &1 - zEe^{-(q+\gamma c)A_i+\gamma C_i} = 0 \qquad (30)\\ &= 1 - z\widehat{a}(q+c\gamma)\widehat{f_C}(-\gamma) = 1 - z\widehat{a}(q+c\gamma)\frac{\mu}{\mu-\gamma}. \end{aligned}$$

Applying the optional stopping theorem, using $\sum_{i=0}^{N_\tau} A_i = \tau, \sum_{i=0}^{N_\tau}(cA_i - C_i) = U_\tau - u$, and the *independence of* τ, U_τ, yields:

$$\begin{aligned} E[M_0] &= 1 = E[z^{N_\tau}e^{-q\tau-\gamma(U_\tau-u)}\mathbf{1}_{\{\}}\tau < \infty]\\ &= e^{\gamma u}\widehat{\Psi}(q,z;u)\frac{\mu}{\mu-\gamma} \Longrightarrow \Psi(q,z;u) = (1-\frac{\gamma}{\mu})e^{-\gamma u}, \end{aligned}$$

and (25) follows by taking $z = 1$.

For (27), one applies again the optional stopping, but with the time n starting from 1. With A_i and C_i independent, this yields

$$E[M_1] = z\,\widehat{a}_0\,(q+c\gamma)\,\frac{\mu}{\mu-\gamma} = e^{\gamma u}\widehat{\Psi}(q,z;u)\frac{\mu}{\mu-\gamma}$$

and the case A_i and C_i dependent yields a slightly more complicated result. ■

Acknowledgements: We take pleasure in thanking H. Albrecher, B. Avanzi, L. Breuer, E. Frostig, J. Ivanovs, R. Loeffen, M.Pistorius and T. Rolski for useful discussions.

References

1. Abate, J., Choudhury, G.L., and Whitt, W.: Calculating the M/G/1 busy-period density and lifo waiting-time distribution by direct numerical transform inversion. *Operations Research Letters*, **18** (1995) 113–119.
2. Abate, J. and Whitt, W.: Approximations for the M/M/1 busy-period distribution. *Queueing Theory and its Applications, Liber Amicorum for JW Cohen*, (1988) 149–191.
3. Abate, J. and Whitt, W.: Solving probability transform functional equations for numerical inversion. *Operations Research Letters*, **12** (1992) 275–281.
4. Albrecher, H., Hartinger, J., and Thonhauser, S.: On exact solutions for dividend strategies of threshold and linear barrier type in a sparre andersen model. *Astin Bulletin*, **37** (2007) 203–233.
5. Albrecher, H. and Ivanovs, J.: Strikingly simple identities relating exit problems for lévy processes under continuous and poisson observations, *arXiv preprint arXiv:1507.03848*, (2015).
6. Albrecher, H., Ivanovs, J., and Zhou, X.: Exit identities for levy processes observed at poisson arrival times. *arXiv preprint arXiv:1403.2854*, (2014).
7. Ambagaspitiya, R.S.: Ultimate ruin probability in the sparre andersen model with dependent claim sizes and claim occurrence times. *Insurance: Mathematics and Economics*, **44** (2009) 464–472.
8. Andersen, E.S.: On sums of symmetrically dependent random variables, *Scandinavian Actuarial Journal*, (1953) 123–138.
9. Asmussen, S. and Albrecher, H. *Ruin Probabilities*, vol. 14, World Scientific, 2010.
10. Avanzi, B.: Strategies for dividend distribution: A review. *North American Actuarial Journal*, **13** (2009) 217–251.
11. Avram, F., Badescu, A., Pistorius, M., and Rabehasaina, L.: On a class of dependent sparre andersen risk models and applications to central branch networks. *Work in progress*, (2015).
12. Avram, F., Kyprianou, A.E., and Pistorius, M.R.: Exit problems for spectrally negative lévy processes and applications to (canadized) russian options. *The Annals of Applied Probability*, **14** (2004) 215–238.
13. Avram, F., Minca, A., and Pistorius, M.: On the management of central branch risk networks. *Work in progress*, (2015).
14. Avram, F., Palmowski, Z., and Pistorius, M.R.: On the optimal dividend problem for a spectrally negative lévy process. *The Annals of Applied Probability*, (2007) 156–180.
15. Avram, F., Palmowski, Z., and Pistorius, M.R.: Exit problem of a two-dimensional risk process from the quadrant: exact and asymptotic results, *The Annals of Applied Probability*, **18** (2008) 2421–2449.
16. Avram, F., Palmowski, Z., and Pistorius, M.R.: On gerber–shiu functions and optimal dividend distribution for a lévy risk process in the presence of a penalty function. *The Annals of Applied Probability*, **25** (2015) 1868–1935.
17. Azcue, P. and Muler, N.: Optimal reinsurance and dividend distribution policies in the cramér-lundberg model. *Mathematical Finance*, **15** (2005) 261–308.
18. Badescu, A., Cheung, E., and Rabehasaina, L.: A two-dimensional risk model with proportional reinsurance. *Journal of Applied Probability*, **48** (2011) 749–765.
19. Bean, N.G., Kontoleon, N., and Taylor, P.G.: Markovian trees: properties and algorithms. *Annals of Operations Research*, **160** (2008), 31–50.

20. Blanc, J.: On the numerical inversion of busy-period related transforms, *Operations Research Letters*, **30** (2002) 33–42.
21. Borovkov, K.A. and Dickson, D.: On the ruin time distribution for a sparre andersen process with exponential claim sizes. *Insurance: Mathematics and Economics*, **42** (2008) 1104–1108.
22. Chan, W.S., Yang, H., and Zhang, L.: Some results on ruin probabilities in a two-dimensional risk model. *Insurance: Mathematics and Economics*, **32** (2003) 345–358.
23. Chung, S.: *Saddlepoint Approximation to Functional Equations*, PhD thesis, Colorado State University, 2010.
24. Conolly, B.: The busy period in relation to the single-server queueing system with general independent arrivals and erlangian service-time. *Journal of the Royal Statistical Society. Series B (Methodological)*, (1960) 89–96.
25. Consul, P.C. and Famoye, F.: *Lagrangian Probability Distributions*, Springer, 2006.
26. Cuyt, A., Petersen, V.B., Verdonk, B., Waadeland, H., and Jones, W.B.: *Handbook of Continued Fractions for Special Functions*, Springer, 2008.
27. Feller, W.: *An Introduction to Probability Theory and its Applications. Vol. II.*: Second edition, John Wiley & Sons Inc., New York, 1971.
28. Finch, P.: On the busy period in the queueing system Gi/G/1. *Journal of the Australian Mathematical Society*, **2** (1961) 217–228.
29. Frolov, G. and Kitaev, M.: Improvement of accuracy in numerical methods for inverting laplace transforms based on the post-widder formula. *Computers & Mathematics with Applications*, **36** (1998), 23–34.
30. Frolov, G. and Kitaev, M.: A problem of numerical inversion of implicitly defined laplace transforms. *Computers & Mathematics with Applications*, **36** (1998) 35–44.
31. Frostig, E.: Upper bounds on the expected time to ruin and on the expected recovery time. *Advances in Applied Probability*, (2004) 377–397.
32. Frostig, E.: On ruin probability for a risk process perturbed by a lévy process with no negative jumps. *Stochastic Models*, **24** (2008) 288–313.
33. Frostig, E., Pitts, S.M., and Politis, K.: The time to ruin and the number of claims until ruin for phase-type claims. *Insurance: Mathematics and Economics*, **51** (2012) 19–25.
34. Geiger, J. and Kersting, G.: Depthfirst search of random trees, and poisson point processes. In *Classical and Modern Branching Processes*, Springer, 1997, 111–126.
35. Gerber, H.U.: Games of economic survival with discrete-and continuous-income processes. *Operations research*, **20** (1972) 37–45.
36. Good, I.: The lagrange distributions and branching processes. *SIAM Journal on Applied Mathematics*, **28** (1975) 270–275.
37. Good, I.J.: Generalizations to several variables of lagrange's expansion, with applications to stochastic processes. In *Mathematical Proceedings of the Cambridge Philosophical Society*, **56**, Cambridge Univ. Press, 1960, 367–380.
38. Hautphenne, S., Latouche, G., and Remiche, M.A.: Transient features for markovian binary trees. In *Proceedings of the Fourth International ICST Conference on Performance Evaluation Methodologies and Tools*, ICST (Institute for Computer Sciences, Social-Informatics and Telecommunications Engineering), 2009, 18.
39. Hautphenne, S., Latouche, G., and Remiche, M.A.: Algorithmic approach to the extinction probability of branching processes. *Methodology and Computing in Applied Probability*, **13** (2011) 171–192.

40. Ivanovs, J.: *One-sided Markov Additive Processes and Related Exit Problems*, PhD thesis, University of Amsterdam, 2011.
41. Ivanovs, J.: A note on killing with applications in risk theory. *Insurance: Mathematics and Economics*, **52** (2013) 29–34.
42. Ivanovs, J.: Spectrally-negative Markov additive processes 1.0. https://sites.google.com/site/jevgenijsivanovs/files, 2013. Mathematica 8.0 package.
43. Ivanovs, J.: Sparre-Andersen identity: there is more to it. *arXiv preprint arXiv:1501.04542*, (2015).
44. Ivanovs, J. and Palmowski, Z.: Occupation densities in solving exit problems for markov additive processes and their reflections. *Stochastic Processes and their Applications*, **122** (2012) 3342–3360.
45. Ivanovs, J. and Palmowski, Z.: Occupation densities in solving exit problems for markov additive processes and their reflections, *Stochastic Processes and their Applications*, **122** (2012), 3342–3360.
46. Kendall, D.G.: Some problems in the theory of queues. *Journal of the Royal Statistical Society. Series B (Methodological)*, (1951) 151–185.
47. Kuznetsov, A., Kyprianou, A.E., and Rivero, V.: The theory of scale functions for spectrally negative lévy processes. In *Lévy Matters II*, Springer, 2013, 97–186.
48. Kyprianou, A.E. and Palmowski, Z.: Fluctuations of spectrally negative markov additive processes. In *Séminaire de probabilités XLI*, Springer, 2008, 121–135.
49. Lambert, A.: The contour of splitting trees is a lévy process. *The Annals of Probability*, **38** (2010) 348–395.
50. Lambert, A., Simatos, F., and Zwart, B.: Scaling limits via excursion theory: interplay between crump–mode–jagers branching processes and processor-sharing queues. *The Annals of Applied Probability*, **23** (2013) 2357–2381.
51. Landriault, D., Renaud, J.F., and Zhou, X.: An insurance risk model with parisian implementation delays. *Methodology and Computing in Applied Probability*, (2013) 1–25.
52. Landriault, D., Shi, T., and Willmot, G.E.: Joint densities involving the time to ruin in the sparre andersen risk model under exponential assumptions, *Insurance: Mathematics and Economics*, **49** (2011) 371–379.
53. Le Gall, J.F. and Le Jan, Y.: Branching processes in lévy processes: the exploration process. *Annals of probability*, (1998) 213–252.
54. Li, B., Wu, R., and Song, M.: A renewal jump-diffusion process with threshold dividend strategy. *Journal of Computational and Applied Mathematics*, **228** (2009) 41–55.
55. Limic, V.: A lifo queue in heavy traffic. *The Annals of Applied Probability*, (2001) 301–331.
56. Loeffen, R.: On optimality of the barrier strategy in de finettis dividend problem for spectrally negative lévy processes. *The Annals of Applied Probability*, **18** (2008) 1669–1680.
57. Loeffen, R.L. and Renaud, J.F.: De finetti's optimal dividends problem with an affine penalty function at ruin. *Insurance: Mathematics and Economics*, **46** (2010) 98–108.
58. Løkka, A. and Zervos, M.: Optimal dividend and issuance of equity policies in the presence of proportional costs. *Insurance: Mathematics and Economics*, **42** (2008) 954–961.
59. McMillan, B. and Riordan, J.: A moving single server problem. *The Annals of Mathematical Statistics*, (1957) 471–478.

60. Mishkoy, G., Benderschi, O., Giordano, S., and Bejan, A.I.: On numerical algorithms for solving multidimensional analogs of the kendall functional equation. *Buletinul Academiei de Stiinte a Republicii Moldova. Matematica*, (2008) 118–121.
61. Neuts, M.: *The Queue with Poisson Input and General Service Times, Treated as a Branching Process*, Defense Technical Information Center, 1966.
62. Otter, R.: The multiplicative process. *The Annals of Mathematical Statistics*, (1949) 206–224.
63. Pollaczek, F.: Sur la rpartition des priodes d'occupation ininterrompue d'un guichet. *Compt. Rend. Acad. Sci. Paris*, **234** (1952), 2042-2044.
64. Renaud, J.F.: On the time spent in the red by a refracted lévy risk process. *Journal of Applied Probability*, **51** (2014) 1171–1188.
65. Rolski, T., Schmidli, H., Schmidt, V., and Teugels, J.: *Stochastic Processes for Insurance and Finance*, vol. 505, John Wiley & Sons, 2009.
66. Shreve, S.E., Lehoczky, J.P., and Gaver, D.P.: Optimal consumption for general diffusions with absorbing and reflecting barriers. *SIAM Journal on Control and Optimization*, **22** (1984) 55–75.
67. Takács, L.: Transient behavior of single-server queueing processes with Erlang input. *Transactions of the American Mathematical Society*, (1961) 1–28.
68. Takács, L. *Introduction to the Theory of Queues*, vol. 584, Oxford University Press New York, 1962.
69. Takács, L.: The stochastic law of the busy period for a single server queue with poisson input. *Journal of Mathematical Analysis and Applications*, **6** (1963) 33–42.
70. Viskov, O.V.: Some comments on branching processes. *Mathematical Notes of the Academy of Sciences of the USSR*, **8** (1970) 701–705.
71. Wishart, D.: A queueing system with chi-square service-time distribution. *The Annals of Mathematical Statistics*, **27** (1956) 768–779.
72. Yano, K.: Functional limit theorems for processes pieced together from excursions. *arXiv preprint arXiv:1309.2652*, (2013).
73. Zhang, Z., Yang, H., and Yang, H.: On a sparre andersen risk model perturbed by a spectrally negative levy process. *Scandinavian Actuarial Journal*, **2013** (2013) 213–239.

Control of Parallel Non-Observable Queues: Asymptotic Equivalence and Optimality of Periodic Policies

Jonatha Anselmi[1,2], Bruno Gaujal[3,4,5] and Tommaso Nesti[1,6]

Basque Center for Applied Mathematics (BCAM), Al. de Mazarredo 14, 48009 Bilbao, Basque Country, Spain[1]

INRIA Bordeaux Sud Ouest, 200 av. de la Vieille Tour, 33405 Talence, France[2]
jonatha.anselmi@inria.fr

Univ. Grenoble Alpes, LIG, F-38000 Grenoble, France[3]

CNRS (French National Center for Scientific Research), LIG F-38000 Grenoble, France[4]

INRIA[5]
bruno.gaujal@inria.fr

Centrum Wiskunde & Informatica, Group Stochastics, Amsterdam, The Netherlands[6]
t.nesti@cwi.nl

Abstract. We focus on a queueing system composed of a dispatcher that routes jobs to one among a set of non-observable, parallel queues. There is neither job splitting nor job replication: an arriving job is sent to exactly one queue to receive some processing. The fundamental problem is to understand which policy should the dispatcher implement to minimize the stationary mean waiting time of the incoming jobs. We present a structural property that holds in the classic scaling of the system where the network demand (arrival rate of jobs) grows proportionally with the number of queues. Assuming that each queue of type r is replicated k times, we consider a set of policies that are periodic with period $k\sum_r p_r$ and such that exactly p_r jobs are sent in a period to each queue of type r. When $k \to \infty$, our main result shows that all the policies in this set are *equivalent*, in the sense that they yield the same mean stationary waiting time, and *optimal*, in the sense that no other policy having the same aggregate arrival rate to *all* queues of a given type can do better in minimizing the stationary mean waiting time. This property holds in a strong probabilistic sense. Furthermore, the limiting mean waiting time achieved by our policies is a convex function of the arrival rate in each queue, which facilitates the development of a further optimization aimed at solving the fundamental problem above for large systems.

Keywords: Parallel queues, periodic policies, asymptotic equivalence, asymptotic optimality, convex optimization.
AMS 2000 subject classification: Primary 60K25

1 Introduction

In computer and communication networks, the access of jobs to resources (web servers, network links, etc.) is usually regulated by a dispatcher. A fundamental problem is which algorithm should the dispatcher implement to minimize the mean delay experienced by jobs. There is a vast literature on this subject and the structure of the optimal algorithm strongly depends on *i)* the information available to the dispatcher, *ii)* the topology of the network and *iii)* how jobs are processed by resources. We are interested in a scenario where:

- The dispatcher has *static* information of the system;
- The network topology is parallel;
- Resources process jobs according to the first-come-first-served discipline.

Static information means that the dispatcher knows the probability distributions of job sizes and inter-arrival times but cannot observe the dynamic state of resources such as the current number of jobs in their queues. This scenario can be of interest in the context of volunteer computing, cloud computing, web server farms, etc.; see, e.g., [21, 25, 26] respectively.

In this framework, the problems of finding an algorithm, or *policy*, that minimizes the mean stationary delay and of determining the minimum mean stationary delay are both considered difficult; see, e.g., [1, 2] for an overview. A policy can be defined as a function that maps a natural number n, corresponding to the n-th job arriving to the dispatcher, to a probability mass function P_n over the set of resources. When the n-th job arrives, the dispatcher sends it to resource i with probability $P_n(i)$. Unfortunately, the problem of finding an optimal policy is intractable and for this reason two extreme families of policies have received particular attention in the literature: *probabilistic* policies, obtained when P_n is constant (in n), and *deterministic* policies, obtained when P_n puts the whole mass on a single resource.

When dealing with probabilistic policies, the difficulty of the problem is simplified by the fact that the arrival process at each resource is a renewal process, provided that the same holds for the arrival process at the dispatcher. This allows one to decompose the problem and, using the theory of the mean waiting time of the single GI/GI/1 queue, to immediately reduce it to a relatively simple optimization problem. This problem is usually convex and there exist efficient numerical procedures for their solution; e.g., [9, 14, 15, 31, 33, 35].

Contrariwise, when dealing with deterministic policies, one of the main difficulties is that the arrival process at each resource is hardly ever a renewal process. This prevents one from decomposing the problem and directly using the classic theory of the single queue as it has been done for probabilistic policies. Given this difficulty, researchers divided this problem in two subproblems:

i) In the first subproblem, the optimal deterministic policy is searched among all the deterministic policies ensuring that the long-term fractions of jobs to be sent to each resource is kept fixed (denote such fractions by vector p);

ii) In the second subproblem, the output of the first subproblem is employed to develop a further optimization over p.

In this paper, we focus on the first subproblem and, under some system scaling, we identify a set of policies that are optimal. This result is used to reduce the second subproblem to the solution of a convex optimization.

One of the folk theorem of queueing theory says that determinism in the inter-arrival times minimizes the waiting time of the single queue [19, 24]. In view of this classic insight and fixing fractions p, it is not surprising that an optimal policy tries to make the arrival process at each resource as *regular* (or less variable) as possible. Thus, our stochastic scheduling problem can be essentially converted into a problem in word combinatorics. If the dispatcher must ensure fractions p, the main result known in the literature is that *balanced sequences* are optimal admission sequences [1, 20]. However, balanced sequences of given rates p are known to exist in very few particular cases. These cases are captured by Fraenkel's conjecture, which is still open to the best of our knowledge [37]; see also [2, Chapter 2], which contains an overview of which rates p are balanceable.

Matter of fact, the problem of finding an optimal deterministic policy is still considered difficult [2, 3, 8, 11, 22, 23, 39]. The only exceptions are when resources are stochastically equivalent, where round-robin[1] is known to be optimal in a strong sense [28], or when the dispatcher routes jobs to two resources, where balanced sequences can be always constructed no matter the value of rates p [20]. In presence of more than two queues, we stress that balanced sequences with given rates p do not exist in general. This non-existence makes the problem difficult and one still wonders which structure should an optimal policy have when p is not balanceable. When the routing is performed to two resources, jobs join the dispatcher following a Poisson process and service times have an exponential distribution, the optimal rates p as function of the inter-arrival and service times have a fractal structure, see [17, Figure 8]. This puts further light on the complexity of the problem even in a simple scenario.

While deterministic policies are believed to be more difficult to study than probabilistic policies (deterministic and probabilistic in the sense described above), they can achieve a significantly better performance [3]. This holds also for the variance of the waiting time because, as discussed above, the arrival process at each resource is much more regular in the deterministic case, especially if there are several resources as we show in this paper. A particular class of deterministic policies, namely *billiard* policies, have been recently implemented in the context of large volunteer and cloud computing to improve the performance of real applications such as SETI@home [25, 26].

1.1 Contribution

In the framework described above, we are interested in deriving structural properties of deterministic policies when the system size is large. We study a scaling of the system where the arrival rate of jobs, λk, grows to infinity proportionally with the number of resources (queues in the following), Rk, while keeping the network load (or utilization) fixed. This scaling is often used in the queueing community. Specifically, there are R types of queues, and k is the number of queues in each type, i.e., the parameter that we will let grow to infinity. Beyond issues related to the tractability of the problem, this type of scaling is motivated by the fact that the size of real systems is large and that replication of resources is commonly used to increase system reliability.

First, with respect to a class of periodic policies, we define the random variable of the waiting time of each incoming job. This is done using Lindley's equation [27] and a suitable initial randomization. Using such randomization, we can adapt the framework developed by Loynes in [29] to our setting where jobs are sent to a set of parallel

[1] Round-robin sends the n-th job to resource $(n \bmod R) + 1$, where R is the total number of resources.

queues. In particular, Theorem 1 shows the monotone convergence in distribution of the waiting time of each incoming job. Then, with respect to a given vector $p \in \mathbb{N}^R$, we define a certain subset of policies that are periodic with period $k\sum_r p_r$ and such that exactly p_r jobs are sent in a period to each queue of type r. While further details will be developed in Section 3.1, this set is meant to imply that queues of a given type are visited in a round-robin manner and that arrivals are "well distributed" among the different queue types. When $k \to \infty$, our main result states that all the policies in this set are *equivalent*, in the sense that they yield the same mean stationary waiting time, and *optimal*, in the sense that no other policy having the same aggregate arrival rate to *all* queues of a given type can do better in minimizing the mean stationary waiting time. In particular, we show that the stationary waiting time converges both in distribution and in expectation to the stationary waiting time of a system of independent D/GI/1 queues whose parameters only depend on p, λ and the distribution of the service times. This is shown in Theorems 2 and 3, respectively.

The main idea underlying our proof stands in analyzing the sequence of stationary waiting times along appropriate subsequences. Along these subsequences, it is possible to extract a pattern for the arrival process of each queue that is common to all members of the subsequence. Such pattern is exploited to establish monotonicity properties in the language of stochastic orderings. These properties hold for the considered subsequences only: they do not hold true along any arbitrary subsequence and counterexamples can be given. These properties will imply the uniform integrability of the sequence of stationary waiting times and will allow us to work on expected values.

Summarizing, fixing the proportions p of jobs to send to each queue type and given k large, our results state that all the policies belonging to the set that we identify in Section 3.1 yield the same asymptotic performance and are asymptotically optimal. Furthermore, using known properties of the D/GI/1 queue, we obtain that the stationary mean waiting time obtained in our limit is a convex function of p. This reduces the complexity of subproblem *ii)* above because it boils down to the solution of a convex optimization problem.

This paper is organized as follows. Section 2 introduces the model under investigation and provides a characterization of the stationary waiting time (Theorem 1); Section 3 introduces a class of policies and presents our main results (Theorems 2 and 3); Section 4 is devoted to proofs; finally, Section 5 draws the conclusions of this paper. A version of this paper also appeared in [5].

2 Parallel Queueing Model

We consider a queueing system composed of R types of queues (or resources, servers) working in parallel. Each queue of type r is replicated k times, for all $r = 1, \dots, R$, so there are kR queues in total. Parameter k is a scaling factor and we will let it grow to infinity. The service discipline of each queue is first-come-first-served (FCFS) and the buffer size of each queue is infinite. A stream of jobs (or customers) joins the queues through a dispatcher. The dispatcher routes each incoming job to a queue according to some policy and instantaneously. There is neither job splitting nor job replication: an arriving job is sent to exactly one queue to receive some processing. Figure 1 illustrates the structure of the queueing model under investigation. In the following, indices r, κ, n will be implicitely assumed to range from 1 to R, from 1 to k, in $\mathbb{N}$, respectively.

All the random variables that follow will be considered belonging to a fixed underlying probability triple $(\Omega, \mathcal{F}, \Pr)$.

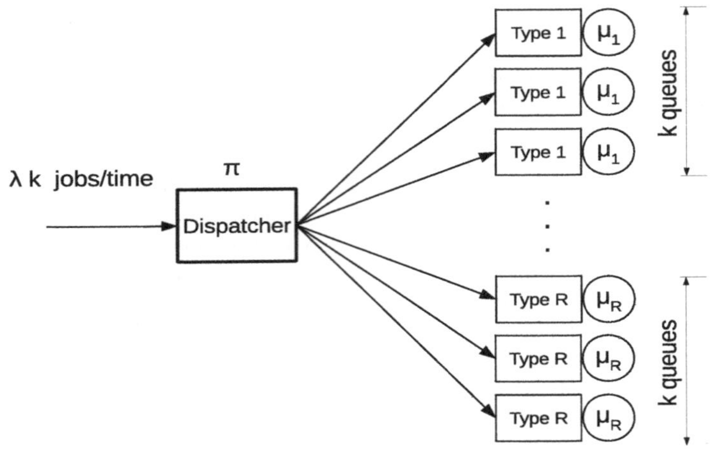

Fig. 1. Structure of the parallel queueing model under investigation.

Let $(T_n^{(k)})_{n\in\mathbb{N}}$ and $(S_{n,\kappa,r}^{(k)})_{n\in\mathbb{N}}$ be given sequences of i.i.d. random variables in $\mathbb{R}_+$[2]. These sequences are all assumed to be independent each other. Quantity $T_n^{(k)}$ is interpreted as the inter-arrival time between the n-th and the $(n+1)$-th jobs arriving to the dispatcher. Quantity $S_{n,\kappa,r}^{(k)}$ is interpreted as the service times of the n-th job arriving at the κ-th queue of type r. We assume that $S_{1,\kappa,r}^{(k)} =_{st} S_{1,1,r}^{(k)}$ and that $\mathbb{E}S_{n,\kappa,r}^{(k)} = \mu_r^{-1}$. For the arrival process at the dispatcher, we will refer to the following cases.

Case 1. The process $(T_n^{(k)})_{n\in\mathbb{N}}$ is a renewal process with rate λk and such that $Var\, T_n^{(k)} = o(1/k)$.

Case 2. The process $(T_n^{(k)})_{n\in\mathbb{N}}$ is a Poisson process with rate λk.

Case 3. The process $(T_n^{(k)})_{n\in\mathbb{N}}$ is constant with rate λk, i.e., $T_n^{(k)} = (\lambda k)^{-1}$.

It is clear that Cases 2 and 3 are both more restrictive than Case 1.

Let $\|\cdot\|$ denote the L_1-norm.

Let $q \overset{\text{def}}{=} (q_{r,\kappa}) \in \mathbb{Q}_+^R \times \mathbb{Q}_+^k$ be such that $\|q\| = 1$. Quantity $q_{r,\kappa}$ will be interpreted as the proportion of jobs sent to queue (r,κ).

Let $n^* \overset{\text{def}}{=} n^*(q) \overset{\text{def}}{=} \min\{n \in \mathbb{Z}_+ : n\,q \in \mathbb{Z}_+^R \times \mathbb{Z}_+^k\}$. Since q is a vector of rational numbers, $n^* < \infty$.

Let V be a discrete random variable with values in $\{1,\dots,n^*\}$ such that $\Pr(V = i) = 1/n^*$, for all $i = 1,\dots,n^*$. We assume that V is independent of any other random variable.

Let $\mathcal{A}_q(k)$ be the set of all functions $\pi : \mathbb{N} \to \{1,\dots,R\} \times \{1,\dots,k\}$ such that for all r and κ

$$q_{r,\kappa} = \frac{1}{n^*} \sum_{n=V}^{n^*+V-1} \mathbf{1}_{\{\pi(n)=(r,\kappa)\}} \text{ and } \pi(n) = \pi(n+n^*) \quad (1)$$

[2] For any $E \subseteq \mathbb{R}$, we let $E_+ \overset{\text{def}}{=} \{x \in E : x > 0\}$.

for all n, where $\mathbf{1}_E$ denotes the indicator function of event E. Thus, these functions are periodic with period n^* and $n^* q_{r,\kappa}$ is the number of jobs sent in a period to queue (r,κ). We refer to each element $\pi \in \mathcal{A}_q(k)$ as a *policy* (or a q-policy) operated by the dispatcher, and it is interpreted as follows: $\pi(n) = (r,\kappa)$ means that the n-th job arriving to the dispatcher is sent to the k-th queue of type r if $n \geq V$, otherwise it means that the n-th job is discarded. Thus, the outcome of random variable V gives the index of the first job that is actually served by some queue. In other words, $\pi(V)$ is the first queue that serves some job.

Let $(T^{(k)}_{n,\kappa,r}(\pi))_{n\in\mathbb{N}}$ be the sequence of inter-arrival times that are induced by policy π at the κ-th queue of type r (under any of the cases above). By construction, $T^{(k)}_{n,\kappa,r}(\pi)$ is the sum of a deterministic number of inter-arrival times seen at the dispatcher. The arrival process $(T^{(k)}_{n,\kappa,r}(\pi))_{n\in\mathbb{N}}$ can be made stationary if it is allowed a shift in time and a suitable randomization (independent of V and of any other random variable) for the inter-arrival time of the first arrival of each queue. We assume for now that this has been done (details will be given at the beginning of Section 4). As done in [29] and according to [16, p. 456], this implies that we can extend the stationary process $(T^{(k)}_{n,\kappa,r}(\pi))_{n\in\mathbb{N}}$ to form a stationary process $(T^{(k)}_{n,\kappa,r}(\pi))_{n\in\mathbb{Z}}$ (clearly, the same holds for the process $(S^{(k)}_{n,\kappa,r})_{n\in\mathbb{N}}$).

The waiting time of the n-th job arriving to the κ-th queue of type r induced by a policy $\pi \in \mathcal{A}(k)$ is denoted by $W^{(k)}_{n,r,\kappa}(\pi)$. It is the time between its arrival at the dispatcher (or equivalently at the queue) and the start of its service, and it is defined as follows: for $n = 0$, $W^{(k)}_{n,r,\kappa}(\pi) = 0$ and for $n > 0$,

$$W^{(k)}_{n,r,\kappa}(\pi) \stackrel{\text{def}}{=} \left(W^{(k)}_{n-1,r,\kappa}(\pi) + S^{(k)}_{n,\kappa,r} - T^{(k)}_{n,\kappa,r}(\pi)\right)^+, \tag{2}$$

where $x^+ \stackrel{\text{def}}{=} \max\{x, 0\}$. Equation (2) is known as Lindley's recursion [27]. The assumption that $W^{(k)}_{0,r,\kappa}(\pi) = 0$ serves to avoid technicalities[3]. It is known that the sequence of random variables $(W^{(k)}_{n,r,\kappa}(\pi))_{n\in\mathbb{N}}$ converges in distribution to the random variable

$$W^{(k)}_{r,\kappa}(\pi) \stackrel{\text{def}}{=} \left(\sup_{n\geq 0} \sum_{n'=0}^{n} S^{(k)}_{-n',\kappa,r} - T^{(k)}_{-n',\kappa,r}(\pi)\right)^+ \tag{3}$$

and that $W^{(k)}_{n,r,\kappa}(\pi) \leq_{st} W^{(k)}_{n+1,r,\kappa}(\pi)$, where $\leq_{st}$ denote the usual stochastic order; see [29]. We refer to $W^{(k)}_{r,\kappa}(\pi)$ as the *stationary* waiting time of jobs at the κ-th queue of type r.

Given π, let $f : \mathbb{N} \to \mathbb{N} \times \{1,\ldots,R\} \times \{1,\ldots,k\}$ be a mapping with the following meaning: $f(n) = (n', r, \kappa)$ means that the n-th job arriving to the dispatcher is the n'-th customer joining queue (r,κ). If $n < V$, no job is sent to any queue and thus we assume $f(n) = 0$. Note that $f(n)$ is a deterministic function of random variable V. Since V is uniform over $\{1,\ldots,n^*\}$, for all $n \geq n^*$ we have

$$\Pr(f_2(n) = r, f_3(n) = \kappa) = q_{r,\kappa}, \tag{4}$$

where f_j refers to the j-th component of f. With this notation, quantity $W^{(k)}_{f(n)}(\pi)$ is the waiting time of the n-th job arriving to the dispatcher induced by a policy $\pi \in \mathcal{A}_q(k)$, for all $n \geq n^*$.

---

[3] Using a standard coupling argument and that each queue will empty in finite time almost surely, what follows can be generalized easily to the case where $W^{(k)}_{0,r,\kappa}(\pi) \geq 0$.

Now, let $(Q_{r,\kappa})_{\forall r,\kappa}$ denote a partition of set $\{1,\ldots,n^*\}$. The subsets $Q_{r,\kappa}$, for all r and k, are thus disjoint and we further require that the number of points in $Q_{r,\kappa}$ is $n^*\, q_{r,\kappa}$. Let

$$W^{(k)}(\pi) \stackrel{\text{def}}{=} \sum_{r=1}^{R}\sum_{\kappa=1}^{k} \mathbf{1}_{\{V\in Q_{r,\kappa}\}} W_{r,\kappa}^{(k)}(\pi). \tag{5}$$

Since V is independent of the $W_{r,\kappa}^{(k)}(\pi)$'s, the distribution of $W^{(k)}(\pi)$ is the finite mixture of the $W_{r,\kappa}^{(k)}(\pi)$'s with weights q. Next theorem says that $W^{(k)}(\pi)$ can be interpreted as the right random variable describing the stationary waiting time of jobs achieved with policy $\pi \in \mathcal{A}_q(k)$. It is proven by adapting the framework developed by Loynes in [29]. In the remainder of the paper, convergence in distribution (in probability) is denoted by $\xrightarrow{d}$ (respectively, $\xrightarrow{\Pr}$).

Theorem 1. *Let Case 1 hold. Let q be such that $q_{r,\kappa}\lambda < \mu_r$ for all r,κ and $\pi \in \mathcal{A}_q(k)$. Then,*

$$W_{f(n)}^{(k)}(\pi) \leq_{st} W_{f(n+1)}^{(k)}(\pi) \tag{6}$$

and

$$W_{f(n)}^{(k)}(\pi) \xrightarrow[n\to\infty]{d} W^{(k)}(\pi). \tag{7}$$

By using the monotone convergence theorem, Theorem 1 implies that also the moments of $W_{f(n)}^{(k)}(\pi)$ converge to the moments of $W^{(k)}(\pi)$, provided that they are finite.

Finally, for a given $p \in \mathbb{R}_+^R$, we define the auxiliary random variable $W_r(p)$, which corresponds to the stationary waiting time of a D/GI/1 queue with inter-arrival times $(T_{n,r}(p))_{n\in\mathbb{N}}$ where $T_{n,r}(p) = \|p\|/(p_r\lambda)$ and service times $(S_{n,1,r})_{n\in\mathbb{N}}$, for all r. Let also $\overline{V} \stackrel{\text{def}}{=} \overline{V}(p)$ be a random variable with values in $\{1,\ldots,R\}$ independent of any other random variable and such that $\Pr(\overline{V}=r) = p_r/\|p\|$, for all r, and let

$$W(p) \stackrel{\text{def}}{=} \sum_{r=1}^{R} \mathbf{1}_{\{\overline{V}=r\}} W_r(p). \tag{8}$$

Note that $\mathbb{E}W(p) = \sum_r \frac{p_r}{\|p\|}\mathbb{E}W_r(p)$ can be interpreted as the mean waiting time of R independent D/GI/1 queues averaged over weights $p/\|p\|$.

We will be interested in establishing convergence results for $W^{(k)}$ when $k \to \infty$. With respect to some policies to be defined, we will show forms of convergence to $W(p)$.

2.1 Discussion

Some remarks about the model above and Theorem 1 follow:

- To agree on a definition for $\mathbb{E}W^{(k)}$ or other moments of $W^{(k)}$, one does not necessarily need to know the distribution of $W^{(k)}$, e.g., one can achieve that using Cesaro sums [2]. Matter of fact, existing works agree on the structure of $\mathbb{E}W^{(k)}$ *without* constructing the distribution of $W^{(k)}$. On the other hand, our approach needs to know the distribution of $W^{(k)}$ (and thus Theorem 1) because we can prove convergence results for $\mathbb{E}W^{(k)}$ *only* through a distributional convergence argument, that is [12, Theorem 3.5, pp. 31]. In particular, to prove $\mathbb{E}W^{(k)}$ converges to $\mathbb{E}W(p)$, we will prove convergence in distribution of $W^{(k)}$ to $W(p)$ and then the uniform integrability of the sequence of the $W^{(k)}$'s. This is the reason why we need the characterization of the distribution of the stationary waiting time $W^{(k)}$, which we give in Theorem 1 and prove through classical arguments.

- Though several works focused on finding policies that minimize the expected value $\mathbb{E}W^{(k)}$ (see the introduction), the analysis of $\mathbb{E}W^{(k)}$ in our scaling where $k \to \infty$ seems new.
- We assume that q is a vector of rational numbers. From a practical standpoint, this is not a loss of generality for obvious reasons. As an additional remark to support this assumption, we note that it has been proven that in several cases the q-vector that minimize $\min_{\pi \in \mathcal{A}_q(k)} \mathbb{E}W^{(k)}(\pi)$ is indeed rational [2, Theorem 32, p.136]; see also [38].
  Our approach needs this assumption to prove (7). In the case where $\pi(n)$ is not periodic, a case that we do not consider here, and the limit

$$\lim_{n\to\infty} \frac{1}{n} \sum_{n'=1}^{n} \mathbf{1}_{\{\pi(n')=(r,\kappa)\}}$$

  does not exist, it would be interesting to know whether some convergence in distribution of $W^{(k)}_{f(n)}(\pi)$ occurs. A totally different argument will be needed here.
- The fact that in Case 1 we require that $\operatorname{Var} T_n^{(k)} = o(k)$ is not a loss of generality for Theorem 1, as we do not let $k \to \infty$ there. Case 1 covers the case where $T_n^{(k)}$ has a (Gk, α)-phase-type distribution where both G and α are not functions of k.
- We may have assumed that each queue of type r was replicated $kz_r + o(k)$ times instead of just k times. This is essentially equivalent to our setting and our proofs can be easily adapted to this case though we do not do it for simplicity.

3 Main Results

For $p \in \mathbb{Z}_+^R$, let

$$\mathcal{P}_p(k) \stackrel{\text{def}}{=} \left\{ q \in \mathbb{Q}_+^R \times \mathbb{Q}_+^k : \sum_{\kappa=1}^{k} \frac{q_{r,\kappa}}{k} = \frac{p_r}{\|p\|}, \forall r \right\} \tag{9}$$

and

$$\overline{\mathcal{A}}_p(k) \stackrel{\text{def}}{=} \bigcup_{q \in \mathcal{P}_p(k)} \mathcal{A}_q(k). \tag{10}$$

The set $\overline{\mathcal{A}}_p(k)$ is interpreted as the set of all periodic policies for which $\lambda k \frac{p_r}{\|p\|}$ is the aggregate mean arrival rate of jobs to all type-r queues. We consider the following problem.

Problem 1. Let $p \in \mathbb{Z}_+^R$ be given. Determine the optimizers and the optimal objective function value of

$$\min_{\pi \in \overline{\mathcal{A}}_p(k)} \mathbb{E}W^{(k)}(\pi). \tag{11}$$

As discussed in the introduction, this is considered as a difficult problem. We are interested in establishing structural properties of Problem 1 when k is large. In the following, we define a certain class of policies and then present our main results.

3.1 A Class of Periodic Policies

For $p \in \mathbb{Z}_+^R$, we define $\mathcal{C}_p^{(k)}$ as the subset of all policies $\pi \in \overline{\mathcal{A}}_p(k)$ that satisfy the following properties:

- The sequence $(\pi_1(n))_{n\in\mathbb{N}}$ has period $\|p\|$ and p_r is the number of jobs sent to type-r queues per period, for $1 \le r \le R$.
- Let $n_1, n_2, \ldots$ be the subsequence of all jobs that are sent to queues of type r. Then, $\pi_2(n_1) = 1$, and $\pi(n_{j+1}) = \pi(n_j) + (0,1)$ if $\pi_2(n_j) < k$ and $\pi(n_{j+1}) = (r,1)$ otherwise.

The first property specifies the periodicity of π with respect the types of queues, while the second with respect to queues. Thus, the second property says that queues of type r are accessed by jobs in a round-robin (or cyclic) order starting from queue 1, and we need it to ensure that the cardinality of $\mathcal{C}_p^{(k)}$, i.e., $|\mathcal{C}_p^{(k)}|$, does not vary with k. One can verify that

$$|\mathcal{C}_p^{(k)}| = \frac{\|p\|!}{\prod_r (p_r!)}. \tag{12}$$

These policies can be implemented in a distributed manner. More precisely, one can think that there are two tiers of dispatchers: the dispatcher in the first tier schedules jobs *inter*-group, while the dispatchers in the second tier, R in total, schedule jobs *intra*-group and implement round-robin.

With respect to a sequence of policies $(\pi^{(k)})_{k\in\mathbb{N}}$, where $\pi^{(k)} \in \mathcal{C}_p^{(k)}$, we will show (Lemma 2)

$$T_{n,\kappa,r}^{(k)}(\pi^{(k)}) \xrightarrow[k\to\infty]{\Pr} \frac{\|p\|}{p_r \lambda}, \qquad \forall n. \tag{13}$$

This means that the finite dimensional distributions of the arrival process at each queue will be 'close' to the deterministic process, when k is large, which implies that some form of convergence to $W(p)$ should occur in view of the continuity of the stationary waiting time [13].

3.2 Asymptotic Equivalence and Optimality

Next theorem proves a first form of convergence of $W^{(k)}(\pi^{(k)})$ to $W(p)$.

Theorem 2. *Let Case 1 hold. Let $p \in \mathbb{Z}_+^R$ be such that $\lambda \frac{p_r}{\|p\|} < \mu_r$ for all r. Let also an arbitrary sequence $(\pi^{(k)})_{k\in\mathbb{N}}$ be given where $\pi^{(k)} \in \mathcal{C}_p^{(k)}$ for all k. Then*

$$W^{(k)}(\pi^{(k)}) \xrightarrow[k\to\infty]{d} W(p). \tag{14}$$

Theorem 2 is not enough to claim that $\mathbb{E}W^{(k)}(\pi^{(k)})$ converges as well. Convergence of the first moment is important from an operational standpoint, as in practice one desires to optimize over $\mathbb{E}W^{(k)}$ or $\operatorname{Var} W^{(k)}$. Under some additional assumptions, next theorem states that also the expected value and the variance of $W^{(k)}$ converge. Furthermore, it states that all the policies in set $\mathcal{C}_p^{(k)}$ are both asymptotically equivalent and asymptotically optimal, with respect to the criterion in Problem 1.

Theorem 3. *Let Case 2 or 3 hold. Let $p \in \mathbb{Z}_+^R$ be such that $\lambda \frac{p_r}{\|p\|} < \mu_r$ for all r. Let also an arbitrary sequence $(\pi^{(k)})_{k\in\mathbb{N}}$ be given where $\pi^{(k)} \in \mathcal{C}_p^{(k)}$ for all k. If $\mathbb{E}[(S_{1,\kappa,r}^{(k)})^3] < \infty$, then*

$$\underline{\lim}_{k\to\infty} \min_{\pi\in\overline{\mathcal{A}}_p(k)} \mathbb{E}W^{(k)}(\pi) = \lim_{k\to\infty} \mathbb{E}W^{(k)}(\pi^{(k)}) \tag{15}$$

$$\lim_{k\to\infty} \mathbb{E}W^{(k)}(\pi^{(k)}) = \mathbb{E}W(p). \tag{16}$$

Furthermore, if $\mathbb{E}[(S_{1,\kappa,r}^{(k)})^5] < \infty$, *then*

$$\lim_{k\to\infty} \operatorname{Var} W^{(k)}(\pi^{(k)}) = \sum_r \tfrac{p_r}{\|p\|} \left(\operatorname{Var} W_r(p) + (\mathbb{E}W_r(p) - \mathbb{E}W(p))^2\right). \tag{17}$$

Provided that k is large, thus, no other policy in $\overline{\mathcal{A}}_p(k) \setminus \mathcal{C}_p^{(k)}$ can do better than any $\pi^{(k)} \in \mathcal{C}_p^{(k)}$ to minimize $\mathbb{E}W^{(k)}$. It also explicits the limiting value of $\mathbb{E}W^{(k)}(\pi^{(k)})$, which is $\mathbb{E}W(p)$. It is known that $\mathbb{E}W(p)$ is a convex function in p (e.g., [31]).

To prove Theorem 3, we use the well-known fact that [6]

$$W_r(p) \leq_{icx} W_{r,\kappa}^{(k)}(\pi) \tag{18}$$

for any $\pi \in \overline{\mathcal{A}}_p(k)$, where $\leq_{icx}$ denotes the increasing-convex order (see, e.g., [34, 36] for their definition). Using this lower bound first and then that the waiting time of the D/GI/1 queue is convex increasing in its arrival rate, it is not difficult to show that

$$\mathbb{E}W(p) \leq \mathbb{E}W^{(k)}(\pi). \tag{19}$$

Then, we prove that the sequence $\mathbb{E}W^{(k)}(\pi^{(k)})$ is upper bounded by a sequence that converges to the lower bound in (19). An observation here is that the lower bound (18) holds under conditions that are weaker than those assumed in this paper; see [24]. For instance, it is possible to extend (19) (and thus Theorem 3) to the case where i) policies are not periodic, ii) the fractions of jobs to send in each queue are not necessarily rational numbers, and iii) policies are randomized [32], that is the case where $\pi(n)$ is any probability mass function over the set of queues. We do not investigate these extensions in further detail.

4 Appendix

In this section, we develop proofs for Theorems 1, 2 and 3. Before doing this, we fix some additional notation and show how it is possible to make the arrival process at each queue stationary with respect to any policy in $\pi^{(k)} \in \mathcal{A}_q(k)$, $q \in \mathbb{Q}_+^{kR}$ such that $\|q\| = 1$ (as assumed in Section 2).

Let us consider the κ-th queue of type r and its arrival process $(T_{n,\kappa,r}^{(k)})_{n\in\mathbb{N}}$. Each inter-arrival time clearly depends on the policy $\pi^{(k)}$ implemented by the dispatcher, i.e., $T_{n,\kappa,r}^{(k)} = T_{n,\kappa,r}^{(k)}(\pi^{(k)})$, though in the following we drop such dependence for notational simplicity. Since $\pi^{(k)}$ is periodic by construction with period n^* and $n^* q_{r,\kappa}$ jobs have to be sent within a cycle to the queue identified by the couple (r,κ), the sequence $(T_{n,\kappa,r}^{(k)})_{n\in\mathbb{N}}$ is composed of a repeated pattern of $n^* q_{r,\kappa}$ inter-arrival times, that we can write as

$$A_{1,\kappa,r}^{(k)}, A_{2,\kappa,r}^{(k)}, \ldots, A_{n^* q_{r,\kappa},\kappa,r}^{(k)}, \tag{20}$$

where each quantity $A_{j,\kappa,r}^{(k)}$, $j = 1, \ldots, n^* q_{r,\kappa}$, is the sum of a deterministic number (that depends on $\pi^{(k)}$) of inter-arrival times to the dispatcher. We denote such number by $a_{j,r,\kappa}^{(k)}$. Thus,

$$A_{j,\kappa,r}^{(k)} =_{st} \sum_{n=1}^{a_{j,r,\kappa}^{(k)}} T_n^{(k)}, \tag{21}$$

where $=_{st}$ denotes equality in distribution.

If $\pi^{(k)} \in \mathcal{C}_p^{(k)}$ for some $p \in \mathbb{Z}_+^R$, then we notice that $a_{j,r,\kappa}^{(k)}$ does not vary with κ because by symmetry the arrival processes of all queues of a given type are equal, in distribution, up to a shift in time. In this case, the arrival process at *any* queue of type r becomes a sequence composed of a repeated pattern of p_r inter-arrival times that we can write as

$$A_{1,\kappa,r}^{(k)}, A_{2,\kappa,r}^{(k)}, \ldots, A_{p_r,\kappa,r}^{(k)}. \tag{22}$$

Therefore, when $\pi^{(k)} \in \mathcal{C}_p^{(k)}$, we will just write $a_{j,r}^{(k)}$ instead of $a_{j,r,\kappa}^{(k)}$ and it is also clear that

$$\sum_{j=1}^{p_r} a_{j,r}^{(k)} = k\|p\| \tag{23}$$

and that

$$\mathbb{E}T_{n,\kappa,r}^{(k)} = \frac{1}{p_r}\sum_{j=1}^{p_r} a_{j,r}^{(k)}\,\mathbb{E}T_n^{(k)} = \frac{1}{p_r}\sum_{j=1}^{p_r} \frac{a_{j,r}^{(k)}}{k\lambda} = \frac{\|p\|}{p_r\lambda}. \tag{24}$$

Now, we want to make $(T_{n,\kappa,r}^{(k)})_{n\in\mathbb{N}}$ stationary. This can be done as follows by randomizing over the first inter-arrival time of queue (r,κ). Now, let us consider the auxiliary random variables $U_{r,\kappa}$, for all r and κ, which we assume independent each other and of any other random variable and having a uniform distribution in $[0,1]$. Then, we take

$$T_{1,r,\kappa}^{(k)} \stackrel{\text{def}}{=} \sum_{j=1}^{n^* q_{r,\kappa}} A_{j,r,\kappa}^{(k)} \mathbf{1}_{\left\{U_{r,\kappa}\in\left[\frac{j-1}{n^* q_{r,\kappa}}, \frac{j}{n^* q_{r,\kappa}}\right]\right\}} \tag{25}$$

Therefore, if $T_{1,r,\kappa}^{(k)} = A_{j,r,\kappa}^{(k)}$, for some $j < n^* q_{r,\kappa}$, then $T_{2,r,\kappa}^{(k)} = A_{j+1,r,\kappa}^{(k)}$ and so forth according to the pattern (20). Defined in this manner, one can see that $(T_{n,r,\kappa}^{(k)})_{n\in\mathbb{N}}$ is stationary as desired. At this point, the issue is the following. Consider two queues, say (r,κ) and (r',κ'), and suppose that $T_{1,r,\kappa}^{(k)} = A_{j,r,\kappa}^{(k)}$ and $T_{1,r',\kappa'}^{(k)} = A_{j',r',\kappa'}^{(k)}$. We should check whether policy $\pi^{(k)}$ is actually able to induce arrival processes at the queues equal (samplepath-wise) to the ones built above through (25). One can easily see that this can be done with a possible shift of time for the arrival process at the queues and possibly discarding a finite number of jobs. This is allowed because these operations do not change the stationary behavior.

In the remainder, we will use stochastic orderings. We will denote by $\leq_{st}$, $\leq_{cx}$ and $\leq_{icx}$, the *usual stochastic order*, the *convex order* and the *increasing convex order*, respectively; we point to, e.g., [34, 36] for their definition.

We will also refer to the following lemma, which can be easily proven.

Lemma 1. *Let N be a finite positive integer and suppose $(f_{k,n})_{(k,n)\in\mathbb{N}\times\{1,\ldots,N\}}$ is a semi-infinite array of numbers such that for some constant c, $\lim_{k\to\infty} f_{k,n} = c$, for $n \in \{1,\ldots,N\}$. Then, for any sequence $(n_k)_{k\in\mathbb{N}}$ with values in $\{1,\ldots,N\}$, $\lim_{k\to\infty} f_{k,n_k} = c$.*

We now give proofs for our results, i.e., Theorems 1, 2 and 3.

4.1 Proof of Theorem 1

Let random variables V_1 and V_2 be given such that $V_1 =_{st} V$ and $V_2 = V_1 - 1$ if $V_1 > 1$ otherwise $V_2 = n^*$. We prove (6) through Strassen's theorem building a coupling $(\tilde{W}_{f(n)}^{(k)}(\pi), \tilde{W}_{f(n+1)}^{(k)}(\pi))$ of $W_{f(n)}^{(k)}(\pi)$ and $W_{f(n+1)}^{(k)}(\pi)$ through V_1 and V_2 ensuring that

$\tilde{W}^{(k)}_{f(n)}(\pi) \le \tilde{W}^{(k)}_{f(n+1)}(\pi)$. This is done as follows. First, let $\tilde{W}^{(k)}_{f(n)}(\pi)$ and $\tilde{W}^{(k)}_{f(n+1)}(\pi)$ be the Loynes waiting times (see [29, p. 501]) obtained when the first queue to serve a job is given by the outcome of V_1 and V_2, respectively; thus, $\tilde{W}^{(k)}_{f(n)}(\pi)$ is the waiting time at time 0 with n' jobs in the past at queue (r,κ), provided that $f(n) = (n', r, \kappa)$. Then, we let the (Loynes) waiting times be driven by the same realizations of the random inter-arrival and service times.

We now prove (7). Since π is periodic with period n^*, we first observe that π returns the same queue along subsequence $(n\,n^* + i)_{n\in\mathbb{N}}$, for all $i = 1, \ldots, n^*$. Similarly, also the second and third component of $f(n\,n^*+i)$ do not change along these subsequences, though they are not known in advance because they depend on the outcome of random variable V, see (4). Thus, for all $i = n^* + 1, \ldots, 2n^*$, by construction we have

$$\Pr(f_2(n\,n^* + i) = r, f_3(n\,n^* + i) = \kappa) = \Pr(f_2(i) = r, f_3(i) = \kappa) = q_{r,\kappa}, \tag{26}$$

and we get

$$\lim_{n\to\infty} \Pr(W^{(k)}_{f(nn^*+i)}(\pi) \le t) \tag{27a}$$

$$= \lim_{n\to\infty} \sum_{r,\kappa} q_{r,\kappa} \Pr(W^{(k)}_{n,r,\kappa}(\pi) \le t \,|\, (f_2(i), f_3(i)) = (r,\kappa)) \tag{27b}$$

$$= \sum_{r,\kappa} q_{r,\kappa} \Pr(W^{(k)}_{r,\kappa}(\pi) \le t) \tag{27c}$$

$$= \Pr(W^{(k)}(\pi) \le t). \tag{27d}$$

In (27b), we have conditioned on $f(i)$. In (27c), we have used that $W^{(k)}_{n,r,\kappa}(\pi)$ converges in distribution to $W^{(k)}_{r,\kappa}(\pi)$; see [29]. In (27d), we have used the definition of $W^{(k)}(\pi)$. Now, since the limit in (27d) does not depend on i, the proof is concluded by applying Lemma 1 once noted that n^* is a finite positive integer. ■

4.2 Proof of Theorem 2

We first observe that we can prove this theorem under some assumption on the sequence $(\pi^{(k)})_{k\in\mathbb{N}}$. Given $\pi^{(1)} \in \mathcal{C}_p^{(1)}$, we require that for all k:

$$\pi_1^{(k)}(n) = \pi_1^{(1)}(n), \qquad \forall n. \tag{28}$$

One may refer to these sequences as the 'natural' scaling of policy $\pi^{(1)}$: in the two-tier interpretation of our policies, (28) means that the dispatcher at the first tier implements the same policy, to queue types, when k grows.

These sequences will be assumed along this proof. If this theorem holds for these sequences, then it also holds for all the sequences in view of Lemma 1 and of the fact that the cardinality of $\mathcal{C}_p^{(k)}$ does not vary with k.

For $m \in \mathbb{N}$, let

$$k_m \stackrel{\text{def}}{=} m\,\text{lcm}(p), \tag{29}$$

where lcm(p) denotes the least common multiple of $p_1, \ldots, p_R$. The subsequences $(k_m + i)_{m\in\mathbb{N}}$, for all $i = 1, \ldots, \text{lcm}(p)$, play a key role in our proof of Theorem 3. Along these subsequences, next fact holds true and follows by construction of the policies in set $\mathcal{C}_p^{(k)}$: it is a direct consequence of the fact that queues of the same type are visited in a round-robin manner.

Fact 1 *For $m > 1$, $j = 1, \dots, p_r$ and $i \in \mathbb{N}$, $a_{j,r}^{(p_r m+i)} = a_{j,r}^{(p_r(m-1)+i)} + \|p\|$.*

<u>Proof.</u> By construction, we have $\pi_2^{(p_r m+i)}(V) = 1$ and $\pi^{(p_r m+i)}(V) = \pi^{(p_r(m-1)+i)}(V)$, which in some sense couples the arrival processes at queues of the $(p_r m + i)$-th and $(p_r(m-1)+i)$-th systems.

Without loss of generality, let us assume that $(r, 1)$ is the queue that receives the first job.

Now, Fact 1 holds true because $(p_r m + i) - (p_r(m-1) + i) = p_r$ jobs must be sent to some queues of type r of the $(p_r m + i)$-th system in the time interval $[V + a_{j,r}^{(p_r(m-1)+i)}, V + a_{j,r}^{(p_r m+i)} - 1]$, for $j = 1$, and the number of arrivals at the dispatcher in that interval is exactly $\|p\|$ by construction of the policies in $\mathcal{C}_p^{(k)}$ for any k (see subsection 3.1)

This argument applies to all the other queues because we have considered periodic policies. ∎

We show Fact 1 in the following example, to help understanding its meaning and proof.

Example 1. Assume $R = 2$, $p = (3, 2)$, $r = 1$, $i = 1$, $m = 2$, $j = 1$. Assume also that $\pi \in \mathcal{C}_p^{(1)}$ is such that $(\pi_1(V + n))_{n=0,\dots,4} = (1, 1, 2, 1, 2)$. Then, the sequence of queues to be visited for both systems $p_r m + i$ and $p_r(m-1) + i$ is given in Table 1, where we can see that the decomposition in Fact 1 holds.

$$p_r m + i:\quad \overbrace{(1,1)\ (1,2)\ (2,1)\ (1,3)\ (2,2)\ (1,4)\ \underbrace{(1,5)\ (2,3)\ (1,6)\ (2,4)\ (1,7)}_{\|p\|}}^{a_{j,r}^{(p_r m+i)}}\ (1,1)\ \dots$$

$$p_r(m-1) + i:\quad \underbrace{(1,1)\ (1,2)\ (2,1)\ (1,3)\ (2,2)\ (1,4)}_{a_{j,r}^{(p_r(m-1)+i)}}\ (1,1)\ (2,3)\ (1,2)\ (2,4)\ (1,3)\ (1,4)\ \dots$$

Table 1. Illustrative example for the decomposition in Fact 1.

Since Fact 1 holds for any $m > 1$, we can make the replacement $m \to m\frac{\mathrm{lcm}(p)}{p_r}$ for which we obtain

$$a_{j,r}^{(k_m+i)} = a_{j,r}^{(k_m - p_r + i)} + \|p\|. \tag{30}$$

Unfolding this recursion, for $m \in \mathbb{N}$, $i = 1, \dots, \mathrm{lcm}(p)$, we get

$$a_{j,r}^{(k_m+i)} = a_{j,r}^{(k_m - 2p_r + i)} + 2\|p\| = \dots = a_{j,r}^{(k_{m-1}+i)} + \frac{\mathrm{lcm}(p)}{p_r}\|p\| = a_{j,r}^{(i)} + m\frac{\|p\|}{p_r}\mathrm{lcm}(p) \tag{31}$$

and therefore

$$\frac{a_{j,r}^{(k_m+i)}}{k_m + i} = \frac{a_{j,r}^{(i)} + m\frac{\|p\|}{p_r}\mathrm{lcm}(p)}{m \cdot \mathrm{lcm}(p) + i} \xrightarrow[m\to\infty]{} \frac{\|p\|}{p_r}. \tag{32}$$

As a technical observation, in (31) we note why we require index i to range in $\{1, \dots, \mathrm{lcm}(p)\}$: it 'closes' the recursion.

Since (32) holds for all $i = 1, \dots, \text{lcm}(p)$, by using Lemma 1 we obtain

$$\lim_{k\to\infty} \frac{a_{j,r}^{(k)}}{k} = \frac{\|p\|}{p_r}. \tag{33}$$

As a comment, we note here that (33) could be proven without Fact 1 and what has followed. However, we stress that we will need Fact 1 later anyhow, as it will play a crucial role in the proof of our main result Theorem 3 (see Lemma 3.iii).

Lemma 2. *Under the hypotheses of Theorem 2,* $T_{n,r,\kappa}^{(k)} \to \frac{\|p\|}{\lambda p_r}$ *in probability, as* $k \to \infty$.

Proof. For all $\epsilon > 0$,

$$\begin{aligned}
\Pr\left(|T_{n,r,\kappa}^{(k)} - \tfrac{\|p\|}{\lambda p_r}| \geq \epsilon\right) &= \Pr\left(|T_{n,r,\kappa}^{(k)} - \mathbb{E}T_{n,r,\kappa}^{(k)}| \geq \epsilon\right) && \text{(34a)}\\
&\leq \frac{1}{\epsilon^2} \operatorname{Var} T_{n,r,\kappa}^{(k)} && \text{(34b)}\\
&= \frac{1}{\epsilon^2}\left(\mathbb{E}(\operatorname{Var} T_{n,r,\kappa}^{(k)}|U_{r,\kappa}) + \operatorname{Var}\mathbb{E}(T_{n,r,\kappa}^{(k)}|U_{r,\kappa})\right) && \text{(34c)}\\
&= \frac{1}{\epsilon^2}\left(\frac{1}{p_r}\sum_{j=1}^{p_r} \operatorname{Var} A_{j,r,\kappa}^{(k)} + (\mathbb{E}A_{j,r,\kappa}^{(k)} - \mathbb{E}T_{n,r,\kappa}^{(k)})^2\right) && \text{(34d)}\\
&= \frac{1}{\epsilon^2}\left(\frac{1}{p_r}\sum_{j=1}^{p_r} a_{j,r}^{(k)} \operatorname{Var} T_1^{(k)} + \left(\frac{a_{j,r}^{(k)}}{\lambda k} - \frac{\|p\|}{\lambda p_r}\right)^2\right) && \text{(34e)}\\
&\xrightarrow[k\to\infty]{} 0. && \text{(34f)}
\end{aligned}$$

In (34b), (34c) and (34e), we have used Chebyshev's inequality, the law of total variance and that the $T_n^{(k)}$'s are i.i.d., respectively. In (34f), we have used (33) and that $a_{j,r}^{(k)} \operatorname{Var} T_1^{(k)} \to 0$ because $\operatorname{Var} T_1^{(k)} = o(k)$ and $a_{j,r}^{(k)} \leq k\|p\|$. ■

Since $T_{n,r,\kappa}^{(k)}$ converges in probability for each n, also the finite dimensional distributions of the process $(T_{n,r,\kappa}^{(k)})_{n\in\mathbb{N}}$ converge to the one of the constant process with rate $\frac{\lambda p_r}{\|p\|}$. Together with the fact that $\mathbb{E}T_{n,r,\kappa}^{(k)} = \lim_{k\to\infty} \mathbb{E}T_{n,r,\kappa}^{(k)} = \frac{\|p\|}{\lambda p_r}$, we can use the continuity of the stationary waiting time (see [13, Theorem 22]) to establish that

$$W_{r,\kappa}^{(k)}(\pi^{(k)}) \xrightarrow[k\to\infty]{d} W_r(p). \tag{35}$$

Using (35) and that Cesaro sums converge if each addend converges, we obtain

$$\begin{aligned}
\lim_{k\to\infty} \Pr(W^{(k)}(\pi^{(k)}) \leq t) &= \lim_{k\to\infty} \sum_{r=1}^{R} \frac{p_r}{\|p\|k} \sum_{\kappa=1}^{k} \Pr(W_{r,\kappa}^{(k)}(\pi^{(k)}) \leq t) && \text{(36a)}\\
&= \sum_{r=1}^{R} \frac{p_r}{\|p\|} \Pr(W_r(p) \leq t) = \Pr(W(p) \leq t) && \text{(36b)}
\end{aligned}$$

as desired. ■

4.3 Proof of Theorem 3

Proof of (15) and (16). Given that $\mathcal{C}_p^{(k)} \subseteq \overline{\mathcal{A}}_p(k)$, (15) and (16) hold true if

$$\mathbb{E}W(p) \le \mathbb{E}W^{(k)}(\pi) \tag{37}$$

for all $\pi \in \overline{\mathcal{A}}_p(k)$ and

$$\lim_{k\to\infty} \mathbb{E}W^{(k)}(\pi^{(k)}) \le \mathbb{E}W(p). \tag{38}$$

Let $\mathbb{E}W_{r,\kappa}(x)$ be the mean waiting time of a D/GI/1 queue with arrival rate λx and i.i.d. service times having the same distribution of $S_{1,r,\kappa}$. Inequality (37) is a fairly direct application of known results: for all $\pi \in \overline{\mathcal{A}}_p(k)$,

$$\begin{aligned}
\mathbb{E}W^{(k)}(\pi) &= \sum_{r,\kappa} q_{r,\kappa}\mathbb{E}W_{r,\kappa}^{(k)}(\pi) && \text{(39a)}\\
&\ge \sum_{r,\kappa} q_{r,\kappa}\mathbb{E}W_{r,\kappa}(kq_{r,\kappa}) && \text{(39b)}\\
&\ge \sum_{r,\kappa} \frac{p_r}{k\|p\|}\mathbb{E}W_{r,\kappa}(\tfrac{p_r}{\|p\|}) && \text{(39c)}\\
&= \sum_{r} \frac{p_r}{\|p\|}\mathbb{E}W_r(p) = \mathbb{E}W(p). && \text{(39d)}
\end{aligned}$$

In (39b), we have used the lower bound in [24]. In (39c), we have used Karamata's inequality once noticing that i) $\mathbb{E}W_{r,\kappa}(x) = \mathbb{E}W_{r,1}(x)$, ii) the majorization $(\frac{p_r}{\|p\|},\dots,\frac{p_r}{\|p\|}) \prec (kq_{r,1},\dots,kq_{r,k})$ holds, and iii) the mean waiting time of a D/GI/1 queue is convex increasing in the arrival rate (see, e.g., [31, Theorem 5], [18]), which means that $q_{r,\kappa}\mathbb{E}W_{r,\kappa}(kq_{r,\kappa})$ is convex in $q_{r,\kappa}$.

We now prove (38). As in the proof of Theorem 2, this can be done assuming that (28) holds. Thus, the sequences (28) will be assumed along this proof.

The remainder of the proof basically works as follows. First, we bound the waiting times of our G/GI/1 queues through the waiting times of suitable GI/GI/1 queues. Second, we show that the sequence of such waiting times converges in distribution to $W(p)$. Then, we show that the waiting times of such GI/GI/1 queues are non-increasing in the $\le_{icx}$-sense along the sequences $k_m + i$, for all $i = 1,\dots,\mathrm{lcm}(p)$, which allows us to conclude that the sequence is uniformly integrable; this is the point where we will use Fact 1. Finally, we use [12, Theorem 3.5, pp. 31] to conclude that also the sequence of the expected values converges to $\mathbb{E}W(p)$.

Associated to each queue of type r, we define an auxiliary random variable, $\overline{T}_r^{(k)}$, such that

$$\overline{T}_r^{(k)} =_{st} \sum_{n=1}^{\min_{j=1,\dots,p_r} a_{j,r}^{(k)}} T_n^{(k)}. \tag{40}$$

Next lemma provides properties satisfied by $\overline{T}_r^{(k)}$ that will be used later. We recall that $k_m = m\,\mathrm{lcm}(p)$, see (29).

Lemma 3. *Under the hypotheses of Theorem 3, the following properties hold:*

i) We have

$$\lim_{k\to\infty} \mathbb{E}\overline{T}_r^{(k)} = \lim_{k\to\infty} \min_{j=1,\dots,p_r} \frac{a_{j,r}^{(k)}}{\lambda k} = \frac{\|p\|}{\lambda p_r}. \tag{41}$$

ii) $\overline{T}_r^{(k)} \to \frac{\|p\|}{\lambda p_r}$ *in probability, as* $k \to \infty$.
iii) For all $i = 1, \dots, \mathrm{lcm}(p)$,

$$-\overline{T}_r^{(k_{m+1}+i)} \leq_{icx} -\overline{T}_r^{(k_m+i)}. \tag{42}$$

Proof. i) This is an immediate consequence of (33).

ii) For all $i = 1, \dots, \mathrm{lcm}(p)$, let $j_i^* \in \arg\min_{j=1,\dots,p_r} \frac{a_{j,r}^{(i)}}{i}$. Then, from (32), we get

$$j_i^* \in \arg\min_{j=1,\dots,p_r} \frac{a_{j,r}^{(k_m+i)}}{k_m+i}, \quad \forall m > 1. \tag{43}$$

In view of Lemma 1, the convergence in ii) holds if we can show that

$$\overline{T}_r^{(k_m+i)} \xrightarrow[m\to\infty]{\Pr} \frac{\|p\|}{\lambda p_r},$$

for all $i = 1, \dots, \mathrm{lcm}(p)$. Given (43), this amounts to show that

$$\sum_{n=1}^{a_{j_i^*,r}^{(k_m+i)}} T_n^{(k_m+i)} \xrightarrow[m\to\infty]{\Pr} \frac{\|p\|}{\lambda p_r}, \quad \forall i = 1, \dots, \mathrm{lcm}(p). \tag{44}$$

We prove the former by showing (the stronger statement) that $\sum_{n=1}^{a_{j,r}^{(k)}} T_n^{(k)} \xrightarrow[k\to\infty]{\Pr} \frac{\|p\|}{\lambda p_r}$, for all j. Now, using that $\{|X - c| > 2\epsilon\} \subseteq \{|X - \mathbb{E}X| > \epsilon\} \cup \{|\mathbb{E}X - c| > \epsilon\}$ for a random variable X, we have

$$\Pr\Big(|\sum_{n=1}^{a_{j,r}^{(k)}} T_n^{(k)} - \tfrac{\|p\|}{p_r\lambda}| \geq \epsilon\Big) \leq \Pr\Big(|\sum_{n=1}^{a_{j,r}^{(k)}} T_n^{(k)} - \mathbb{E}\sum_{n=1}^{a_{j,r}^{(k)}} T_n^{(k)}| \geq \epsilon\Big) +$$
$$\Pr\Big(|\tfrac{\|p\|}{p_r\lambda} - \mathbb{E}\sum_{n=1}^{a_{j,r}^{(k)}} T_n^{(k)}| \geq \epsilon\Big).$$

The second term in the right-hand side of former inequality tends to zero as $k \to \infty$ by (33). The following shows that also the first term goes to zero:

$$\Pr\Big(|\sum_{n=1}^{a_{j,r}^{(k)}} T_n^{(k)} - \mathbb{E}\sum_{n=1}^{a_{j,r}^{(k)}} T_n^{(k)}| \geq \epsilon\Big) \leq \frac{1}{\epsilon^2}\sum_{n=1}^{a_{j,r}^{(k)}} \mathrm{Var}\, T_n^{(k)} \tag{46a}$$

$$\leq \frac{1}{\epsilon^2} a_{j,r}^{(k)} o(1/k) \xrightarrow[k\to\infty]{} 0. \tag{46b}$$

In (46a), we have used Chebyshev's inequality and that the $T_n^{(k)}$'s are independent. In (46b), we have used that $a_{j,r}^{(k)} \leq k\|p\|$.

iii) We use that $X \leq_{icx} Y$ if and only if there exists an other random variable Z such that $X \leq_{st} Z$ and $Z \leq_{cx} Y$ [30].

Using (31) and that $\frac{a_{j_i^*,r}^{(k_m+i)}}{(k_m+i)} \leq \frac{\|p\|}{p_r}$ for all m (by (32)), the first observation is that

$$a_{j_i^*,r}^{(k_{m+1}+i)} \geq a_{j_i^*,r}^{(k_m+i)} + \mathrm{lcm}(p)\frac{a_{j_i^*,r}^{(k_m+i)}}{k_m+i}, \tag{47}$$

where j_i^* is defined above in point i). Thus, we have

$$-\overline{T}_r^{(k_{m+1}+i)} =_{st} - \sum_{n=1}^{a_{j_i^*,r}^{(k_{m+1}+i)}} T_n^{(k_{m+1}+i)} \tag{48a}$$

$$\leq_{st} - \sum_{n=1}^{a_{j_i^*,r}^{(k_m+i)} + \mathrm{lcm}(p)\frac{a_{j_i^*,r}^{(k_m+i)}}{k_m+i}} T_n^{(k_{m+1}+i)} \stackrel{\text{def}}{=} -Z. \tag{48b}$$

Now, it remains to show that $-Z \leq_{cx} -\overline{T}_r^{(k_m+i)}$, which is equivalent to show that

$$Z \leq_{cx} \overline{T}_r^{(k_m+i)}, \tag{49}$$

see [34, Theorem 3.A.12]. Since

$$\mathbb{E}Z = \frac{1}{\lambda} \frac{a_{j_i^*,r}^{(k_m+i)} + \mathrm{lcm}(p)\frac{a_{j_i^*,r}^{(k_m+i)}}{k_m+i}}{k_m+i+\mathrm{lcm}(p)} = \frac{1}{\lambda}\frac{a_{j_i^*,r}^{(k_m+i)}}{k_m+i} = \mathbb{E}\overline{T}_r^{(k_m+i)}, \tag{50}$$

(49) holds trivially under Case 3. Now, let Case 2 hold. Noticing that both $\overline{T}_r^{(k_m+i)}$ and Z have Erlang distributions with the same mean, to prove (49) is enough to show that $\mathrm{Var}\, Z \leq \mathrm{Var}\, \overline{T}_r^{(k_m+i)}$; see [36, p. 14]. We have

$$\lambda^2 \,\mathrm{Var}\, Z = \frac{a_{j_i^*,r}^{(k_m+i)} + \mathrm{lcm}(p)\frac{a_{j_i^*,r}^{(k_m+i)}}{k_m+i}}{(k_{m+1}+i)^2} \tag{51a}$$

$$= \frac{a_{j_i^*,r}^{(k_m+i)}}{(k_m+i)^2} \frac{(k_m+i)(k_m+i+\mathrm{lcm}(p))}{(k_m+i+\mathrm{lcm}(p))^2} \tag{51b}$$

$$\leq \frac{a_{j_i^*,r}^{(k_m+i)}}{(k_m+i)^2} = \lambda^2 \,\mathrm{Var}\, \overline{T}_r^{(k_m+i)} \tag{51c}$$

as desired. ∎

We now present an argument that allows us to uniformly bound the second moment of $W_{r,\kappa}^{(k)}$.

Let $\delta_r \stackrel{\text{def}}{=} \frac{1}{2}\left(\frac{\|p\|}{p_r\lambda} - \frac{1}{\mu_r}\right)$ and

$$k^* \stackrel{\text{def}}{=} \min\left\{k > 0 : 0 \leq \frac{\|p\|}{p_r\lambda} - \min_{j=1,\dots,p_r} \frac{a_{j,r}^{(k')}}{k'\lambda} \leq \delta_r, \quad \forall k' \geq k\right\}. \tag{52}$$

Lemma 4. $k^* < \infty$.

Proof. This is immediate because $\delta_r > 0$ by hypothesis, $\lim_{k'\to\infty} \min_{j=1,\dots,p_r} \frac{a_{j,r}^{(k')}}{k'\lambda} = \frac{\|p\|}{p_r\lambda}$ (see (41)), and $\frac{\|p\|}{p_r\lambda} \geq \min_{j=1,\dots,p_r} \frac{a_{j,r}^{(k)}}{k\lambda}$ for all k. ∎

Let $\overline{W}_r^{(k)}$ denote the stationary waiting time of a GI/GI/1 queue with (i.i.d.) inter-arrival times $(\overline{T}_{n,r}^{(k)})_{n\in\mathbb{N}}$ where $\overline{T}_{n,r}^{(k)} =_{st} \overline{T}_r^{(k)}$ (see (40)) and service times $(S_{n,\kappa,r}^{(k)})_{n\in\mathbb{N}}$.

By coupling the $\overline{T}^{(k)}_{n,r}$'s and the $T^{(k)}_{n,\kappa,r}$'s in the obvious manner, one can easily see that $(\overline{T}^{(k)}_{1,r},\ldots,\overline{T}^{(k)}_{n,r}) \leq (T^{(k)}_{1,\kappa,r},\ldots,T^{(k)}_{n,\kappa,r})$[4] and therefore we have $(\overline{T}^{(k)}_{1,r},\ldots,\overline{T}^{(k)}_{n,r}) \leq_{st} (T^{(k)}_{1,\kappa,r},\ldots,T^{(k)}_{n,\kappa,r})$. Using, e.g., [7, pp. 217, 220], this implies

$$W^{(k)}_{r,\kappa} \leq_{st} \overline{W}^{(k)}_{r}. \tag{53}$$

Furthermore, given that $-\overline{T}^{(k_{m+1}+i)}_{n,r} \leq_{icx} -\overline{T}^{(k_m+i)}_{n,r}$ for all $i = 1,\ldots,\mathrm{lcm}(p)$ (by Lemma 3) and that the $(\overline{T}^{(k)}_{n,r})_{n\in\mathbb{N}}$ are independent, we can use [6, p. 337] to establish that

$$\overline{W}^{(k_{m+1}+i)}_{r} \leq_{icx} \overline{W}^{(k_m+i)}_{r}, \tag{54}$$

for all $i = 1,\ldots,\mathrm{lcm}(p)$. Therefore, given $\mathbb{E}\left((W^{(k)}_{r,\kappa})^2\right) < \infty$ for all k and Lemma 4, we can uniformly bound the second moment of $W^{(k)}_{r,\kappa}$ as follows

$$\begin{aligned}
\sup_{k\geq k^*} \mathbb{E}\left((W^{(k)}_{r,\kappa})^2\right) &\leq \sup_{k\geq k^*} \mathbb{E}\left((\overline{W}^{(k)}_{r})^2\right) && \text{(55a)}\\
&= \max_{i=1,\ldots,\mathrm{lcm}(p)} \sup_{m:k_m+i\geq k^*} \mathbb{E}\left((\overline{W}^{(k_m+i)}_{r})^2\right) && \text{(55b)}\\
&= \max_{i=1,\ldots,\mathrm{lcm}(p)} \mathbb{E}\left((\overline{W}^{(k_{m_i^*}+i)}_{r})^2\right) && \text{(55c)}\\
&< \infty, && \text{(55d)}
\end{aligned}$$

where $m_i^* \stackrel{\text{def}}{=} \min\{m : k_m + i \geq k^*\}$. In (55a) and (55c), we have used (53) and (54), respectively. In (55d), we have used that

$$\mathbb{E}\overline{T}^{(k)}_{n,r} = \min_{j=1,\ldots,p_r} \frac{a^{(k)}_{j,r}}{k\lambda} \geq \frac{\|p\|}{p_r\lambda} - \delta_r = \frac{1}{2}\frac{\|p\|}{p_r\lambda} + \frac{1}{2}\frac{1}{\mu_r} > \frac{1}{\mu_r}, \quad \forall k \geq k^*, \tag{56}$$

i.e. the ergodicity condition, and that the third moment of service times is finite, which imply that the second moment of $\overline{W}^{(k)}_{r}$ is finite [6, pg. 270].

Now, using the continuity of the stationary waiting time of GI/GI/1 queues [6, Corollary X.6.4] and part i) and ii) of Lemma 3, we have $\overline{W}^{(k)}_{r} \xrightarrow[k\to\infty]{d} W_r(p)$, and given the uniform integrability (55) we have that also the expected values converge [12, Theorem 3.5, pp. 31], i.e.,

$$\lim_{k\to\infty} \mathbb{E}\overline{W}^{(k)}_{r} = \mathbb{E}W_r(p). \tag{57}$$

With the above relations, we can conclude the proof of (38)

$$\begin{aligned}
\lim_{k\to\infty} \mathbb{E}W^{(k)}(\pi^{(k)}) &= \lim_{k\to\infty} \sum_r \sum_\kappa \frac{p_r}{\|p\|k} \mathbb{E}W^{(k)}_{r,\kappa}(\pi^{(k)}) && \text{(58a)}\\
&\leq \lim_{k\to\infty} \sum_r \frac{p_r}{\|p\|} \mathbb{E}\overline{W}^{(k)}_{r} && \text{(58b)}\\
&= \sum_r \frac{p_r}{\|p\|} \mathbb{E}W_r(p). && \text{(58c)}
\end{aligned}$$

[4] Given $x, y \in \mathbb{R}^d$, here $x \leq y$ means $x_i \leq y_i$ for all $i = 1,\ldots,d$.

This concludes the proof of (15) and (16). ■

Proof of (17). Let Q be a discrete random variable with values in $\{1,\dots,R\}\times\{1,\dots,k\}$ such that $\Pr(Q=(r,\kappa))=\frac{p_r}{\|p\|k}$. We assume that this random variable is independent of any other random variable. By definition of $W^{(k)}(\pi^{(k)})$ and using the law of total variance, we obtain

$$\operatorname{Var} W^{(k)}(\pi^{(k)}) = \mathbb{E}(\operatorname{Var} W^{(k)}(\pi^{(k)})|Q) + \operatorname{Var}\mathbb{E}(W^{(k)}(\pi^{(k)})|Q) \tag{59a}$$

$$= \sum_{r,\kappa} \frac{p_r}{\|p\|k}\left(\operatorname{Var} W^{(k)}_{r,\kappa}(\pi^{(k)}) + \left(\mathbb{E}W^{(k)}_{r,\kappa}(\pi^{(k)}) - \mathbb{E}W^{(k)}(\pi^{(k)})\right)^2\right). \tag{59b}$$

When $k\to\infty$, we have already established that $\mathbb{E}W^{(k)}(\pi^{(k)})\to\mathbb{E}W(p)$ and that $\mathbb{E}W_r(p)\le\mathbb{E}W^{(k)}_{r,\kappa}(\pi^{(k)})\le\mathbb{E}\overline{W}^{(k)}_{r,\kappa}\to\mathbb{E}W_r(p)$. Therefore, it only remains to show that the second moment of $W^{(k)}_{r,\kappa}(\pi^{(k)})$ converge to $\mathbb{E}[W_r(p)^2]$. This is done by using the same argument above for the convergence of the first moment. Hence, using the continuity of the waiting time and of the square function [6, Corollary X.6.4] and part i) and ii) of Lemma 3, we obtain $(\overline{W}^{(k)}_r)^2 \xrightarrow{d} W_r(p)^2$, as $k\to\infty$. Furthermore, the second moment of $(\overline{W}^{(k)}_r)^2$ is finite because the fifth moment of the service times is finite [6, pg. 270] and (54) ensures that the sequence $(\overline{W}^{(k)}_r)^2$ is uniformly integrable because it is non-increasing along subsequences $(k_m+i)_{m\in\mathbb{N}}$, for all $i=1,\dots,\operatorname{lcm}(p)$. Thus, $\mathbb{E}[(\overline{W}^{(k)}_r)^2]\to\mathbb{E}[W_r(p)^2]$. Together with (18), as desired we obtain

$$\lim_{k\to\infty}\mathbb{E}[(W^{(k)}_{r,\kappa}(\pi^{(k)}))^2]=\mathbb{E}[W_r(p)^2]. \tag{60}$$

This concludes the proof of (17). ■

Theorem 3 is proved.

5 Conclusions

We have derived structural properties concerning a known problem in the literature of stochastic scheduling, that is Problem 1. Fixing the proportion of jobs to send on each queue, p, we have identified a class of periodic policies and have proven that all the policies in this class are asymptotically equivalent and optimal. The limiting mean waiting time achieved by these policies, $\mathbb{E}W(p)$ (see (8)), is expressed in terms of a linear combination of independent D/GI/1 queues and has the convenient property of being convex in p. We believe that these structural properties provide researchers and practitioners with new means about the considered problem. For instance, one consequence of these results is that the problem of computing the optimal proportions of jobs to send to each queue, which is considered a difficult problem (see the introduction), boils down, asymptotically, to the solution of an optimization problem of the form:

$$\min \mathbb{E}W(p) \quad \text{s.t.:} \quad p\in\mathcal{S},$$

for $\mathcal{S}$ compact and convex, and we stress that $\mathbb{E}W(p)$ is a convex function of p. Using a classic result in convex optimization, this means that a polynomial number of evaluations of the objective function $\mathbb{E}W(p)$ are sufficient to converge to an optimizer of the problem. Given that each objective evaluation is efficient [9, 14, 15, 31, 33, 35], this lets us conclude that we have significantly reduced much of the difficulty of Problem 1. In the case where service times have an exponential distribution, $\mathbb{E}W(p)$ admits a very simple characterization because it is the weighted mean waiting time of R D/M/1 queues [4, 10].

References

1. Altman, E., Gaujal, B., and Hordijk, A.: Balanced sequences and optimal routing. *J. ACM*, **47** (2000) 752–775.
2. Altman, E., Gaujal, B., and Hordijk, A.: Multimodularity, convexity, and optimization properties. *Math. Oper. Res.*, **25** (2000) 324–347.
3. Anselmi, J., and Gaujal, B.: Optimal routing in parallel, non-observable queues and the price of anarchy revisited. In *International Teletraffic Congress* (2010) 1–8.
4. Anselmi, J., and Gaujal, B.: The price of forgetting in parallel and non-observable queues. *Perform. Eval.*, **68** (2011) 1291–1311.
5. Anselmi, J., Gaujal, B., and Nesti, T.: Control of parallel non-observable queues: asymptotic equivalence and optimality of periodic policies. *Stochastic Systems*, to appear.
6. Asmussen, S. *Applied Probability and Queues.* Springer, New York, 1987.
7. Baccelli, F. and Brémaud, P. *Elements of Queueing Theory: Palm Martingale Calculus and Stochastic Recurrences*, Stochastic Modelling and Applied Probability, Series Volume 26, Springer-Verlag Berlin Heidelberg, 2003.
8. Bar-Noy, A., Bhatia, R., Naor, J., and Schieber, B.: Minimizing service and operation costs of periodic scheduling (extended abstract). In *Soda* (H. J. Karloff, ed.), ACM/SIAM, 1998, 11–20.
9. Bell C. H. and Stidham S.: Individual versus social optimization in the allocation of customers to alternative servers. *Management Science*, **29** (1983) 831–839.
10. Bhat, U. N. *An Introduction to Queueing Theory: Modeling and Analysis in Applications*, Birkhauser, Boston, 2008.
11. Bhulai, S., Farenhorst-Yuan, T., Heidergott, B., and van der Laan, D.: Optimal balanced control for call centers. *Annals OR*, **201** (2012) 39–62.
12. Billingsley, P. *Convergence of Probability Measures.* Wiley series in probability and statistics, New York, 1999.
13. Borovkov, A.: *Stochastic Processes in Queueing Theory.* Applications of mathematics, Springer-Verlag, 1976.
14. Borst, S. C.: Optimal probabilistic allocation of customer types to servers. *ACM Sigmetrics '95/Performance '95, ACM*, New York, NY, USA, 1995, 116–125,
15. Combé M. B. and Boxma O. J.: Optimization of static traffic allocation policies. *Theor. Comput. Sci.*, **125** (1994) 17–43.
16. Doob, J.: *Stochastic Processes.* Wiley Publications in Statistics. John Wiley & Sons, 1953.
17. Gaujal, B., Hyon, E., and Jean-Marie, A.: Optimal routing in two parallel queues with exponential service times. *Discrete Event Dynamic Systems*, **16** (2006) 71–107.
18. Gun, L., Jean-Marie, A., Makowski, A. M., and Tedijanto, T. *Convexity Results for Parallel Queues with Bernoulli Routing.* Technical Report, TR 1990-52. University of Maryland, USA, 1990.

19. Hajek, B.: The proof of a folk theorem on queuing delay with applications to routing in networks. *J. ACM*, **30** (1983) 834–851.
20. Hajek, B.: Extremal splitting of point processes. *Math. Oper. Res.*, **10** (1986) 543–556.
21. Harchol-Balter, M., Scheller-Wolf, A., and Young, A. R.: Surprising results on task assignment in server farms with high-variability workloads. In *Sigmetrics/Performance*, ACM, 2009, 287–298.
22. Hordijk, A., Koole, G. M., and Loeve, J. A.: Analysis of a customer assignment model with no state information. *Probability in the Engineering and Informational Sciences*, **8** (1994) 419–429.
23. Hordijk, A. and van der Laan, D.: Periodic routing to parallel queues and billiard sequences. *Mathematical Methods of Operations Research*, **59** (2004) 173–192.
24. Humblet P.: *Determinism Minimizes Waiting Time in Queues*. LIDS-P-. Laboratory for Information and Decision Systems, 1982.
25. Javadi, B., Kondo, D., Vincent, J.-M., and Anderson, D. P.: Discovering statistical models of availability in large distributed systems: An empirical study of seti@home. *IEEE Trans. Parallel Distrib. Syst.*, **22** (2011) 1896–1903.
26. Javadi, B., Thulasiraman, P., and Buyya, R.: Cloud resource provisioning to extend the capacity of local resources in the presence of failures. In *HPCC-ICESS*, IEEE Computer Society, 2012, 311–319.
27. Lindley, D. V.: The theory of queues with a single server. *Mathematical Proceedings of the Cambridge Philosophical Society*, **48** (1952) 277–289.
28. Liu, Z. and Righter, R.: Optimal load balancing on distributed homogeneous unreliable processors. *Journal of Operations Research*, **46** (1998) 563–573.
29. Loynes, R: The stability of a queue with nonindependent interarrival and service times. *Mathematical Proceedings of the Cambridge Philosophical Society*, **58** (1962) 497–520.
30. Makowski, A.: On an elementary characterization of the increasing convex ordering, with an application. *Journal of Applied Probability*, **31** (1994) 834–840.
31. Neely, M. J. and Modiano, E.: Convexity in queues with general inputs. *IEEE Transactions on Information Theory*, **51** (2005) 706–714.
32. Puterman, M. L.: *Markov Decision Processes: Discrete Stochastic Dynamic Programming*. John Wiley & Sons, Inc., New York, NY, USA, 1st edition, 1994.
33. Sethuraman, J. and Squillante, M. S.: Optimal stochastic scheduling in multiclass parallel queues. *Sigmetrics '99*, ACM, New York, NY, USA, 1999, 93–102,
34. Shaked, M. and Shanthikumar, J. G.: *Stochastic Orders and their Applications*. Academic Press, 1994.
35. Shanthikumar, J. G. and Xu, S. H.: Asymptotically optimal routing and service rate allocation in a multiserver queueing system. *Operations Research*, **45** (1997) 464–469.
36. Stoyan, D. and Daley, D. *Comparison Methods for Queues and Other Stochastic Models*. Wiley series in probability and mathematical statistics: Applied probability and statistics, Wiley, 1983.
37. Tijdeman, R.: Fraenkel's conjecture for six sequences. *Discrete Mathematics*, **222** (2000) 223–234.
38. van der Laan, D. *The Structure and Performance of Optimal Routing Sequences*. Universiteit Leiden, 2003.
39. van der Laan, D.: Routing jobs to servers with deterministic service times. *Math. Oper. Res.*, **30** (2005) 195–224.

Average Cost Optimisation for a Power Supply Management Model*

Ksenia Chernysh

Heriot-Watt University
School of Mathematical and Computer Sciences, Edinburgh, UK
xeniachernysh@gmail.com

Abstract. We present a power supply management model and a stochastic optimisation problem related to it. With certain assumptions, we show that a solution to the problem exists, and is unique. We describe a number of properties of its behaviour.
Keywords: continuous time stochastic control, continuous-time Markov process, Poisson process, average cost optimality, stochastic fluid programs, power supply management, convex cost functional.
AMS 2000 subject classification: Primary 93E20, Secondary 60J25

1 Introduction

In this paper, we introduce a stochastic optimisation problem related to power supply management. We formulate a mathematical model and develop techniques for its analysis. Optimisation problems related to power generation have been actively studied in recent times, for example see [6, 9, 13, 16].

The model developed is inspired by wind power and by a collection of data related to this provided by National Grid. The data represents the errors in wind power prediction. We work under the assumption that these errors occur at time instants modelled by a Poisson process with a fixed intensity. We show under this assumption, our optimisation model falls into the class of Stochastic Fluid Programs (SFPs), a class of models introduced by N.Bäuerle in [1, 2].

There is vast literature on the optimal stochastic control. We refer the reader to textbooks [3, 4, 11, 15] for the standard theory and to [14] for recent developments. We consider that the state space of the controlled process is continuous and actions are applied in continuous time. There are a number of models in a similar setting, such as Piecewise-Deterministic Markov Processes (PDMP) [7, 8], Markov Decision Drift Processes(MDDPs) [10] (predecessor of PDMPs, PDMP generalises the setting for MDDPs), and Stochastic Fluid Programs [2, 1].

The dynamics of the controlled process in all these models are given by two objects:

- a flow function, between the jump times of the underlying environment, modelled by a continuous-time Markov process,

* The author is grateful to the EPSRC grant EP/I017954/1 and National Grid for providing the data. We thank Dr. Alexey Piunovskiy for organising the very interesting workshop and inviting the author to contribute to this volume. The author is thankful to her PhD supervisors Prof. S. Foss and Dr. J.R.Cruise, and to Dr. S. Zachary for all the discussions on the topic.

– a probability transition kernel that defines the value of the process at the jump times.

The setting of continuous control for PDMPs assumes that the controller can instantly change the underlying Markov process (by speeding it up or slowing it down), the transition kernel, and the flow function. To the best of our knowledge, there is no general theory on the average-cost optimisation for PDMPs in the case of the flow function control.

By contrast, for SFPs the underlying Markov process is uncontrolled, while the flow function is controlled. The formalism of SFPs is the most appropriate to us, because it allows us to prove results for the power supply model without a pre-described flow function, taking into consideration a wider set of possible solutions.

There are two main contributions of this paper. First, we find a class of real-world energy control systems, where the optimisation task is of great importance, and introduce a new mathematical model for it. Under the assumption that the inter-arrival times of the error process are exponentially distributed, the cost functional can be given by a cost rate function, so that the total cost will be the integral of the cost rate function with respect to time. However, without this assumption, the structure of the cost functional is more complicated. To the best of our knowledge, such optimisation problems with similar cost functionals have not been studied in literature. Second, we apply results from [1] to show the existence of the optimal control. By assuming the cost functional is strictly convex we prove uniqueness, and describe the behaviour of the optimal solution. We comment on how to find the optimal solution in a smaller class of controls and discuss a number of properties of the solution.

The rest of this paper is organised as follows. We present the power supply management problem in Section 2. We formulate a mathematical model in Section 2.1. We summarise the main definitions and results on the theory of SFPs in Section 3. and explain why our model may be regarded as an SFP in Section 3.2. Section 4 presents our main results on the explicit form and the uniqueness of the solution and explores a number of properties of the solution.

2 The Power Supply Problem

For simplicity, we consider only four types of power supply in this paper: conventional generation, renewable generation, imported power and local storage. We summarize here the main properties which are of interest for this work.

– Conventional power plants require time to ramp up and down. Therefore, they must be scheduled in advance. If there is a shortage of power supply in the system, conventional generation cannot immediately provide the needed power.
– Imported power is more expensive than the conventional generation. However in the event of a power shortage, imported power can be supplied immediately via interconnectors. Therefore, regardless of its high price, it is still purchased to prevent blackouts.
– The uncertainty in the wind power prediction brings difficulty to the management of the power supply. We suppose that all wind power plants are switched on and uncontrollable. The entire amount of wind power produced is supplied to the system automatically. Therefore we are interested in the so-called *net demand*, which is the difference between the total demand and power produced by wind farms. In this paper, the net demand is treated as a stochastic process.

- We suppose that the storage may be used for covering small shortages.

System operators, National Grid, predict the net demand. The power system acts autonomously to cover the mean of the predicted net demand, but it is unable to deal with the forecast errors. Therefore, in the event of a large power shortage imported power is purchased.

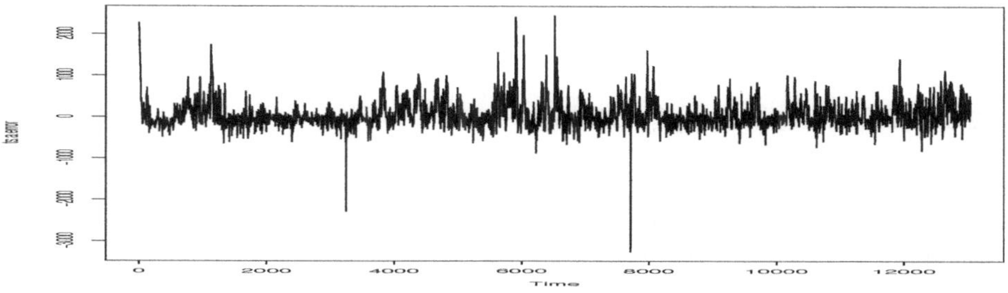

Fig. 1. Errors in MW in wind power prediction, for 1.5 year period.

Figure 1 represents forecast errors (prediction minus actual wind generation) made in the period from April 2011 to October 2012. The prediction was made 4 hours in advance, the data was kindly provided by Dr. A. Richards, National Grid.

When the error is negative there is no need to provide additional power. This is true since at the moment the power supply was scheduled, the prediction of the wind power was less optimistic than the actual outcome. Therefore demand is satisfied.

When the error is positive, one faces a situation where the provided power supply fails to meet demand. This shortage in the power supply must be covered immediately which can be done with the help of imported power.

We assume that the smaller forecast errors may be covered by the local storage, therefore we restrict our interest to only the larger errors. Figure 2 is obtained from the previous one by removing the shortfalls below the level 1000MW.

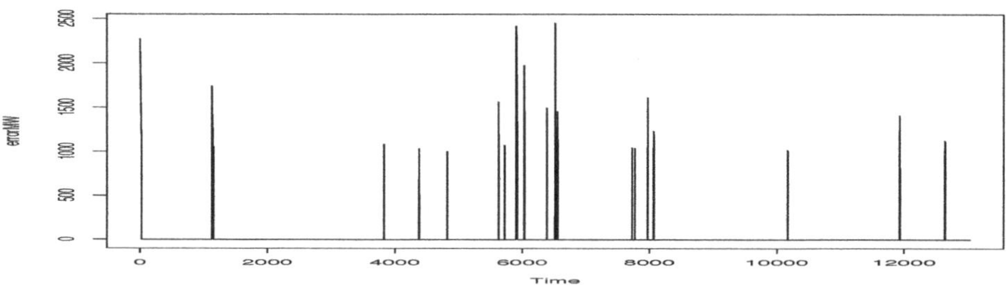

Fig. 2. Only positive errors which are higher than 1000MW, for 1.5 year period.

Hence we model the error process as a collection of spikes (errors) of random heights with random times between two consecutive shortfalls.

The expenditure for running this system is the total price for nuclear power purchased at shortage moments to balance the system. To reduce the total expenditure we come up with a different idea. We assume that we are allowed to provide additional conventional generation. The produced power can be used to satisfy the sudden unpredicted changes in the net demand and possibly to reduce the expenditure connected with the expensive imported power. The total cost will then be comprised of the cost for imported power at shortfall moments and the cost for the additional conventional generation.

2.1 The Power Supply Model

We assume that the sequence of shortfall times $\{T_n\}_{n\in\mathbb{Z}}$ forms a Poisson process $\nu_t = \max\{n : T_n \leq t\}, t \geq 0$ with a constant intensity $\lambda > 0$. Suppose that the shortfalls might be of k different types and let $\mathcal{Y} = \{1, 2, \ldots, k\}$. For instance, one can take $k = 2$ and consider an example where the shortfalls smaller than 1500MW are of the type 1 and the larger ones are of the type 2.

Assume that knowledge of the shortfall type at time instant T_n , denoted by Y_n, becomes available one time period in advance at moment T_{n-1}. In this paper, we assume that $\{Y_n\}$ is an irreducible Markov chain with the state space $\mathcal{Y}$. Notation $Y(t)$ is reserved for the corresponding continuous time Markov process.

A positive-valued process $X(t)$ represents the level of additional conventional generation we choose to provide. The only restriction is that for all $t, \delta > 0$ the increments of the process $X(t)$ must satisfy inequalities

$$-B\delta \leq X(t+\delta) - X(t) \leq U\delta,$$

where $U, B > 0$ are known positive constants, *ramp constraints*. Thus, the rate of increase/decrease of conventional generation must be bounded.

There are given two functions $g : \mathbb{R}^+ \to \mathbb{R}^+$ and $f : \mathbb{R}^+ \times \mathcal{Y} \to \mathbb{R}^+$. We refer to them as *the running cost* and *the terminal charges* respectively.

The amount charged over the time period $[0, T]$ is defined as

$$L(X(s), [0,T]) = \int_0^T g(X(s))ds + \sum_{k=1}^{\nu_T} f(X(T_k), Y(T_k)). \tag{1}$$

The following example of functions f and g is considered as canonical. Suppose that $c_1 > 0$ is the unit cost for conventional generation. Let $g(x) = c_1 x$, be the running cost of the system. Assume that ξ_n is a random variable, which stands for the error size at time T_n. If the state of the Markov chain is $Y_n = y$ then the shortfall size ξ_n has a certain distribution F_y. The expenditure the controller faces on average is

$$f(X(T_n), y) = c_2 \mathbb{E} \max(\xi_n - X(T_n), 0),$$

where $c_2 > c_1$ is the unit cost for nuclear power.

3 Stochastic Fluid Programs

3.1 Definition

We discuss the class of models, stochastic fluid programs (SFPs), introduced in Bäuerle [1, 2]. Let $\mathcal{Y}$ be a finite set and Q be the generator of a Markov process $Y(t)$ that takes values in $\mathcal{Y}$. Denote by T_n the jump times of the process $Y(t)$. We call $Y(t)$ the environment process. We assume that $Y(t)$ is a uniformised Markov process (see [17] for the definition), and $\{T_{n+1} - T_n\}_n$ are exponentially distributed with a common parameter λ. Let $S \subset \mathbb{R}^n$ be a closed set and $\mathcal{B}(S)$ be the Borel σ-algebra on S. Then $E = S \times \mathcal{Y}$ is the state space of the system. A compact and convex set $\mathcal{U} \subset \mathbb{R}^k$ is the action set of the system. For each $y \in \mathcal{Y}$ there is a linear function $b^y : \mathcal{U} \to \mathbb{R}^n$.

We define the controlled process $X(t)$ now. Value $X(0)$ can be an arbitrary element of S. Suppose that at time t the action $u_t \in \mathcal{U}$ is taken, then the left derivative of the controlled process $X(t)$ is equal to $b^y(u_t)$, so that

$$X(t) = X(0) + \int_0^t b^y(u_v)dv.$$

Any measurable function $u : E \times \mathbb{R}^+ \to \mathcal{U}$ is called *an open-loop control.* Assume that $Y(v) = y$ for $v \in [T_n, T_{n+1}]$ and $X(T_n) = x$, then the dynamics of the process $X(t)$ until time T_{n+1} under the open-loop control u are

$$X(t) = X(T_n) + \int_0^{t-T_n} b^y(u(x, y, v))dv, \tag{2}$$

This is the same as to say that action $u(x, y, v)$ is taken at time $v + T_n$. We consider only *admissible open-loop controls*, such that the process $X(t)$ never leaves the set S. It can be given precisely by

$$D(x, y) = \{u : \forall t \geq 0 \quad X(t) \in S, \text{ if } X(0) = x, Y(0) = y\}.$$

Definition 1. *A policy $\pi = (u_n)$ is a sequence of admissible open-loop controls.*

Suppose that $(X(T_n), Y(T_n)) = (x, y)$, then action $u_n(x, y, v - T_n)$ is applied at the moment v. Since the exponential random variable $T_{n+1} - T_n$ can take arbitrarily large values, the controls $u_n(x, y, t)$ are defined for all $t \in \mathbb{R}^+$. This explains why they are called open-loop controls. Clearly, the constructed process has a left derivative at each point and is continuous.

For a policy π we define probability measures $\mathbb{P}^\pi_{x,y}$ and expectations $\mathbb{E}^\pi_{x,y}$ in the following way:

$$\mathbb{P}^\pi_{x,y}(X(0) = x, Y(0) = y) = 1;$$
$$\mathbb{P}^\pi_{x,y}(T_{n+1} - T_n > t | T_0, \dots, T_n, X(T_0), \dots, X(T_n), Y(T_0), \dots, Y(T_n)) = \exp(-\lambda t);$$
$$\mathbb{P}^\pi_{x,y}((X(T_{n+1}), Y(T_{n+1})) \in B \times \{y_0\} | T_0, X(T_0), Y(T_0), \dots, T_n, X(T_n), Y(T_n))$$
$$= \mathbb{P}(Y(T_{n+1}) = y_0 | Y(T_n)) \times \mathbb{I}\left(\left[X(T_n) + \int_0^{T_{n+1}-T_n} b^y(u(x, y, v))dv\right] \in B\right),$$

for $y_0 \in \mathcal{Y}$ and $B \in \mathcal{B}(S)$. We write $\mathbb{P}^u_{x,y}$ and $\mathbb{E}^u_{x,y}$, when we talk about the one step-transition under open-loop control u.

An important subclass of open-loop controls is called *the class of feedback controls.* Feedback controls formalise an idea of having "Markovian" controls, when an action taken depends only on the state (x, y) and neither on the history of the process nor, in particular, on the time.

Definition 2. *Suppose that function $\psi : S \times \mathcal{Y} \to U$ is given. Then the corresponding process $X_\psi(t)$ with $X(0) = x$ is defined as*

$$X_\psi(t) = x + \int_0^t b^{Y(s)}(\psi(X(s), Y(s)))ds$$

We call the rule ψ feedback policy.

As one can see, the class of feedback policies is a proper subclass of the class of open-loop controls.

Finally, we introduce *the cost rate function*

$$c : S \times \mathcal{Y} \times \mathcal{U} \to \mathbb{R}^+,$$

which is assumed to be convex and continuous in x and u for all $y \in \mathcal{Y}$.

Definition 3. *Collection $(E, \mathcal{U}, b, Q, c)$ defines a stochastic fluid program.*

Definition 4. *Set $\pi_s = u_{\eta_s}(s - \eta_s)$.*

1. *Suppose that $\beta > 0$ is a discount parameter. The expected β-discounted cost over an infinite horizon under policy π starting at (x, y) is*

$$V_\pi^\beta(x, y) = \mathbb{E}_{x,y}^\pi \left[\int_0^\infty e^{-\beta t} c(X(s), Y(s), \pi_s) ds \right]$$

2. *For a fixed initial value (x, y) the average cost functional under policy π is defined by*

$$L_\pi(x, y) = \overline{\lim_{t \to \infty}} \frac{1}{t} \mathbb{E}_{x,y}^\pi \int_0^t c(X(s), Y(s), \pi_s).$$

3. *The minimal (β)-discounted and the average costs are*

$$V^\beta(x, y) = \inf_\pi V_\pi^\beta(x, y) \quad \text{and} \quad L(x, y) = \inf_\pi L_\pi(x, y).$$

Then the policy π is optimal in both cases if it attains the minimum.

We use notation $C(x, y, u)$ for the expected cost between two jumps under control u. We can write

$$C(x, y, u) = \int_0^\infty \exp(-(\lambda^{-1} + \beta)t) c(\phi_t(x, y, u), y, u(t)) dt$$

for $\beta \geq 0$. Here the case $\beta = 0$ corresponds to the time-average cost functional.

3.2 Our Model is an SFP

Suppose the state space $S = \mathbb{R}^+$ and the environment state $\mathcal{Y}$ is finite. Process $X(t)$ stands for the level of additional conventional generation provided, whilst the process $Y(t)$ stands for the type of the next shortfall. Let $b^y(x) = x$. The presence of the ramp constraints yields to the following definition of the action set as $\mathcal{U} = [-B, U]$.

The next lemma helps us to show that the cost functional introduced in Section 2.1 can be simplified.

Lemma 1. *Suppose that τ is an exponential random variable with parameter λ then for any function $g \geq 0$ and stochastic process $X(t)$ holds*

$$\mathbb{E}\int_0^\tau g(X(t))dt = \mathbb{E}\frac{g(X(\tau))}{\lambda}.$$

Proof. Recall that the exponential distribution is the only probability distribution for which the tail probability and the density $p(t)$ are proportional

$$\lambda\mathbb{P}\{\tau \geq t\} = p(t).$$

Using this and the change of the integration order one has

$$\begin{aligned}\mathbb{E}\int_0^\tau g(X(t))dt &= \int_0^\infty \int_0^\tau \lambda g(X(t))dt\exp(-\lambda\tau)d\tau = \\ &= \int_0^\infty \left(\int_t^\infty \lambda\exp(-\lambda\tau)d\tau\right) g(X(t))dt = \\ &= \int_0^\infty g(X(t))\exp(-\lambda t) = \mathbb{E}\frac{g(X(\tau))}{\lambda}.\end{aligned}$$

■

Remark 1. The proposition still holds if the nonnegativity of g is replaced with the absolute integrability.

Suppose that the error process $Y(t)$ is a uniformised Markov process, so that $T_{n+1} - T_n$ are i.i.d. exponentially distributed with parameter λ. As before $\nu_t = \max\{n : T_n \leq t\}$.

Suppose that the process $X(t)$ does not leave a compact set K a.s.. The time-average version of the cost functional (1) can be written as

$$\begin{aligned}\overline{\lim_{t\to\infty}} \frac{\mathbb{E}L(X(\cdot),[0,t])}{t} &= \overline{\lim_{t\to\infty}}\, \mathbb{E}\frac{\int_0^t g(X(t))dt + \sum_{i=1}^{\nu_t} f(X(T_i), Y_{i-1})}{t} \\ &= \overline{\lim_{t\to\infty}}\, \mathbb{E}\frac{\int_0^t g(X(t)) + \lambda f(X(t), Y(t))dt}{t} - \mathbb{E}\frac{\int_{\nu_t}^t \lambda f(X(t), Y(t))dt}{t} \\ &= \overline{\lim_{t\to\infty}}\, \mathbb{E}\frac{\int_0^t g(X(t)) + \lambda f(X(t), Y(t))dt}{t}.\end{aligned}$$

The last equality holds since

$$0 \leq \mathbb{E}\frac{\int_{\nu_t}^t \lambda f(X(t), Y(t))dt}{t} \leq \frac{\mathbb{E}(T_{n+1} - T_n)\max_{x\in K, y\in\mathcal{Y}} f(x,y)}{t} \leq \frac{\max_{x\in K, y\in\mathcal{Y}} f(x,y)}{\lambda t}.$$

By a similar reasoning

$$\begin{aligned}\overline{\lim_{t\to\infty}} \frac{\mathbb{E}L(X(\cdot),[0,t])}{t} &= \overline{\lim_{t\to\infty}}\, \mathbb{E}\frac{\sum_{i=1}^{\nu_t} \lambda^{-1} g(X(T_i)) + f(X(T_i), Y_{i-1}) + \int_{\nu_t}^t \lambda g(X(t))dt}{t} \\ &= \overline{\lim_{t\to\infty}}\, \mathbb{E}\frac{\sum_{i=1}^{\nu_t} \lambda^{-1} g(X(T_i)) + f(X(T_i), Y_{i-1})}{t}.\end{aligned}$$

The similar relations hold for the infinite-time discounted cost functional as well. Consequently, one may take functions

$$\tilde{g}(x) = 0 \text{ and } \tilde{f}(x,y) = \lambda^{-1}g(x) + f(x,y)$$

and consider a new model, where the charges apply only at points T_n. Or similarly, by letting

$$\widehat{g}(x,y) = g(x) + \lambda f(x,y) \text{ and } \widehat{f}(x,y) = 0$$

we reduce it to a model with the running cost only. The collection $(S \times \mathcal{Y}, \mathcal{U}, b, Q, \widehat{g})$ is an SFP. In case of time-average and discounted cost optimisation our problem may be regarded as this SFP.

3.3 Main Results and Assumptions for SFPs

This section summarises the results from Bäuerle [1, 2], which are relevant to our work. We start by stating assumptions and results for the β-discounted optimality. We proceed further by providing additional assumptions to formulate Theorem 2, which states the existence of the optimal solution and provides the average cost optimality equation (ACOE). We discuss Assumptions 2 and slightly modify them.

The following assumption is needed to verify the existence of a β-discounted optimal policy in Bäuerle [2].

Assumption 1. *For $\beta > 0$ there exists a policy π^β such that $V^\beta_{\pi^\beta}(x,y) < \infty$ for all $(x,y) \in E$.*

The following theorem states the existence of the optimal policy for $\beta > 0$ and provides the β-discounted cost optimality equation.

Theorem 1 (Theorem 4 and Lemma 5 in Bäuerle [2]). *Suppose that Assumption 1 holds. Then*

- *V^β is the minimal solution of*

$$V(x,y) = \min_{a \in D(x,y)} \left[C(x,y,a) + \mathbb{E}^a_{x,y} \exp(-(\beta+\lambda)T_1) V(\phi(X(T_1)), Y(T_1)\right]. \quad (3)$$

- *There exists a minimiser $f^\beta : S \times \mathcal{Y} \times \mathbb{R}^+ \to \mathcal{U}$ of (3) and the stationary policy $(f^\beta, f^\beta, f^\beta, \ldots)$ is β-discounted optimal.*
- *Solution $V^\beta(x,y)$ is convex in x for all y.*

Throughout literature, the equation (3) is generally referred to as the *Bellman equation.*

Fix $\kappa \in E$ and put

$$h^\beta(x,y) = V^\beta(x,y) - V^\beta(\kappa).$$

Assumption 2. *1. For all $\beta > 0$ there exists a policy π^β such that $V^\beta_{\pi^\beta}(x,y) < \infty$ for all $(x,y) \in E$.*

2. *There exists a policy π such that $L_\pi(x,y) < \infty$ for all $(x,y) \in E$.*
3. *There exist constants $l \in \mathbb{R}$ and $\hat{\beta} > 0$ and upper semicontinuous function $M : E \to \mathbb{R}^+$ such that for all $(x,y) \in E$ and $0 \le \beta \le \hat{\beta}$ holds*

$$l \le h^\beta(x,y) \le M(x,y)$$

and for all $(x,y) \in E$ and $u \in D((x,y))$

$$\mathbb{E}^u_{x,y} M(X(T_1), Y(T_1)) < \infty.$$

Theorem 2 (Theorems 4,5, in Bäuerle [1]). *Suppose that $S = \mathbb{R}^n$, or $S = (\mathbb{R}^+)^n$. Suppose that assumptions 2 hold. For $S = (\mathbb{R}^+)^n$ assume further that*

$$V^{\beta}_{\pi^\beta}(x, y) \quad \text{is increasing in } y.$$

Then the following is true.

1. *There exists a constant $\rho \geq 0$ and a convex function $h : E \to \mathbb{R}$ such that the average optimality equation holds for all $x \in E$*

$$\lambda\rho + h(x, y) = \inf_{u \in D((x,y))} \left[C(x, y, u) + \mathbb{E}^u_{x,y} h(X(T_1, Y(T_1))\right]. \tag{4}$$

2. *There exists a function $f^0 : S \times \mathcal{Y} \times \mathbb{R}^+ \to \mathbb{R}$, such that infimum of the right-hand side of equation (4) for (x, y) attains in it. There also exists a sequence $\beta_n \to 0$ such that*

$$f^0(x, y) = \lim_{n\to\infty} f^{\beta_n}(x, y). \tag{5}$$

3. *Suppose that $f_0(x, y, t)$ can be given by a feedback control ψ_0 and either $c(x, y, u)$ does not depend on y or the set of discontinuity points of mapping*

$$t \to \psi(X_\psi(t), Y(t))$$

   is of measure zero.
   Then f_0 is an average optimal policy and ρ is the minimal average cost.

The theorem may be proven without the additional assumption on the monotonicity of function V and it is done in Chernysh [5]. We use the latter version of the theorem for our purposes.

Remark 2. Controls $f^\beta(x, y)$ are functions of t. Their convergence in formula (5) is understood in the Young topology (see Davis [8]).

4 Results

In this subsection we assume that $S \subset \mathbb{R}$. We assume the action space U is convex and compact, hence it is a closed connected subset of $\mathbb{R}^n$. The images $b^y(\mathcal{U})$ are one-dimensional convex compact sets for all $y \in \mathcal{Y}$, thus they are closed intervals. For each $y \in \mathcal{Y}$ point 0 either belongs to it or the whole interval lies to one or the other side of 0. For our convenience we need the following.

Assumption 3. *We assume that $0 \in b^y(U)$ for all $y \in \mathcal{Y}$, so the control process $X(t)$ is allowed to move up and down, and also stop, regardless of the state of the environment.*

Under Assumption 3 $b^y(\mathcal{U}) = [-B(y), U(y)]$.

Define function $\bar{k}(x, y, z, t) : \mathbb{R}^+ \times \mathcal{Y} \times \mathbb{R}^+ \times \mathbb{R}^+ \to \mathbb{R}^+$ as

$$\bar{k}(x, y, z, t) = \mathbb{I}_{\{x\leq z\}} \min(x + U(y)t, z) + \mathbb{I}_{\{x>z\}} \max(x - B(y)t, z). \tag{6}$$

This is a function, with trajectory, which starts at x, goes to z, at speed equal to $U(y)$ if $x < z$ or $-B(y)$ if $x > z$, and stops at level z when it reaches it.

In order to prove our main results we need an auxiliary lemma.

Lemma 2. *Suppose that $f : \mathbb{R} \to \mathbb{R}$ is a strictly convex function and it attains its minimum at point $a \in \mathbb{R}$. Assume further, that P is a probability measure on $\mathbb{R}^+$, absolutely continuous with respect to Lebesque measure. Consider a class of functions $X : \mathbb{R}^+ \to \mathbb{R}$, where $X(0) = x_0$ and X satisfies the following assumptions with positive constants U and B:*

$$-B\delta \leq (X(t+\delta) - X(t)) \leq U\delta.$$

Then the optimization problem in the class described above

$$L(X(t), x_0) = \int_0^\infty f(X(t))dP(t) \to \min_{X(t)}$$

has a unique solution given by

$$X^*(t) = \mathbb{I}(x_0 > a)\max(x_0 - Bt, a) + (1 - \mathbb{I}(x_0 > a)\min(x_0 + Ut, a).$$

Proof. For the entire proof suppose that $x_0 < a$. The opposite case is treated in a similar way. Firstly, note that if $X(t)$ is the optimal policy, then for all $t \geq 0$ holds $X(t) \leq a$. Consider a new policy $X_1(t) = \min(X(t), a)$ then the set $\{t : f(X_1(t)) < f(X(t)))\}$ has positive measure due to the presence of constraints U and B. Therefore,

$$L(X_1(t), x_0) < L(X(t), x_0).$$

Hence, the policy crossing level a cannot be optimal.

The second step is to observe that if we have two controls $X_1(t) < X_2(t) \leq a$ for all $t \geq 0$ then clearly

$$L(X_2(t), x_0) < L(X_1(t), x_0).$$

This shows that at each point t the optimal control $X(t)$ should be as close to the optimal point a as possible. The fastest movement towards a is linear with speed U. The dynamics are formalised by function $X^*(t)$. ■

We start with a solution of the problem for the discounted optimality problem.

Theorem 3. *Suppose that $S = \mathbb{R}^+$, function $c(x, y, u)$ does not depend on u and is strictly convex in x, and Assumptions 1 and 3 hold for $\beta > 0$. Assume that the initial value is known $(X(0), Y(0)) = (x_0, y_0)$.*

Then there exists a unique function $l^\beta(y) : \mathcal{Y} \to \mathbb{R}^+$ such that the trajectory of the corresponding β-discounted optimal process $X(t)$ can be found recursively in the following manner.

- *$X(0) = x_0$;*
- *for $t \in [0, T_1]$ let $X(t) = \bar{k}(x_0, y_0, t, l^\beta(y_0))$;*
- *for $t \in [T_n, T_{n+1}]$ let $X(t) = \bar{k}(X(T_n), Y(T_n), t - T_n, l^\beta(Y(T_n))$.*

Remark 3. If Assumption 3 fails then none of the controls given by formula (6) are admissible.

Remark 4. The sample path of the process $X(t)$ may be written as

$$X(t) = \bar{k}(X_{\eta_t}, t - \eta_t, l^\beta(Y_{\eta_t})),$$

where as before $\eta_t = \max\{n : T_n < t\}$.

Remark 5. There exist a set of levels $\{l^\beta(y)\}_{y \in \mathcal{Y}}$, such that whenever the environment is in the state y, the trajectory starts moving towards the level $l(y)$ at the maximal possible speed and stops if it reaches this level.

Remark 6. The other possibility to formulate the theorem is to explicitly give the control function $u_0(x, y, t)$ which is done in the proof of the theorem.

Proof of Theorem 3. Due to Theorem 1 there exist a constant $\rho \geq 0$, a convex function $h : E \to \mathbb{R}$ and an open-loop control u_0, which satisfy the discounted optimality equation (3). Starting with the Bellman equation (3) we may write:

$$\begin{aligned} V(x,y) &= \min_{a \in D(x,y)} \left(C(x,y,a) + \mathbb{E}^a_{x,y} \exp(-(\beta+\lambda)T_1) V(\phi(X(T_1)), Y(T_1)\right) \\ &= \min_{a \in D(x,y)} \left(\int_0^\infty c(X(t),y) \exp(-(\beta+\lambda)t)dt + \right. \\ &\left. + \mathbb{E}^a_{x,y} \exp(-(\beta+\lambda)T_1) V(X(T_1), Y(T_1)) \right) \\ &= \min_{a \in D(x,y)} \int_0^\infty \left(c(X(t),y) + \right. \\ &+ \left. \sum_{y' \in \mathcal{Y}, y' \neq y} \frac{-q_{yy'}}{q_{yy}} (\beta+\lambda) V(X(t), y')) \right) \exp(-(\beta+\lambda)t)dt. \end{aligned} \tag{7}$$

The function

$$c(x,y) + \sum_{y' \neq y} \frac{-q_{yy'}}{q_{yy}} (\beta+\lambda) V(x,y')) =: H(x,y) \tag{8}$$

is strictly convex in x due to convexity of $V(x,y)$ for all y and strict convexity of $c(x,y)$. Therefore there exist points $l^\beta(y)$, where the minimum of function (8) attains. Then $l^\beta(y) \leq \sup_y \arg\min_x c(x,y)$.

Applying Lemma 2 to $H(x,y)$ and $dP(t) = \exp(-(\beta+\lambda)t)dt$ one has

$$X^*(t) = \bar{k}(x, y, t, l^\beta(y)),$$

which provides the minimum in the right-hand side of formula (7).

Therefore, in terms of open-loop controls, we can write

$$f^\beta(x,y,t) = \begin{cases} \begin{cases} U(y) & \text{if } t < \frac{l^\beta(y) - x}{U(y)} \\ 0 & \text{if } t \geq \frac{l^\beta(y) - x}{U(y)} \end{cases} & \text{if } x < l^\beta(y), \\ \\ \begin{cases} -B(y) & \text{if } t < \frac{x - l^\beta(y)}{B(y)} \\ 0 & \text{if } t \geq \frac{x - l^\beta(y)}{B(y)} \end{cases} & \text{if } x \geq l^\beta(y). \end{cases}$$

■

A similar result holds for the time-average optimality.

Theorem 4. *Assume that $S = \mathbb{R}^+$, and $c(x, y, u)$ does not depend on u. Suppose that Assumptions 2 and 3 hold. Suppose that the process starts from the initial point (x_0, y_0).*

Then there exists a function $l^0(y) : \mathcal{Y} \to \mathbb{R}^+$ such that the trajectory of the corresponding time-average optimal process $X(t)$ can be constructed recursively in the following manner.

– $X(0) = x_0$;

– *for* $t \in [0, T_1]$ *let* $X(t) = \bar{k}(x_0, y_0, t, l^0(y_0))$;
– *for* $t \in [T_n, T_{n+1}]$ *let* $X(t) = \bar{k}(X(T_n), Y(T_n), t - T_n, l^0(Y(T_n)))$.

Proof. The proof is similar to the proof of Theorem 3 with only a slight modification related to a different shape of the Bellman equation. Under assumptions 2 there exists a constant ρ and a convex function $h(x, y)$ satisfying the equation (4), by Theorem 2. Hence, starting with the ACOE one can write

$$\lambda\rho + h(x, y) = \inf_u \left[\int_0^\infty c(X(t), y) \exp(-\lambda^{-1} t) dt + \int_E h(x', y') P(x, y, u, dx', dy') \right] =$$
$$\inf_u \int_0^\infty \left(c(X(t), y) + \sum_{y' \in \mathcal{Y}, y' \neq y} \frac{q_{y_0 y'}}{q_{y_0 y_0}} \lambda h(X(t), y') \right) \exp(-\lambda^{-1} t) dt.$$

Let

$$H(x, y) := \frac{c(x, y)}{\lambda} + \sum_{y' \neq y} q_{yy'} h(x, y').$$

We can proceed along the lines of the proof of Theorem 3 to obtain the result. ■

We have shown in Theorems 3 and 4 that the trajectories of the optimal control processes $X(t)$ can be fully described by the levels $\{l^\beta(y)\}$. Therefore, one may restrict the area of search from the set of all open-loop controls to the set of models with levels $l_1, \dots, l_n$, which are treated here as variables.

For small values of n it is possible to obtain an explicit equation for the stationary distribution of process $X(t)$ and, hence, for the cost functional. Then it is possible to get the exact values $(l_i)_{i=1}^n$ by solving the corresponding equations numerically. The technique is explained in Chernysh [5].

The next corollary shows that the sequential limit in formula (5) can be replaced by the continuous limit and that the convergence of the optimal controls is equivalent to the convergence of the optimal levels.

Corollary 1. *Under the assumptions of Theorem 4 for all* (x, y), *we have*

1.
$$f^0(x, y) = \lim_{\beta \to 0} f^\beta(x, y),$$

2.
$$l(y) = \lim_{\beta \to 0} l^\beta(y).$$

The proof of Corollary 1 can be found in Chernysh [5].

The statement of Theorem 4 may be made stronger by weakening the assumptions.

Theorem 5. *Suppose* $\mathcal{Y}$ *is finite and there exists a policy* π *such that for all* (x, y) *holds* $L_\pi(x, y) < \infty$.

Then assumptions 2 hold.

Proof. We first show that $V_\beta^\pi(x, y) < \infty$ for small β. Corollary (1c) on page 183 of Widder [18] in our notation says that for any policy π and any initial state (x, y) holds

$$\overline{\lim_{\beta \to 0}}\, \beta V_\pi^\beta(x, y) \leq L_\pi(x, y).$$

By the definition of the upper limit for $\varepsilon = 1$ there exists $\bar{\beta}$ such that

$$\sup_{\beta \le \bar{\beta}} \beta V_\beta^\pi \le L_\pi(x) + 1,$$

and therefore,

$$V_\beta^\pi \le \frac{L_\pi(x) + 1}{\beta} < \infty \text{ for } \beta \le \bar{\beta}.$$

Therefore, due to Theorem 3 the levels $\{l^\beta(y)\}$ exist for small β.

Now we want to find a function $M(x,y)$ and a constant $L > 0$. Let $x_0 = \max_y \arg\min_x c(x,y) < \infty$ and $y_0(\beta) = \arg\min_y V^\beta(x_0, y)$. Suppose that π is an arbitrary policy and $X_\pi(t)$ is the controlled process corresponding to it, then a process defined by $X_\pi^*(t) = \min(X_\pi(t), x_0)$ gives a better value locally and, therefore, it holds for β-optimality too. Hence, we obtain that the optimal policy should not lie above level x_0. This yields that $l^\beta(y) \le x_0$ for all $\beta > 0$ and $y \in \mathcal{Y}$. Now note that if $X(0) = x_1 \le x_0$ then for any $t \ge 0$ holds $X(t) \le x_0$, so the process never leaves compact $[0, x_0]$ once it reaches it.

Let $h^\beta(x,y) = V^\beta(x,y) - V^\beta(x_0, y_0(\beta))$.

Consider the process with initial value fixed at (x_1, y_1). The rest of the proof is split into two parts for $x_1 \le x_0$ and $x_1 > x_0$. We write $(X_i(t), Y_i(t))$ for the process that starts at (x_i, y_i) for $i = 0, 1$, assuming that $y_0 = y_0(\beta)$.

Suppose that $x_1 \le x_0$. Random time τ_1 is defined as

$$\tau_1 = \min\{t : Y_1(t) = Y_0(t)\}$$

Time τ_1 is a.s. finite and, moreover, $\mathbb{E}\tau_1 < \infty$ (see Lindvall [12] pp 25, 35-36). Denote by τ_2 random time such that

$$\tau_2 = \min\{t : X_1(t + \tau_1) = X_0(t + \tau_1)\}.$$

We now show that $\mathbb{E}\tau_2 < \infty$. Let $w = \min_y(\min(U(y), B(y))$. As before sequence $\{T_k\}$ denotes jumping times of the embedded Markov chain of the process $Y(t)$. If any of $T_k - T_{k-1} \ge \frac{x_0}{w}$ then $X_1(T_k) = X_0(T_k)$, because there is enough time to reach level $l^\beta(Y(T_{k-1}))$ from any point $x \in [0, x_0]$. Since $T_k - T_{k-1}$ is an exponential random variable with parameter λ we have

$$\mathbb{P}\left\{T_k - T_{k-1} \ge \frac{x_0}{w}\right\} = \exp\left(\frac{-\lambda x_0}{w}\right) := q^*.$$

There exists a random variable $\chi \sim \text{Geom}(q^*)$ such that $\tau_2 \le \frac{x_0}{w}\chi$, thus

$$\mathbb{E}\tau_2 \le \frac{x_0}{w}\mathbb{E}\chi = \frac{x_0}{w} \times \frac{1}{q^*} < \infty.$$

Therefore,

$$|h^\beta(x_1, y_1)| = |V^\beta(x_1, y_1) - V^\beta(x_0, y_0)| \le 2\mathbb{E}(\tau_1 + \tau_2) \max_{x \in [0, x_0], y} c(x,y) =: L.$$

Let $M(x,y) = L$ for $x_1 \le x_0$.

Now suppose that $x_0 > x_1$. Then

$$h^\beta(x_1, y_1) = \Big(V^\beta(x_1, y_1) - V^\beta(x_0, y_1)\Big) + \Big(V^\beta(x_0, y_1) - V^\beta(x_0, y_0(\beta))\Big) \ge 0.$$

The first bracket is positive due to convexity of functions V^β and to the fact that $x_1 > x_0 \geq l^\beta(y_1)$, where the function attains its minimum. The second bracket is positive due to the definition of y_0.

We now show the upper bound. Notice that for $t \leq \frac{x_1 - x_0}{\min_y B(y)}$, the process $X_1(t)$ decreases to x_0. Let $\tau_1^* = \max(\tau_1, \frac{x_1 - x_0}{\min_y B(y)})$, then for $t > \tau_1^*$ holds $Y_1(t) = Y_0(t)$ and $X_1(t) \leq x_0$. Let $\tau_2^* = \min\{t : X_1(t + \tau_1^*) = X_0(t + \tau_1^*)\}$. Analogously to the previous case we get $\mathbb{E}\tau_2^* < \infty$. Hence we have

$$|h^\beta(x_1, y_1)| = \leq \max\left(\mathbb{E}\tau_1, \frac{x_1 - x_0}{\min_y B(y)}\right) \max_y c(x_1, y) +$$
$$+ \mathbb{E}\tau_2 \max_{x \in [0, x_0], y} c(x, y) =: M(x_1, y_1) < \infty.$$

Therefore we have constructed L and $M(x, y)$ as needed. ■

4.1 On the Continuity with Respect to the Parameters

In this subsection we analyse dependence of the optimal solution given in Theorem 4 on the system parameters. We consider only dependence on the ramp constraints U and B here, when both values tend to infinity.

In this subsection we assume that the optimal levels $l^0(y)$ exist, and that the ramp constraints do not depend on the states of the process $Y(t)$. Now we compare models with different values of the ramp constraints U and B, however assuming that the matrix Q is fixed. Then the optimal levels $l^0(y)$ might be seen as functions of U and B, which we denote by $l^0_{U,B}(y)$. Let $m(y) = \arg\min_x c(x, y)$, and that $m(i) \leq m(j)$ if $1 \leq i \leq j \leq n$. Let $\mathbf{m} = (m(1), m(2), \ldots, m(n))$.

We use the following notation in this section.

- For the set of levels $\ell = \{l(y)\}_{y \in \mathcal{Y}}$ we write $X^\ell_{U,B}(t)$ for the corresponding process with the levels ℓ and the ramp constraints U and B.
- The optimal policy for the ramp constraints U and B is $X^{\ell^0_{U,B}}_{U,B}(t)$, where

$$\ell^0_{U,B} = (l^0_{U,B}(1), l^0_{U,B}(2), \ldots, l^0_{U,B}(n))$$

  is the set of the optimal levels. We denote it by $X_{U,B}$, for the sake of notational simplicity.
- Suppose that $U = B = \infty$, then for $t \in [T_n, T_{n+1}]$ the trajectory $X^\ell_{\infty,\infty}(t)$ is defined by

$$X^\ell_{\infty,\infty}(t) = l(Y_{T_n}).$$

  Note that the trajectory is not continuous.
- Let $\widetilde{X} = X^{\mathbf{m}}_{\infty,\infty}$.
- Let $\Delta(U, B) = (m(n) - m(1)) \max(1/U, 1/B)$.
  Note that for any pair $m(1) \leq x_1, x_2 \leq m(n)$, the time to reach level x_2 starting at level x_1 is not bigger than $\Delta(U, B)$, if the ramp constraints are equal to U and B.

Consider the limiting case where $U = B = \infty$. In this situation there are no ramp constraints, so the trajectories are no longer necessarily continuous, and any positive function on $\mathbb{R}^+$ may be considered as a possible trajectory. Then the trajectory $\widetilde{X}$

is average-optimal in the class of positive functions on $\mathbb{R}^+$ as it minimises the cost functional at every point:

$$\forall x, y \quad c(x,y) \geq c(m(y),y) \quad \Longrightarrow$$
$$\forall X(t), y \quad \mathbb{E}\int_0^{\tau} c(X(s),y)ds \geq \mathbb{E}\int_0^{\tau} c(\widetilde{X}(s),y)ds.$$

Intuition suggests that the higher the possible speed (ramp constraint), the faster it is possible to move to the optimal levels $l^0(y)$, thus the optimal levels $l^0_{U,B}(y)$ should approach $m(y)$ as $U, B \to \infty$. This is the content of the following theorem.

Theorem 6. *Suppose that $U, B \to \infty$ then $l^0_{U,B}(y) \to m(y)$, for all $y \in \mathcal{Y}$.*
Moreover, there exists $\widetilde{C} > 0$ such that, for all $y \in \mathcal{Y}$ holds

$$|l^0_{U,B}(y) - m(y)| \leq \widetilde{C}\left(\max\left(\frac{1}{U}, \frac{1}{B}\right)\right).$$

To prove the theorem we need a simple auxiliary lemma.

Lemma 3. *1. There exists a constant $\widehat{C}$ such that*

$$L(X^{\mathbf{m}}_{U,B}) - L(\widetilde{X}) \leq \widehat{C}\Delta(U,B).$$

In particular,

$$L(X^{\mathbf{m}}_{U,B}) \to L(\widetilde{X}) \quad as \quad U, B \to \infty.$$

2. Suppose that a policy X is defined by the set of levels $\{l(y)\}_{y\in\mathcal{Y}}$ and there exists y_0 such that

$$|l(y_0) - m(y_0)| \geq \delta.$$

Then

$$L(X) - L(\widetilde{X}) \geq \mathbb{P}\{Y(t) = y_0\} \times \delta \times \lambda\exp(-\Delta(U,B)\lambda). \tag{9}$$

Proof of Lemma 3.

1. Let $M = \max_y \max_{x_1,x_2\in[m(1),m(n)]} |c(x_1,y) - c(x_2,y)|$. Within each time interval $[T_n, T_{n+1}]$ the maximal possible difference between the charges for both policies is

$$\mathbb{E}\int_{T_n}^{T_{n+1}} \left(c(X^{\mathbf{m}}_{U,B}(s),y) - c(\widetilde{X}(s),y)\right) ds \leq \Delta(U,B)M.$$

   Therefore, for the average cost the following inequality holds

$$0 \leq L(X^{\mathbf{m}}_{U,B}) - L(\widetilde{X}) \leq \frac{\Delta(U,B)M}{\lambda},$$

   because $1/\lambda$ is the mean of τ_n. As U and B grow to infinity we have $\frac{\Delta(U,B)M}{\lambda} \to 0$.
2. For the difference between the average cost functionals we have

$$L(X) - L(\widetilde{X}) = \overline{\lim_{k\to\infty}} \frac{\mathbb{E}\int_0^{T_k}\left(c(X(s),Y(s)) - c(\widetilde{X}(s),Y(s))\right)}{\frac{k}{\lambda}}$$
$$\geq \overline{\lim_{k\to\infty}} \frac{\sum_1^k \mathbb{P}\{Y(t) = y_0\}\int_{\Delta(U,B)}^{\infty} \delta \times (s - \Delta(U,B)) \times \lambda\exp(-\lambda s)ds}{\frac{k}{\lambda}}$$
$$= \mathbb{P}\{Y(t) = y\}\delta \times \lambda\exp(-\Delta(U,B)\lambda).$$

■

Now we are ready to prove Theorem 6.

Proof of Theorem 6. We prove the theorem by contradiction.

Suppose there exists y such that the sequence of levels for the average optimal policies $\ell^0(U,B)$ does not converge to $\mathbf{m}$. Then there exist $\delta > 0$ and a subsequence (U_n, B_n), such that

$$|l^0_{U_n,B_n}(y) - m(y)| \geq \delta.$$

Due to the first part of Lemma 3, we can take U_0, B_0 such that for $U > U_0, B > B_0$ the following inequalities hold

$$L(X^{\mathbf{m}}_{U,B}) - L(\widetilde{X}) \leq \frac{\delta\lambda\mathbb{P}\{Y(t)=y\}}{4} \text{ and}$$
$$\exp(-\Delta(U,B)\lambda) \geq \frac{1}{2}.$$

By inequality (9) one has

$$L(X_{U,B} - L(\widetilde{X}) \geq \delta\mathbb{P}\{Y(t)=y\}\lambda\exp(-\Delta(U,B)\lambda) \geq \frac{\delta\lambda\mathbb{P}\{Y(t)=y\}}{2}.$$

Now by the definition of U_0, B_0 and by the triangle inequality one has

$$L(X_{U,B}) - L(X^{\mathbf{m}}_{U,B}) \geq \frac{\delta\lambda\mathbb{P}\{Y(t)=y\}}{4}.$$

Therefore we get a contradiction with the optimality of strategy $X_{U,B}$. Hence the assumption that $l^0_{U,B}(y)$ does not converge to $m(y)$ cannot be true.

Finally, the rate of convergence may be obtained as follows. For $U > U_0$ and $B > B_0$, we have

$$\begin{aligned}
|L(X_{U,B}) - L(\tilde{X})| &\leq |L(X^{\mathbf{m}}_{U,B}) - L(\tilde{X})| = \widehat{C}\Delta(U,B),\\
|L(X_{U,B}) - L(\tilde{X})| &\geq \max_y |l^0_{U,B}(y) - m(y)| \min_y(\mathbb{P}\{Y(t)=y\}) \times \lambda\exp(-\Delta(U,B)\lambda)\\
&\geq \max_y |l^0_{U,B}(y) - m(y)|\widehat{C_1},
\end{aligned}$$

where $\widehat{C_1} > 0$. Hence, $\max_y |l^0_{U,B}(y) - m(y)| \leq \widetilde{C}\Delta(U,B)$, where $\widetilde{C} = \widehat{C} \times \widehat{C_1}^{-1}$. ■

The convergence of the optimal levels $l^0_{U,B}(y)$ to $m(y)$ has a useful application. For large U and B, instead of searching for the optimal levels $l^0_{U,B}(y)$, which is computationally difficult, one may take $m(y)$ as the set of levels for the trajectory, which is nearly optimal. Calculating $m(y)$ is simple, it is nothing more than finding minimums of functions $c(x,y)$.

References

1. Bäuerle, N.: Convex Stochastic Fluid Programs with average cost. *Journal of Mathematical Analysis and Applications,* **259** (2001) 137–156.
2. Bäuerle, N.: Discounted Stochastic Fluid Programs. *Mathematics of Operations Research,* **26** (2001) 401–420.
3. Bertsekas, D. *Dynamic Programming and Optimal Control.* Athena Scientific Belmont, MA, 1995.

4. Bertsekas, D. and Shreve, S. *Stochastic Optimal Control.* Academic Press, NY, 1978.
5. Chernysh. K. *Stochastic Average-cost Control, with Energy-related Applications.* In preparation. PhD thesis, Heriot-Watt University, 2016.
6. Cruise, J., Flatley L., Gibbens, R. and Zachary S.: Optimal control of storage incorporating market impact and with energy applications. arXiv preprint arXiv:1406.3653, 2014.
7. Davis, M.: Piecewise-deterministic Markov processes: A general class of non-diffusion stochastic models. *J.R. Statist. Soc. B,* **46** (1984) 353–388.
8. Davis, M. *Markov Models & Optimization,* Chapman & Hall, London, 1993.
9. Harsha P. and Dahleh, M.: Optimal management and sizing of energy storage under dynamic pricing for the efficient integration of renewable energy. *IEEE Transactions on Power Systems,* **30** (2015) 1164–1181.
10. Hordijk, A. and van der Duyn Schouten, F.: Average optimal policies in Markov Decision Drift Processes with applications to a queueing and a replacement model. *Advances in Applied Probability,* **15** (1983) 274–303.
11. Howard, R. *Dynamic Programming.* Management Science, **12** (1966) 317–348.
12. Lindvall, T. *Lectures on the Coupling Method.* Courier Corporation, 2002.
13. Meyn, S., Barooah, P., Busic, A. and Ehren, J.: Ancillary service to the grid from deferrable loads: the case for intelligent pool pumps in florida. *IEEE 52nd Annual Conference on Decision and Control (CDC),* (2013) 6946–6953.
14. Piunovskiy, A. (ed.) *Modern Trends in Controlled Stochastic Processes.* Luniver Press, 2010.
15. Puterman, M. *Markov Decision Processes: Discrete Stochastic Dynamic Programming.* John Wiley & Sons, 2014.
16. Richmond, N., Jacko P. and Makowski, A.: Optimal planning of slow-ramping power production in energy systems with renewables forecasts and limited storage. *IEEE International Conference on Probabilistic Methods Applied to Power Systems,* (2014) 1–6.
17. Ross, S. *Introduction to Probability Models.* Academic Press, 2014.
18. Widder, D. *The Laplace Transform.* Princeton, 1946.

Lightning Source UK Ltd.
Milton Keynes UK
UKOW06n2003221215

265244UK00001B/74/P

9 781905 986453